T0292394

THE FRONTIERS COLLECTION

Series editors

Avshalom C. Elitzur
Iyar The Israel Institute for Advanced Research, Rehovot, Israel
e-mail: avshalom.elitzur@weizmann.ac.il

Laura Mersini-Houghton
Department of Physics, University of North Carolina, Chapel Hill,
NC 27599-3255, USA
e-mail: mersini@physics.unc.edu

T. Padmanabhan
Inter University Centre for Astronomy and Astrophysics (IUCAA), Pune, India
e-mail: paddy@iucaa.in

Maximilian Schlosshauer
Department of Physics, University of Portland, Portland, OR 97203, USA
e-mail: schlossh@up.edu

Mark P. Silverman
Department of Physics, Trinity College, Hartford, CT 06106, USA
e-mail: mark.silverman@trincoll.edu

Jack A. Tuszynski
Department of Physics, University of Alberta, Edmonton, AB T6G 1Z2, Canada
e-mail: jtus@phys.ualberta.ca

Rüdiger Vaas
Center for Philosophy and Foundations of Science, University of Giessen,
35394 Giessen, Germany
e-mail: ruediger.vaas@t-online.de

THE FRONTIERS COLLECTION

Series Editors
A.C. Elitzur   L. Mersini-Houghton   T. Padmanabhan   M. Schlosshauer
M.P. Silverman   J.A. Tuszynski   R. Vaas

The books in this collection are devoted to challenging and open problems at the forefront of modern science, including related philosophical debates. In contrast to typical research monographs, however, they strive to present their topics in a manner accessible also to scientifically literate non-specialists wishing to gain insight into the deeper implications and fascinating questions involved. Taken as a whole, the series reflects the need for a fundamental and interdisciplinary approach to modern science. Furthermore, it is intended to encourage active scientists in all areas to ponder over important and perhaps controversial issues beyond their own speciality. Extending from quantum physics and relativity to entropy, consciousness and complex systems—the Frontiers Collection will inspire readers to push back the frontiers of their own knowledge.

More information about this series at http://www.springer.com/series/5342

For a full list of published titles, please see back of book or springer.com/series/5342

Klaas Landsman · Ellen van Wolde
Editors

THE CHALLENGE OF CHANCE

A Multidisciplinary Approach from Science
and the Humanities

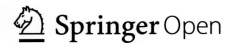 Springer Open

Editors
Klaas Landsman
Faculty of Science
Radboud University
Nijmegen
The Netherlands

Ellen van Wolde
Faculty of Philosophy, Theology and
 Religious Studies
Radboud University
Nijmegen
The Netherlands

ISSN 1612-3018   ISSN 2197-6619   (electronic)
THE FRONTIERS COLLECTION
ISBN 978-3-319-26298-7   ISBN 978-3-319-26300-7   (eBook)
DOI 10.1007/978-3-319-26300-7

Library of Congress Control Number: 2015956118

© The Editor(s) (if applicable) and The Author(s) 2016. This book is published open access.
Open Access This book is distributed under the terms of the Creative Commons Attribution-Noncommercial 2.5 License (http://creativecommons.org/licenses/by-nc/2.5/) which permits any non-commercial use, distribution, and reproduction in any medium, provided the original author(s) and source are credited.
The images or other third party material in this book are included in the work's Creative Commons license, unless indicated otherwise in the credit line; if such material is not included in the work's Creative Commons license and the respective action is not permitted by statutory regulation, users will need to obtain permission from the license holder to duplicate, adapt or reproduce the material.
This work is subject to copyright. All commercial rights are reserved by the Publisher, whether the whole or part of the material is concerned, specifically the rights of translation, reprinting, reuse of illustrations, recitation, broadcasting, reproduction on microfilms or in any other physical way, and transmission or information storage and retrieval, electronic adaptation, computer software, or by similar or dissimilar methodology now known or hereafter developed.
The use of general descriptive names, registered names, trademarks, service marks, etc. in this publication does not imply, even in the absence of a specific statement, that such names are exempt from the relevant protective laws and regulations and therefore free for general use.
The publisher, the authors and the editors are safe to assume that the advice and information in this book are believed to be true and accurate at the date of publication. Neither the publisher nor the authors or the editors give a warranty, express or implied, with respect to the material contained herein or for any errors or omissions that may have been made.

Printed on acid-free paper

This Springer imprint is published by SpringerNature
The registered company is Springer International Publishing AG Switzerland

Contents

Introduction 1
Klaas Landsman, Ellen van Wolde and Noortje ter Berg

**Conceptual and Historical Reflections on Chance (and Related
Concepts)** 9
Christoph H. Lüthy and Carla Rita Palmerino

The Mathematical Foundations of Randomness 49
Sebastiaan A. Terwijn

Randomness and the Madness of Crowds. 67
Utz Weitzel and Stephanie Rosenkranz

Randomness and the Games of Science 91
Jelle J. Goeman

**The Fine-Tuning Argument: Exploring the Improbability of Our
Existence** 111
Klaas Landsman

Chance in the Hebrew Bible: Views in Job and Genesis 1 131
Ellen van Wolde

**Happiness and Invulnerability from Chance: Western and Eastern
Perspectives** 151
Johannes M.M.H. Thijssen and David R. Loy

**The Experience of Coincidence: An Integrated Psychological and
Neurocognitive Perspective** 171
Michiel van Elk, Karl Friston and Harold Bekkering

**When Chance Strikes: Random Mutational Events as a Cause of Birth
Defects and Cancer** 187
Han G. Brunner

Chance, Variation and the Nature of Causality in Ecological Communities. 197
Hans de Kroon and Eelke Jongejans

The Size of History: Coincidence, Counterfactuality and Questions of Scale in History 215
Olivier Hekster

Accidental Harm Under (Roman) Civil Law. 233
Corjo Jansen

Taming Chaos. Chance and Variability in the Language Sciences 249
Roeland van Hout and Pieter Muysken

Biographies. 267

Titles in this Series 273

About the Editors

Klaas Landsman (1963) obtained his Ph.D. in Theoretical High-Energy Physics from the University of Amsterdam in 1989. He was a research fellow at the University of Cambridge from 1989 to 1997, interrupted by an Alexander von Humboldt Fellowship at Hamburg in 1993–1994. He was subsequently a Royal Academy Research Fellow at the University of Amsterdam from 1997 to 2002, and obtained a Pioneer Grant from NWO in 2002. Klaas has held the Chairs in Analysis and subsequently in Mathematical Physics at the Radboud University since 2004, and in 2011 was awarded the Bronze Medal of this university for his outreach work in mathematics. His research is mainly concerned with non-commutative geometry and with the mathematical foundations of quantum theory. The latter topic lies behind his interest in (pure) chance and probability.

Ellen van Wolde (1954) obtained her Ph.D. in Biblical Studies from Radboud University in 1989. She was a professor at the Faculty of Theology of University of Tilburg from 1992 to 2008, and has held the Chairs in Textual Sources of Judaism and Christianity at Radboud University since 2009. In 2005 she was appointed as a member of the KNAW, becoming a member of its Executive Board in 2011. Ellen's research is mainly concerned with the Old Testament Books of Genesis and Job, and with methodological approaches that acknowledge the role culture and language plays in the formation of biblical texts. A related field of interest of hers is the question how chance, bad luck, or coincidence were explained in ancient cultures and religions, especially in so far as these explanations still influence our present views.

Introduction

Klaas Landsman, Ellen van Wolde and Noortje ter Berg

Das Gewebe dieser Welt ist aus Notwendigkeit und Zufall gebildet
(The fabric of reality is built from necessity and chance)
Goethe

Abstract This chapter introduces the theme of the book (i.e., the challenge of chance) and includes brief surveys of the individual chapters.

The collapse of cohesion is one of the features that characterize chance. By sheer accident, or so it seems, something breaks the typical regularity of the natural world, like a comet disrupting the solar system. At a human scale, we find examples like unexpectedly bumping into an old friend, or losing a loved one in an accident. Such (seemingly) random phenomena appear arbitrary; they disrupt our lives and frustrate our human need for logic and meaning. The ensuing feelings of uncertainty and apprehensiveness, in turn, trigger us to search for explanations that will help restore order and normal patterns of cause and effect. In a word, we are challenged by chance, and we have been so at least since antiquity.

How do we respond to such challenges? For thousands of years people have tried to decide whether chance is a fundamental and irreducible phenomenon, i.e. certain events are not caused—they just happen, or whether chance is merely a reflection of our ignorance. Either way, we find the experience of chance hard to deal with. Humans constantly try to understand random phenomena and prefer explanations that (re)install meaning. The question, then, is whether this search for explanation and meaning has succeeded, or, at least, has a fighting chance (sic) to succeed.

This question is more subtle than it appears, since with his revolutionary claim that the universe is necessarily the way it is and yet has no goal, Spinoza cut the thread connecting explanation and purpose. Even necessity was subsequently

K. Landsman (✉)
Faculty of Science, Radboud University, Nijmegen, The Netherlands
e-mail: landsman@math.ru.nl

E. van Wolde
Faculty of Philosophy, Theology and Religious Studies, Radboud University,
Nijmegen, The Netherlands
e-mail: e.vanwolde@ftr.ru.nl

N. ter Berg
Radboud Honours Academy, Radboud University, Nijmegen, The Netherlands

© The Author(s) 2016
K. Landsman and E. van Wolde (eds.), *The Challenge of Chance*,
The Frontiers Collection, DOI 10.1007/978-3-319-26300-7_1

challenged by Darwin's theory of evolution in the 19th century, followed by quantum theory in the 20th, in both of which chance plays a fundamental role. Insult following injury, from Hume and Kant onwards even the causal patterns that permeate traditional science began to be questioned. From Aristotle to the 18th century, natural philosophy had seen these patterns as real, our role being limited to discovering them. Now, however, causality was claimed to be a mere by-product of our subjective need for rules, patterns, and meaning, which eventually led Bertrand Russell to his witticism that causality is "a relic of a bygone age, surviving, like the monarchy, only because it is erroneously supposed to do no harm."

The overall picture was summarized by the chilling words of Physics Nobel Laureate and popular science writer Steven Weinberg: "The more the universe appears comprehensible, the more it also appears pointless." However, he immediately qualified this pessimistic view (quoted from his popular account of the Big Bang entitled *The First Three Minutes*) in the following way: "But if there is no solace in the fruits of our research, there is at least some consolation in the research itself. Men and women are not content to comfort themselves with tales of gods and giants, or to confine their thoughts to the daily affairs of life; they also build telescopes and satellites and accelerators, and sit at their desks for endless hours working out the meaning of the data they gather. The effort to understand the universe is one of the very few things that lifts human life a little above the level of farce, and gives it some of the grace of tragedy."

This effort to understand includes the present book, which complements the excellent interdisciplinary books on chance that have already appeared over the last decades, both at a scholarly[1] and a popular[2] level. By incorporating a wide range of historical and contemporary sciences, the studies presented here allow us to develop a transdisciplinary perspective on chance. Thus our multidisciplinary approach, in which a team of authors explores the issue of chance in the disciplines of philosophy, mathematics, economics, game theory, statistics, physics, theology, neuropsychology, genetics, ecology, history, law, and linguistics, makes us aware of shared insights in these distinct disciplines. Let us first give a short survey of the articles originating in these various disciplines, to conclude with a few thoughts towards a transdisciplinary perspective on chance.

---

[1]See, for example, G. Gigerenzer et al. (eds.), *The Empire of Chance* (Cambridge University Press, 1989), L. Krüger et al., *The Probabilistic Revolution*, Vols. 1, 2 (MIT Press, 1990), I. Hacking, *The Taming of Chance* (Cambridge University Press, 1990), I. Hacking, *The Emergence of Probability* (Cambridge University Press, 2nd ed. 2006), S. Kern, *A Cultural History of Causality* (Princeton University Press, 2004), P. Vogt, *Kontingenz und Zufall: eine Ideen- und Begriffgeschichte* (Akademie-Verlag, 2011).

[2]E.g., N.N. Taleb, *Fooled by Randomness* (Penguin, 2004), W. Poundstone, *Fortune's Formula* (HIll and Wang, 2005), K. Mainzer, *Der kreative Zufall* (C.H. Beck, 2007), N. Silver, *The Signal and the Noise: The Art and Science of Prediction* (Penguin, 2012), D. Hand, *The Improbability Principle* (Bantam Press, 2014).

1 Contents of This Book: Addressing the Challenge

The opening chapter by Lüthy and Palmerino presents a survey of 2500 years of linguistic, philosophical, and scientific reflections on chance, coincidence, fortune, randomness, luck and other related concepts. In particular, they show that any concept of chance could only be understood through the alternative that the particular notion of chance attempted to exclude. And precisely because the alternative that was to be excluded did not have a stable identity, also its anti-pole (i.e. the idea of what chance is) had a variable meaning. For example, 'chance' has been opposed to 'fate', 'providence', 'natural laws', 'determinism,' and 'the knowledge of causes'. This heterogeneous list illustrates what a slippery concept 'chance' really is. The endeavour to pin down and define concepts by contrasting with opposites is a thread that runs throughout this book.

Perhaps the most rigorous way to analyse chance is through pure mathematics. In Terwijn's chapter we are told that even the best efforts in the 20th century to capture randomness mathematically have yielded no single 'true' notion of randomness." Instead, a number of (equivalent) definitions have been proposed that contextualize randomness relative to prior notions such as computability. Accordingly, an object is defined as random if its description cannot be shortened in a computable way, that is, randomness is opposed to computable compressibility. For example, according to this definition, despite the completely irregular distribution of its infinitely many digits the number $\pi = 3.14\ldots$ is not random at all, since instead of giving all these digits we could write a short program to compute them. On the other hand, most real numbers are random in this sense, although, curiously, this fact cannot be proven for any given random number.

Historically, the first application of mathematics to chance was to betting and gambling. Unexpectedly, two centuries later similar methods turned out to lie at the heart of game theory in economics (Weitzel and Rosenkranz). In finance, one typically assumes complete rationality on the part of all actors. In combination with the 'efficient market hypothesis', this would naively seem to imply a deterministic course of events. However, one of the remarkable predictions of game theory is that even on these assumptions the most rational strategy is often a random mixture of a number of alternative possibilities. Of course, this again blurs the alleged demarcation between determinism and chance.

Moving from probability to statistics, Goeman describes how researchers in medical statistics and psychology look for statistical correlations between data in the hope of revealing (publishable) evidence of a chain of cause and effect (for example, to conclude or predict that drinking milk is healthy whereas smoking is not). In a word, statistics is used to 'negotiate' chance. However, as Goeman argues, even ignoring notorious (especially Dutch) cases of scientific fraud, estimates of the unreliability of serious and published clinical studies range from 14 to 89 %, and he makes several proposals to improve this situation.

In the next chapter, Landsman's analysis of the 'fine-tuning argument' bridges the gap between chance in mathematics and physics on the one hand and chance in philosophy and theology on the other. The laws of nature contain parameters that are set at highly specific values for the universe to exist, and for us, humans, to exist in it. The list of possible explanations for this fine-tuning of the universe includes: design by a deity, a 'multi-verse' (so as to increase the probability of the existence of our own universe), 'blind chance', and finally, 'blind necessity'. For Landsman the latter two are the best options but he adds: "The present state of science does not allow us to make such a choice now, and the question even arises whether science will ever be able to make it, except perhaps philosophically."

Contrary to common belief, theological stances from the past were not all deterministic. In the Hebrew Bible, for instance, the book of Job describes the dramatic alternation between fortune and misfortune in a non-deterministic way, as Van Wolde's analysis shows. Job is unaware that God is carrying out an experiment because of a wager with the satan. Job tries to find his own explanations and reasons, but is chastised by God for obscuring "the design by words without knowledge". God's dismissive words reverberate throughout the years of thinking about chance, coincidence, luck, randomness and such concepts. Are these just words without knowledge? Or is it our historical, spatial, and cultural perspective that limits our type of rationality? Van Wolde also discusses this question with respect to the first chapter of the book of Genesis, which for many people, secular or non-secular, is the clearest example of God initiating a cause-driven chain of events. The question, then, is whether this is really the case.

It is a relatively small step from the ancient Near-East to the ancient Greek and Asian worlds. Bringing both philosophical and Buddhist attitudes towards chance into the picture, Thijssen and Loy point out that at first, 'luck' (or 'chance') and 'karma' seem to be opposing concepts. If something happens because of good or bad luck, it is beyond the agent's control whereas, in contrast, karma, implies that agents have a great deal of control (albeit indirect) over what happens. However, both philosophical traditions believe that being invulnerable to bad luck depends upon mental transformation. Western traditions focus more on coping with the emotional effects of bad luck, whereas Eastern traditions concentrate on the agent's motivations. But both aim to change our experience of the world and are still helpful today in our attempts to secure happiness in the face of adversity.

In contrast, the contemporary western approach to chance as an aspect of human life is set in the framework of cognitive neuroscience. van Elk, Friston, and Bekkering discuss the deeply engrained human tendency to give meaning to coincidences. However, it turns out that not only are humans remarkably bad at estimating chances and probabilities, they also tend to perceive a causal nexus between situations even where there is none. In doing so, the original meaning of coincidence is subverted, as it gestures at a perceived connection between events even though we cannot explain the causal mechanism behind it. Following Helmholtz, they argue that the human brain a priori constructs a predictive model of the world, which however may be interrupted or distracted by seemingly random events (neuroscientists typically have a deterministic world picture, so that randomness is never absolute but is only experienced as such). However, it is their very randomness that endows such events with at least subjective explanatory power, in that the brain may conclude that the inexplicable becomes explicable, precisely because it was random.

Medical research has to bridge another chasm, namely from biology and genetics to the feelings of loss when a handicapped child is born 'by chance' to healthy parents. Brunner shows in his study that random genetic mutations that originate at the molecular level can subsequently have either causal or probabilistic consequences for genes, individuals, species, ecosystems, and eventually even for the planet. The example of genetics also raises the question whether random events are beneficial or harmful: on the one hand, random errors of replication during the formation of germ cells can cause birth defects that result in a miscarriage or severe problems for the child and parents. On the other hand, such mutations drive evolution at the level of the species, typically enabling it to improve.

Coincidence also plays a central role in De Kroon and Jongejans' chapter. They counter the statement that "if it's a coincidence, it is not scientific"—a judgment implied in the premises of the previous two chapters. They argue that if 'chance processes' such as a heavy storm occur at the right place and time they could well determine the development of ecosystems and they claim "chance is pervasive in ecological systems." But what is the status of chance here? Qualifying their thesis, the authors argue that chance events typically have a deterministic origin, and that the stochastic nature of their occurrences can often be defined within a range of predictable variation. What remains problematic is the uneasy relationship between the scale-dependence of cause and effect with that of stochasticity.

In his chapter, Hekster tells us that because coincidences are, by definition, not causally related, traditional historians have tended to ignore them. So when is a coincidence just a coincidence, and when does a pattern occur? And why would a historian be interested in 'accidents', 'singular events', or 'contingent circumstances'? Surely, it has been historically decisive that Hitler survived all attempts to kill him (except his own). Yet it is tempting to walk the path of 'what-if history'. But does counterfactual thinking liberate us from a false sense of historical determinism or does it, instead, lead to a view of history as a series of random events? The answer to this question depends entirely on one's sense of the causal forces active in history. A providentialist or determinist will see inevitabilities and necessities. As Hekster argues, much will also depend on how one defines "the intersection between private actions and the public world," where "history develops." At those intersections, coincidences might play an explanatory role, but only if understood in terms of micro-causes related to individual human agency.

In Jansen's article, which deals with 'accidental harm' under Roman law (which has exerted a paramount influence on modern European Law), we once more encounter the Latin word 'casus' with its many meanings, which signifies not just 'accident', but also 'misfortune', 'fate', 'adversity' or 'setback,' which, in the legal context, all amount "to an event resulting in damage which cannot be traced back to another party's fault." For the Roman lawyer, however, 'casus' is not opposed to necessity, but to some state of intentionality. In any case, accidents are seen as purely negative, and the question is simply who is liable for the damage they cause.

Yet at least in Western Europe, after WW II this principle was increasingly countered by the tendency of governments to protect citizens from misfortune, notably by means of a social security system—"from womb to tomb" (Churchill). In recent years such systems seem to be weakening, partly for financial reasons (they are arguably becoming unaffordable), but also under the influence of liberal tendencies to restore the individual's responsibility for whatever happens to him or her.

The chapter by Van Hout and Muysken starts with a rejection of complete generative models of linguistics à la Chomsky, in which chance hardly plays any role and at best represents a lack of knowledge. Instead, they use numerous examples to show that chance, in the sense of language variation, plays a major role at each of the four levels of linguistics: inter-species variability, inter-language variability, variability in the linguistic signal within a given language, and finally inter-individual variability. In each of these four levels, the notion of chance figures as an inherent property; it is a probability mechanism to explain variability. They conclude the final chapter of this book with the comment that random variations in language ultimately originate from the fact that human ways of expressing meaning are far from unique.

2   A Transdisciplinary Perspective on Chance

One of the insights of this collection of articles that struck us as meaningful when looking at chance from such diverse disciplinary perspectives is that two aspects return in many of the contributions, namely the contextuality of chance and its role in explanations.

Contextuality of chance is most clearly seen in scale-dependence, which is found in many biological ecosystems (cf. De Kroon and Jongejans). What seem to be random events at a lower level can produce stability at a higher level. For example, seeds are dispersed at random by the wind, then may germinate into a plant or disappear. Another example is the origins of language variation. Ideas about random origins will be different if studied at the level of species, language in general, different languages, or individual speakers of a given language (Van Hout and Muysken). Random genetic mutations (Brunner) provide yet another case in point. They originate at a molecular level but, subsequently, have causal or probabilistic consequences for genes, individuals, species, ecosystems and thus, ultimately, for the planet as a whole. In history, what seem to be a small-scale state of affairs (such as the legendary beauty of Cleopatra's nose) can have huge consequences for nations and even epochs (Hekster). As a final example, in economics, the (random) individual psychology of a single investor interacts with the rather more deterministic psychology of the 'masses', for example, during the tulip mania in 1637 or the dotcom bubble in the 1990s (Weitzel and Rosenkranz).

Another instance of the contextuality of chance is its perspectival nature. In mathematics (Terwijn), no *absolute* notion of randomness can exist, and in order to properly define the notion, one has to specify *with respect to what* the supposed random objects should be random. Thus a random object is random with respect to a given type of definition, or class of sets. Strikingly, this view is comparable to the theological view presented in literary form in the biblical book of Job (Van Wolde). In the narrated divine speech out of the whirlwind, chance is related to a multifocal view of a universe and interpreted in terms of perspective: God, reflecting on the universe and its inhabitants, states that he does not share the perspective of the stars, weather phenomena, or animals, and that he does not even share the moral convictions of human beings who only want him to share *their* perspective, such as their ideas of justice. Thus what seems to be coincidental at the level of humans (or animals and plants) may be the effect of order at a higher level.

Secondly, throughout history including contemporary science, chance has been used both as an explanation and as the hallmark of an absence of explanation. Thus one may wonder if these apparent antipodes are really as antithetical as they seem. Historiography itself is a prime example. One could argue that Western philosophy would have emerged without Plato, or that there would have been a Scientific Revolution without Newton. But would there have been a communist Russian Revolution without Lenin, or a Holocaust without Hitler? If not, the actual occurrence of these momentous events in history was eventually *caused* by the *random* events of the births of these particular individuals. Similarly, parents with a severely handicapped or stillborn child may feel that their misfortune has no explanation, while their doctor may say it was *caused* by a *random* genetic defect. Appeals to God as the instigator of certain random events play a similar role. In quantum physics it could be claimed that radio-active atoms decay *because* of random events, or it could be said that this decay cannot be explained. The Fine-Tuning Argument brings this dual role of chance to a head, as many contemporary secular scientists seem perfectly happy to attribute the occurrence of life to chance, whereas others regard this as the lack of an 'explanation', and look elsewhere.

The reader is invited to also look at other chapters from these two angles, or indeed from any angle he or she prefers, as chance is an infinitely rich phenomenon that will continue to fascinate humans as long they live. We hope this book will challenge our readers as much as it did the authors.

Acknowledgement This project was initiated by Chris Mollema, who infected all authors with his enthusiasm. The publication of this book would not have been possible without the financial support of the Executive Board of Radboud University. Our thanks also go to our publisher, Springer-Verlag, especially to Angela Lahee, for her unfailing care and attention for this book.

Open Access This chapter is distributed under the terms of the Creative Commons Attribution-Noncommercial 2.5 License (http://creativecommons.org/licenses/by-nc/2.5/) which permits any noncommercial use, distribution, and reproduction in any medium, provided the original author(s) and source are credited. The images or other third party material in this chapter are included in the work's Creative Commons license, unless indicated otherwise in the credit line; if such material is not included in the work's Creative Commons license and the respective action is not permitted by statutory regulation, users will need to obtain permission from the license holder to duplicate, adapt or reproduce the material.

Conceptual and Historical Reflections on Chance (and Related Concepts)

Christoph H. Lüthy and Carla Rita Palmerino

Abstract In everyday language, the use of such words as "chance," "coincidence," "luck," "fortune" or "randomness" strongly overlap. In fact, in some languages, such as German, they coincide in one word (*Zufall*). In others, there is a clear separation between chance events with positive connotations (e.g., "luck," "fortune") and those with bad ones (e.g., "accident," "hazard"). In this essay, we try to sketch the main lines of development of several of these concepts from the ancient Greeks up to modern times, or more precisely, from Democritus and Aristotle up to the world of quantum mechanics. Three elements emerge with particular force. First, "chance," "fortune," "randomness," etc. are in some instances invoked as explanations of events, but in others designate events that occur without an explanations. Second, the meaning of these terms only becomes clear when one understands which alternatives they exclude. Finally, it is conspicuous to see how, after a rigid exclusion of "chance" or "randomness" from the domain of scientific explanation in the early modern period, they were restored to full glory in nineteenth- and twentieth-century biology and physics.

There exists a cluster of words with which we designate events that in some way or another surprise us, either because we didn't expect them, or because they are out of the ordinary, or because they seem inexplicable. "Chance," "coincidence," "randomness," and "luck" are words that belong to this category of surprise. Sure enough, each of them has more technical meanings, particularly when used in

We would like to thank Ellen van Wolde, Klaas Landsman, Jos Uffink and Frederik Bakker for their comments on earlier drafts of this chapter. The section on modern physics is based on a text provided by Klaas Landsman, which we have slightly adjusted.

C.H. Lüthy (✉) · C.R. Palmerino
Center for the History of Philosophy and Science, Faculty of Philosophy,
Theology and Religious Studies, Radboud University, Nijmegen, The Netherlands
e-mail: c.luethy@ftr.ru.nl; c.palmerino@ftr.ru.nl

© The Author(s) 2016
K. Landsman and E. van Wolde (eds.), *The Challenge of Chance*,
The Frontiers Collection, DOI 10.1007/978-3-319-26300-7_2

specific scientific and non-scientific contexts; take, for example, a mathematically precise notion such as Martin-Löf randomness.[1] But as far as everyday language is concerned, our terms strongly overlap. Phrases such as "I met him by chance," "this was an extraordinary coincidence," "I was randomly chosen," or "I was lucky enough to escape" all gesture at the fact that we couldn't have predicted what in fact happened to us or to someone else.

All of these terms are popular, and some are used with great frequency. And yet, it is very difficult to say what exactly they mean. It is impossible to develop either a coherent theory or a single narrative around them. They are simply too soft conceptually, too imprecise, and in fact even contradictory. Most people would probably agree with the Enlightenment philosopher David Hume that "chance, when strictly examined, is a mere negative word, and means not any real power which has anywhere a being in nature." (Hume 1748, Ch. 8.1).

One important reason why it is impossible to give a coherent account of this negative word and of its siblings is that they are used both to offer an explanation and to signal the lack of an explanation! Two examples will suffice to demonstrate this. In the sentence, "She didn't know the game, and that she won was sheer luck," the word "luck" signals the absence of a good explanation (such as routine or skill) to account for the fact that someone won at a game. The logic is quite different in the sentence, "through this lucky coincidence, she managed to win the elections." Here, the "lucky coincidence" offers an abbreviated explanation. The "coincidence" might refer to the fact that Harry, the obvious candidate, had suffered a stroke, and Lucy, his opponent, had on the same day been imprisoned, so that Theodora, whose ambitions had previously seemed implausible, could now win the elections. While in the first sentence the expression "sheer luck" signals the absence of a convincing causal explanation, in the second the expression "lucky coincidence" provides the explanation, while obviously also indicating its unforeseen nature.

Depending on the context, "chance," "coincidence," "randomness," or "luck" do not only indicate the presence or absence of a recognizable causal logic, but they also indicate unknown probabilities, which might or might not be calculable. "Chances are that you won't make it," or "If you are lucky, you might still catch that train," are phrases which imply an embryonic form of probabilistic reasoning of the type "what are the odds that x happens?"

Explanation, lack thereof, or intuited probabilities: it is in this ill-defined, swampy area that the terms we are examining here are located. As a consequence, Madam Fortune, the mythological personification that rules over these swamps, will necessarily also assume multiple roles. At one extreme, she will manifest herself as a divine figure that determines our fate; reference to her will in that case provide a coherent answer for explaining why things that for us had been unpredictable had nevertheless happened. At the other extreme, she is as helplessly exposed to

---

[1] On different mathematical definitions of randomness, see Sebastiaan Terwijn's chapter in this book.

circumstances as we are. A fickle woman placed on the allegorizing weather vane who is swept about by the winds, she is herself the object of unpredictable influences. Explaining an event through fortune characterized in the latter way amounts to empty prattle, as it merely moves unpredictability to a different level.

Despite the elusive and contradictory explanatory value of this cluster of words, there are interesting things than can be said about them. In our first section, we will first try an etymological approach. There, we will encounter a strong presence of falling dice as well as of lots, straws and other literally "aleatoric" objects of gaming and decision making, including the emblematic Wheel of Fortune. But we will also witness a strong and unresolved tension between viewing fortune and chance as a final (possibly divine) explanation for unexpected occurrences, and that of depicting them as merely a higher level of unpredictable randomness.

Our main approach is, however, historical. We will in some detail survey a number of key moments in the history of scientific (or natural philosophical) thought, from the divine fate of Greek tragedy and the chance swerve of Epicurean atoms through the deterministic machine world à la Descartes up to the reintroduction of chance and randomness in scientific theories as diverse as evolutionary theory and quantum physics. In this section, we will see that, as a general rule, philosophy and science have repeatedly tried to drive chance and coincidence out of their domain—unless they could stand for a precise type of causal factor that was required for a specific type of physical explanation—but that, time and again, chance entered anew through the back door.

We will end by concluding that our terms are best understood *ex negativo*. In order to understand what scientists or philosophers of past and present ages mean when they attribute something to chance, coincidence, randomness or luck, it is indispensable to understand what it is that they wish to exclude. Is it necessity, fate, determinism, causal knowledge, regularity, high probability, or something else? Given the obvious vagueness and contradictoriness of the connotations of our original set of words, it will come as no surprise to see that their contraries are just as ill-defined. Still, there is a strong heuristic advantage to this exercise. Being aware of what it is that we wish to exclude, we, the readers of this essay, will at least have some greater clarity of what it is that we implicitly wish to affirm with our underdetermined words.

1 Etymological Prelude

1.1 Dice and Other Falling Objects

We have opened this essay with the observation that "chance," "coincidence," "randomness," and "luck" may possess precise meanings in specific scientific and cultural circumstances, but that in everyday language, their meanings overlap.[2] Let us now add that this overlap is much greater in one language than in another. A particularly striking case is German (and the same is true for Dutch), where the word "Zufall" covers all four English terms: "*eine zufällige Begegnung*" is "a chance encounter"; "*ein seltsamer Zufall*" is translated as "a rare coincidence"; "*ein zufälliger Passant*" would be "a random passer-by"; and "*ein Zufallstreffer*" could be translated as "piece of good luck." Now, *Zufall*, this all-encompassing German word, is an old but literal translation of the Latin *accidens*: "something that falls down" or "upon."

Cadere, the Latin verb for "to fall," stands in fact at the root of several of the words that we are investigating in these pages. To begin with, there is of course the Latin noun *casus*, "the fall," a word that can describe the falling of snow, but also everything else that literally "befalls" us, however improbable it may be. *Casus* is therefore also the Latin word for "chance," "coincidence," or "luck." In Italian, it has retained precisely that meaning: "*Sei per caso in città domani?*" is literally "Are you by chance in town tomorrow?" The English word "case," which barely hides its Latin origin, has lost most of the original significance of *casus*, although the adjective "casual" still retains some of it, as when we speak of a "casual meeting."

What "befalls" us can be pleasant or unpleasant. Whereas *Zufall* is neutral in that respect (an event can be a *glücklicher* or *unglücklicher Zufall*), the Latin *accidens,* of which *Zufall* is a translation, has in many languages assumed a predominantly negative connotation. While the adverb "accidentally" still means "by chance," the noun "accident" has clearly negative connotations. The phrase, "It was an accident," would nowadays never be used with reference to a "fortune" won at the lottery, but most certainly so as to explain why the window is broken. The same negative connotation of "accident" is found in French or Italian, while in German, the oddly inauspicious prefix *un-* in *Unfall* does the same trick. Significantly, the French word *hazard*, which ultimately seems to go back to an Arabic expression relating to the throwing of dice, has had the same double fate as "accident": while *par hazard* is emotionally neutral, simply meaning "by chance," the English "hazard" and "hazardous"—as in "hazardous waste"—are negatively charged.

But the element of "falling" is even more pervasive than that. Just like "case," the English word "chance" also derives in the last instance from the Latin verb *cadere*. It has however made a certain detour, deriving ultimately from *cadentia*, "the ways in which the dice fall," which later became *chéance* in old French. Just like "hazard," which—as mentioned before—may derive from an Arabic word that also refers to the unpredictable way in which the dice fall, "chance" eventually came to designate whatever happens without us being able to determine it. Seen in

---

[2]The following etymological paragraphs were written on the basis of the *Oxford English Dictionary*, 2nd ed. (1989), s.v.; *Historisches Wörterbuch der Philosophie*, voice "Zufall"; *Duden. Deutsches Universalwörterbuch* (2007), s.v.; the *Online Etymological Dictionary*, s.v.; as well as various Latin, Greek, French, German and Italian dictionaries.

this light, Julius Caesar's famous pronouncement, "The die is cast" (*alea iacta est*), which announced his much thought-over decision to cross the Rubicon and start a civil war, would be an oddly inappropriate metaphor, given that Caesar's was everything but a random decision. But in fact, it appears that he spoke the phrase in Greek, citing a line from a comedy by Menander; the Greek phrase *anerrhiphtho kubos* should in fact be translated as *alea iacta esto*, "let the die be thrown," referring not to the decision taken, but instead to the uncertain outcome of the enterprise that was to follow from it (Lewis and Short 1879, s.v. *alea*).

The word "coincidence" derives from the Latin verb *cadere* in a more visible way. A "coincidence" takes place when things "fall" (*cadere*) "together" (*co[n]-*) and "upon" (*-in*) something. The word is not ancient Latin, but medieval, and it seems to have first been used in astrology, where *coincidentia* referred to the joint influence of multiple planets. This genealogy gives us an indication of a basic difference between "chance" and "coincidence": the latter requires more than one thing to happen at the same time. In the sentence, "By chance, I was born into a rich family," you could not replace the first word by "by coincidence." Meeting your neighbour in a far-away vacation location, by contrast, certainly qualifies as a coincidence; after all, you both had to travel there in order for your paths to cross.

These various shades of "falling" are instructive. It is certainly noteworthy how many terms there are in English and other languages that express surprise at a certain event or "occasion" (yet another such word) in terms of a "fall." It is as if the *casus*, "chance" or *Zufall* always fell down from above, literally "out of the blue." The proverbial "stroke of luck" would therefore have to be represented by the gesture of a fast downward arm movement.

More indirectly, the same is true for other words, such as "luck," which—though related to Germanic words for happiness and fortune—seems to have entered English as a gambling term. Like "accident," it might originally have referred to the way in which the dice fall, although this time with a uniquely positive connotation. The "falling" of the *casus* has here been confined to the descent of circumscribed objects on the gambling table. Still, the downward direction has remained intact.

1.2 Fortuna, Wheels and the Lottery

Still related to gambling, but involving quite a different type of movement, is the Wheel of Fortune. In late Antiquity and medieval times, this wheel was the constant attribute of the goddess Fortuna, who was spinning it (either blindfolded, or else maliciously watching) as men and women were literally "rising to fortune" or descending rapidly, "losing their fortune." Whether blind or seeing, Madam Fortune was a puppeteer, we mortals were her puppets. But then, as we have mentioned earlier, she was also regularly depicted in a passive role, herself the victim of unpredictable change. A particularly striking depiction of this latter figure was given by the Roman tragedian Pacuvius, who sketched the following portrait:

"Philosophers proffer the view that Fortune is insane and blind and stupid, /And they teach that she stands on a round, spherical rock: /They assert that, where chance (*fors*) pushes that rock, there Fortuna will fall."[3] Once one realizes that the word *fors*, "chance," stands at the root of the name of the goddess Fortuna, one begins to stare down the mirror cabinet of an infinite regress: we get a situation in which we humans rise and fall, tied to the Wheel of Fortune, while the goddess herself falls from the ball on which she stands, pushed in turn by "chance" (of which one had mistakenly expected her to be the ruler and embodiment).

In his demolition of the pantheon of pagan deities, Saint Augustine in *The City of God* directs his glance also at Fortuna (Book IV, Ch. 18). Why, he asks, is Fortuna traditionally associated with "felicity"—the Romans had initially endowed her with a cornucopia, and had thus viewed her as an exclusively positive figure—although we know that one can also have "bad fortune?" Such an identification doesn't make any sense, according to Augustine. Further, why should Fortuna be considered a goddess, if she can also bring about bad things? Plato tells us clearly that it is the essence of gods to be good; "how, then, is the goddess Fortuna sometimes good and sometimes bad? Is it perhaps that when she is bad, she is not a goddess, but is suddenly transformed into a malignant demon?" (Augustine 1998, 164). And finally, what should we make of the fact that the name of the goddess is also derived from the word *fortuito*, that is, "by accident?" How can she be a goddess if what we ascribe to her happened accidentally? In a few lines, Augustine exposes all the contradictions that reside in the concept of a deified principle of randomness, and all the inner tensions between a principle that should account at the same time for luck, happiness, destiny, the vicissitudes of life and personal success.

It is surprising to see that despite Saint Augustine's debunking, Fortuna was highly popular in the Middle Ages. In the meantime, however, her cornucopia had definitely disappeared for the wheel (Vogt 2011). Fortuna had changed from the positive figure ridiculed by Augustine into a highly ambivalent one. This may come as a surprise, as the idea of the random rise and fall of people (and peoples) is of course profoundly un-Christian, as it contradicts the notion of providence. And yet, it survived, and in fact thrived, in the hands of medieval Christianity. Dante Alighieri eulogizes Fortuna as nothing less than the first creature of God, who rules over the world and makes it spin about according to her occult whims, which are ominously invisible "like the serpent in the grass." With respect to God and to humans, she is "general servant and leader," respectively (*Divina Commedia*, "Inferno," VII.78–84).

It has been argued that the popularity of Fortuna in the Middle Ages is due to the late Roman author Boethius, in whose *Consolation of Philosophy* Fortuna makes a striking appearance, declaring:

---

[3]Pacuvius, ed. O. Ribbeck (1897), vol. 1, vv. 365–375: "Fortunam insanam esse et caecam et brutam perhibent philosophi,/ Saxoque instare in globoso praedicant volubili:/ Id quo saxum inpulerit fors,/ eo cadere Fortunam autumant."

> This is my art, this the game I never cease to play. I turn the wheel that spins. I delight to see the high come down and the low ascend. Mount up, if you wish, but only on condition that you will not think it a hardship to come down when the rules of my game [*ratio ludicri mei*] require it (Boethius 1897, II.2p, trans. modified).

Curiously, while Fortuna goes about her pagan business of causing the rise and fall of people, she seems (at least in this passage), to give us the choice between participating in the "ludicrous game" or abstaining from it. In fact, she quickly recalls to her listener the brutal fall of the Lydian king Croesus. The theme of the fall of kings—and here we are back with the previous etymology, of the *casus* and the "accident"—was popular throughout the Middle Ages. The *Carmina Burana* warns the powerful of the inevitable turning of the wheel: "too high up/ sits the king at the peak/ let him beware of ruin!"[4] In fact, a particularly popular image was that of four kings attached to a wheel, with one ascending (*regnabo*, "I will rule"), one on top (*regno*, "I rule"), one dethroned and descending (*regnavi*, "I have ruled"), and one at the bottom (*sum sine regno*, "I have no kingdom").

However, Boethius' Fortuna does not only seem to give us the choice between taking a ride on her wheel or leaving it, but Boethius himself, in Stoic fashion, recommends that we should seek our tranquillity irrespective of the vicissitudes afflicting our personal lives. Moreover, he suggests that there is a higher, maybe Platonic or else providentially Christian level at which it all makes sense. It has in fact been suggested that the ubiquitous medieval representations of Fortuna should be interpreted through the influence of Boethius (Vollmer 2009). The advantage of this explanation is that it helps us explain how it was possible that the pagan Wheel of Fortune could end up defining the shape and iconographical program of cathedral roses and church interiors (see Fig. 1).

An entirely demythologized, contemporary version of the Wheel of Fortune is the lottery wheel, which is inscribed by numbers corresponding to lottery tickets and a pointer pointing to the rim. The wheel is spun, and when it comes to a standstill, the ticket carrying the number corresponding to the number indicated by the pointer wins. With this device, we have arrived at our last set of terms. Originally, the "lot" was any object—a piece of straw, a chip of wood with a name on it, or, as in so many earlier examples, a die—that was used to determine someone's share, for example in an inheritance. A "lot" of land (and even the trivial "parking lot") still refer to that process of "random allotment" as does the phrase, "what falls to a person by lot." But when we recall the figure of Fortuna spinning her wheel, or deciding the outcome of the draw or the casting of dice, we will understand how "lot" and "lottery" could also come to refer to any "(ill-)fortune" that life has in store for us. The village lottery may assign a lot of land to us; the phrase "It was my lot to be born poor" refers instead to a lottery in which I was not able to buy even my own ticket.

---

[4]*Carmina burana* (1974), song 16: "nimis exaltatus / rex sedet in vertice - / caveat ruinam!".

Fig. 1 Fortuna (1372) depicted on the floor of the Cathedral of Siena. Is the ruler on his throne (*regno*) about to fall (*regnavi*), or is he rather, solidly enthroned, supervising the ascent and descent of the other figures? Strikingly enough, the four philosophers in the corners are all pagan: Euripides ("I have told you, son, to seek fortune through labours," from *Elektra*); Seneca ("A great fortune is a great slavery," from *De consolatione*); Aristotle ("Great fortune makes men more petulant," from the *Politics*); and Epictetus ("Glory not in the gifts of fortune, but in the goods of the souls," from the *Enchiridion*). Their advise has no providential, Christian, or eschatological overtones, but combines classical prudentialism with the topos of *virtus vincit fortunam* ("virtue/determination wins over fortune"): don't seek fortune, but if you do, seek it through hard labour; but be beware that it will negatively affect your character; and anyway, "it's what's inside that counts."

1.3   Randomness and Reckoning with Fortune

With the "randomness" of the lottery's decision-making process, we have arrived at the last word in our etymological survey. There was once an Old Frankish word, **rant*, cognate to the English "running," which eventually became *randir*, "to run fast," in Old French, as well as *randon*, meaning "rush" and "disorder." From the French, it migrated to English, where it became "at random," which originally meant, "at great speed" and hence "without order" and "haphazardly." By 1650, it had acquired one of its current meanings, by referring to events that took place "without definite aim or purpose." Originally, it was actor-bound: an individual was said to act "randomly," that is, without purpose, for example by pointing at someone while blindfolded. The use of the word as an adjective, as well as the identification of "chance events" with "random events," seem to be of more recent date. Of even more recent date are the mathematical theories of randomness, which are an extension of classical probability theory, or the quantum-mechanical randomizers.

These recent developments are interesting from a philosophical perspective. For once one equates randomness with chance, and once chance becomes calculable, as it did over the past three and a half centuries thanks to the mathematical determination of probability, one somehow also domesticates chance, randomness, and possibly even one's lot. Looking at a set of global statistics, one may now state: "The odds were high that I would be born poor." In Boethius' *Consolation* (II.3p), we have Fortune asking defiantly: "Do you wish to count out the score with Fortune?" (*Visne igitur cum fortuna calculum ponere*?). Through the mathematization of probability, we are attempting to do just that: "Reckon with fortune." As several chapters in this book document, this reckoning has taken on high forms of abstraction in various disciplines.

And yet, despite all domestication of chance, luck, fortune, coincidence and randomness in the specialized disciplines, the old meanings have not disappeared. Fortuna may no longer be a deity, but the surprise, the rage, the joy, and the bewilderment that something particular happened to us, of all people, that it had to happen just then and there, has not vanished. Nor have most of the terms and expressions that the Greeks, the Romans and our medieval ancestors used.

Did, then, our etymological exercise tell us anything useful? If we had hoped for a conceptual convergence between the words investigated here, then we were (predictably) deluded. Between the goddess who spins the wheel, the blind and hasty rush forward, life's lottery, the ubiquitous falling of dice, and all other unpredictable coincidences and accidents, there is little that amounts to any overarching notion of how we must "account" for the unforeseen events in life, nature or history. The divergent uses of the words we investigated, and even of single words, is however illuminating. To remind ourselves of the most dramatically ambivalent word, "fortune," we have seen that Fortuna could appear as a goddess of "good fortune," with her cornucopia at the ready; she could be a (still personified) semi-independent cosmic force governing over chance, coincidence and randomness; she could be a way of life that one could choose to follow or else ignore; but "fortune" could also be the well-deserved result of hard work, the danger being, however, that it might corrupt our character.

To be aware of the internal tensions between the various sub-meanings of the words seems to us an important step towards a comprehension of what these words can possibly be intended to achieve. But their full complexity only becomes apparent once one places them into the philosophical and scientific context in which their role in the causal nexus of things was examined. This is what needs to be done next.

2 History

2.1 Greek Origins

Let us therefore turn to an examination of a number of key moments in the intellectual—that is: philosophical and scientific—evolution that our words have
undergone, and the explanatory (or causal) role that was attributed or denied to
them. We must start with ancient Greece, because it is there that our current
terminology takes its origin. It is also there that we find, for the first time in Western
intellectual history, a debate about the status of unexpected events and the way we
must deal with them conceptually.

We have begun our essay with the element of surprise that characterizes the
various terms in question. In ancient Greece, the word that designated an unexpected
turn of events in a human life or in the observed natural world was *tuchê*. In
comedies, tragedies and in works of historiography, *tuchê* is invoked to designate
such unforeseen events, which may derive from the gods or from mere fortune [*tas
tôn theôn tuchaskai to chreôn* (Euripides, *Hercules Furens* 309–11)]. If from the
gods, *tuchê* is of course providential, which means that what to us may seem "by
chance," is instead "by necessity" or "will" at a higher, divine, level. The existence
of such a two-tiered logic explains why Sophocles can speak, in what at first looks
like an oxymoron, of "necessary chance" (*anankaia tuchê*, *Ajax* 485, 803), a combination of words that in other texts is rendered as "fate" (*moira*, *potmos*). But while
the older tragedians Aeschylus and Sophocles seem to have equated *tuchê* with fate,
their younger colleague Euripides was less inclined to attribute all unforeseen events
to a providential plan (Dudley 2012, 137). In *Hecuba* 488–491, a certain Talthybius
wonders, for example, whether it is the gods or rather chance (*tuchê*) that rule over
human affairs, thereby clearly separating the two (Lawrence 2013).

2.2 Aristotle

Distinctions and reflections that in literary works were merely adumbrated were
made most fully explicit in that potent thinker whom Dante called "the master of
those who know," namely Aristotle, whose philosophical and scientific teachings
were to define Western university education until the end of the seventeenth century. In various of his works, we find Aristotle reflecting on the possible role that
chance might play in the natural world and in human affairs. Always an acute
analyst of terminology, he carefully examined various types of chance, distinguishing between *tuchê*, on the one hand, and such related concepts as *to
automaton* (a type of spontaneity), and *eutuchia* (which might be translated as
"good fortune").

Aristotle's most extensive treatment of chance is found in book 2 of his *Physics*.
As is often the case, Aristotle starts his analysis with an historical excursus.

Previous philosophers have failed to give an account of chance, he tells us, which is all the more surprising as some of them have attributed to chance a fundamental role in their physical systems (*Physics* 195b30–196b9). Aristotle here thinks of Empedocles' cosmogony, which relies on air that moves upwards by chance and speaks of the haphazard origin of limbs of animals; but he thinks even more clearly of Democritus, who maintains that "the cosmic order came by chance [...], whereas neither animals nor plants are, or come to be, by chance, but are all caused by Nature or Mind or what else." Aristotle laughs this idea out of court, arguing thus:

> But if this really were so, that very fact ought to give us pause and convince us that the matter needs investigation. For, in addition to the inherently paradoxical nature of such an assertion, we may note that it is exactly in the movements of the heavenly bodies that we never observe what we call casual or accidental variations, whereas in all that these people tell us is exempt from chance such things are common. Of course it ought to be just the other way (Aristotle 1957, 196a25–196b5).

Famously, Aristotle inverts the order: for him, "regular and customary successions," such as those observed in the heavenly motions, must happen by necessity (*ex anankês*), whereas the terrestrial realm is defined by a great degree of randomness. Regular necessity is observed throughout the superlunary sphere, where the sun, the planets and the stars are located and which is defined by one single element, ether, and by constant, circular movements. By contrast, in the sublunary sphere, where the four elements constantly mix and unmix, objects continuously come about and perish again. Here, where we find irregularity and surprising events, we may truly speak of products of chance (*hê tuchê kai to automaton*). Let us here remember that this stark Aristotelian opposition between two cosmological domains, each with its distinct ontological status and its own set of physical laws, was to break down only in the aftermath of Copernicus and Kepler in the course of the seventeenth century.

Aristotle admits that, in our sublunary domain of permanent change, "what we call luck or chance corresponds to some reality" (Aristotle 1957, 196b15–17). At the same time, he rejects the suggestion that *tuchê* should be viewed as a specific type of causality. Instead, chance events should be regarded as accidental, that is to say, concomitant effects of a definite cause: "*Tuchê*," Aristotle writes in his *Physics*, "is a cause only accidentally (*kata symbebêkos*)" (*ibid.*, 197a14f). But what does it mean to be an accidental cause? In his *Metaphysics*, Aristotle defines "accident" as that which happens "neither necessarily, nor usually," adding that there is "no definite cause for an accident, but only a chance, i.e., indefinite cause (*aoriston*)" (Aristotle 1933, 1025a15). If a man goes to the market and "accidentally" meets his debtor, "the reason of his meeting him was that the wanted to go marketing; and so too in all other cases when we allege chance as the cause, there is always some other cause to be found" (Aristotle 1957, 196a1–8). Here, then, we have the typical surprise moment mentioned in our introduction. The man may have wanted to buy cheese and vegetables, but, "as it happened," he encountered his debtor. That the verb *sumbainô*, of which *symbebêkos* ("accident") is the past participle, literally means "to walk together," is most suitable for this specific Aristotelian example, as it provides a quite visual model for what we have earlier defined as a "coincidence": two men walking, each steered by his own intentions, to the market, but "accidentally" ending up in each other's company.

In order to make sense of Aristotle's distinctions, one has to remember that his entire universe, and the causality that is active in it, is everywhere purposeful and goal-driven, so that the explanations he offers tend to be teleological. In such a universe, *tuchê* is an "accident" in the sense that it designates those events that eschew all purposes. In the natural world, a typical class of "accidents" is constituted by monstrous births, which also include female babies and which may be regarded as "failures of purpose in Nature" (Aristotle 1957, 199b4), in the sense that accidental factors hindered the natural development of the seed.[5]

Being "characteristic of the perishable things of the earth" (Aristotle 1937, 641b15), chance manifests itself above all in the domains of biology and of human action. Sometimes, Aristotle in fact wishes to limit the scope of *tuchê* even further, restricting it to rational behaviour. "Neither inanimate things nor brute beasts nor infants can ever accomplish anything by *tuchê*, since they exercise no deliberate choice." By contrast, the larger category, *automaton*, describes cases in which "*any* causal agency incidentally produces a significant result outside its aim" (Aristotle 1957, 197b19–23). Spontaneous generation, in which the presence of warmth can bring about worms or insects in a heap of dung or a warm puddle, is a case in which non-rational agents bring about a meaningful product by a sheer concurrence of circumstances.

If taken in this restrictive meaning, *tuchê* becomes the object of ethical reflection. In his *Eudemian Ethics*, when discussing the cause and the ethical bearing of good luck (*eutuchia*), Aristotle formulates an interesting paradox: we tend to call those persons "fortunate" (*eutuchês*) who "without the aid of reason are usually successful" (Aristotle 1935, 1247b27–28). This is however in contradiction with the accepted definition of chance or fortune (*tuchê*), which implies that something happens neither always nor even regularly (*ibid.*, 1247a31–35). In order to resolve this paradox, Aristotle distinguishes between two types of fortune. The first is due to the aid of a god, whereas the second type of fortune is that of persons who are successful because they instinctively choose for the right course of action. Both sorts of good fortune are "irrational," in the sense that they are not obtained through our conscious choice, but the first is continuous, whereas the second is incidental (*ibid.*, 1248b5–10).

What Aristotle's sundry ethical, physical and biological reflections on chance have in common is an emphasis on the inherent lack of reflection, premeditation or, in short, rationality. Good luck (*eutuchia*), chance (*tuchê*) and spontaneity (*automaton*) are all *paralogos*, unaccountable by reason, either because there is no purpose

---

[5]Both in *Physics* and in the *Generation of Animals*, monsters are regarded as "chance substances"; see Dudley 2012, 171, 175.

involved (as in the case of worms being spontaneously generated in a heap of dung), or because the result of an action was not intended (as the man meeting his debtor on the market square) (*Physics*, 197a10, 18–20 and *Eudemian Ethics*, 127a33–38). It is precisely their undirected, irregular and contingent nature that also renders chance events "unscientific." For Aristotle, "science" (*episteme*) designates a psychological state in which the mind possesses knowledge with regard to the causes of an event. In the case of accidental events, the cause is however "unrecognizable," "indefinite" and "irrational" (*paralogos*) (*Physics*, 197a8–35).

2.3   The Ancient Atomists

So much for Aristotle himself. Let us however return to the atomists he criticized for what he took to be a misguided cosmogony and a misleading causal theory. We recall from above that Aristotle ridiculed Democritus specifically for suggesting that the cosmic order was the product of chance. Interestingly, the doxographer Diogenes Laertius provides a different version of Democritus' convictions, ascribing to him the view that "everything happens according to necessity; for the cause of the coming-into-being of all things is the whirl [that is, the atomic vortex which gave origin to the world], which he calls necessity" (Laertius 1925, IX, 45). Similarly, the only extant fragment of Leucippus, who may have been the inventor of the concept of atom, reads: "Nothing exists at random (*matên*), but everything for a reason (*logos*) and by necessity (*anankê*)" (Kirk et al. 1983, 420).

Why should Aristotle then have attributed to Democritus the view that the world came about by chance (*apo tautomatou*)? According to Edmunds' influential interpretation, he did so to stress the purposeless character of the atomistic cosmos (Edmunds 1972).[6] We recall from above that according to Aristotle's own definition, *automaton* "means an occurrence that is in itself to no purpose" (*Physics* 197b25–30). In other words, what to Leucippus and Democritus was "necessary" and hence the contrary of "chance" would for Aristotle have been its very opposite, namely a blind and therefore unguided and random event. Put differently, what was a deterministic "necessity" to one philosopher was mere "chance" to the other. This is a typical example for the phenomenon that will be discussed in our conclusion: the terms with which we are engaging in this chapter can only be understood if one knows the alternative terms they wish to rule out.

Indeed, as A. A. Long has perceptively pointed out, chance (*tuchê*) is incompatible with necessity (*anankê*) only if the former is taken to indicate events that are the result of sheer contingency and indeterminacy (Long 1977, 67–68). This observation takes us to Epicurus, the first philosopher to have explicitly introduced

---

[6]A similar point is made by Cherniss (1935), 248–49, and Long (1977), 67.

an element of contingency and indeterminacy into the universe. Epicurus in fact criticized previous natural philosophers, including the atomists he followed in his physics, for attributing the origin of the cosmos to necessity and for making man the slave of destiny (Epicurus 1931, *Letter to Pythocles,* 89–90; *Letter to Menoeceus,* 131, 133, 134). He himself hoped to avoid absolute determinism by postulating a *parenklisis,* a spontaneous swerve that atoms suddenly perform, deviating from their rectilinear parallel paths and intermingling as a consequence of these deviations. In his own rendition of Epicurus' theory, Lucretius explained how this swerve, which he called *clinamen,* was responsible for breaking "the bonds of fate and preventing one cause from following from another from infinity" (Lucretius 1924, 2.251).

According to Cicero, the main function of Epicurus' *clinamen* was that of introducing freedom into a universe that would otherwise be fully defined by necessity:

> The reason why Epicurus brought in this theory was his fear lest, if the atom were always carried along by the natural and necessary force of gravity, we should have no freedom whatever, since the movement of the mind was controlled by the movement of the atom. The author of the atomic theory, Democritus, preferred to accept the view that all events are caused by necessity, rather than to deprive the atoms of their natural motions (Cicero 1941, *On Fate,* 23).

While Cicero pitted necessity against freedom, Epicurus himself distinguished between three concepts, namely "necessity," "chance" and "freedom":

> With us lies the chief power in determining events, some of which happen by necessity and some by chance, and some are within our control; for while necessity cannot be called to account, (…) chance is inconstant, but that which is in our control is subject to no master, and to it are naturally attached praise and blame (Epicurus 1926, *Letter to Menoeceus* 133).

In other words, from the ethical point of view, we cannot be blamed for actions that are due to necessity or chance, as both types defy our control. Only those actions that we control are free. But are we in control of the swerves of the atoms in us? How convincing is Lucretius' statement—which incidentally corroborates Cicero's analysis of the *raison d'être* of the swerve—that "what keeps the mind itself from having necessity within it in all actions (…) is the minute swerving of the first beginnings at no fixed place and at no fixed time?" (Lucretius 1947, 2: 288–293). In fact, the debate on whether or not the swerve, which might look like the epitome of randomness, was really meant to offer a plausible account of free will, continues to this day. Scholars presuming Epicurean free will to have been synonymous with "conscious chance" (Bailey in Lucretius 1947, 3: 1287) are opposed by others who think that Epicurean freedom "fits random actions, rather than deliberate and purposive ones" (Furley 1967, 232–233). According to Furley's interpretation, the point of the swerve is merely to allow for a discontinuity in an otherwise deterministic succession of causes, and thereby to assure that the source of a human action can be traced in the agent himself and not in external factors. It does not, however, account for anything like a conscious free action.[7]

If we return our glance to Cicero's analysis of Epicurus, we will find that he opposes Epicurus' worldview not only to that of the older atomist Democritus, but also to that of the Stoic philosopher Chrysippus, according to whom "all things happen by fate and spring from eternal causes governing future events" (Cicero 1941, *On Fate,* 21). Indeed, it would seem that the Greek Stoics held that there exists a rational organizing principle that is found in all things in the world and which determines the course of all events. It is obvious that such a view "leaves no room for alternative developments of the world. There is exactly one course of events (and states) that is in accordance with the rational universal nature" (Bobzien 1998, 31).

While this type of strict determinism might have been compatible with Democritean atomism, it clearly wasn't with Epicurus'. From Plutarch's *On Stoic Self-Contradictions*, we know that Chrysippus derided the argument according to which the soul "takes a swerve of itself and resolves the perplexity." He replied "that the uncaused is altogether non-existent," and warned that "obscure causes insinuate themselves" whenever events appear to happen by chance. In the context of the etymological link between the words "chance" and "hazard" and the throwing of dice, to which we have drawn attention in our previous section, it is interesting to find Chrysippus insisting that dice and scales "cannot fall or incline now one way and now another without the occurrence of some cause" (Plutarch 1976, *On Stoic Self-Contradictions* 1045). The fact that we do not know how the dice will fall does not mean that there is no cause behind their specific fall.

2.4   On Divination and Providence

Let us conclude our section on Antiquity by listening once more to Cicero, and more specifically to his attack on divination, the power to foretell the future. His critique contains important reflections on chance, necessity and the knowledge of the course of nature. As the Latin word *divinatio* clearly indicates, the seer's knowledge of the future is "divinely inspired." This meaning implies that you can only know the future, first, if a god has predetermined it, and secondly, if this god

---

[7]Furley's interpretation was challenged by Fowler (1983) and Purinton (1999), who attribute to Epicurus and Lucretius the view that random swerves are indeed the cause of all voluntary actions. Bobzien (2000) agrees with Furley that the swerve is not responsible for every voluntary action, while O'Keefe (2005) 17, goes as far as to deny that the swerve plays any role in the production of action. While the above-mentioned interpretations are concerned with upward causation (from the atomic to the macroscopic level), David Sedley believes that Epicurus' denial of Democritus' determinism "involves an express assertion of downward causation": volitions are not influenced by, but instead influence atoms' motion (Sedley 1988, 318).

has also revealed his or her plans to the seer. At some point in his *De divinatione*, Cicero criticizes specifically the view that "divination is the foreknowledge and foretelling of events considered as happening by chance [*res fortuitae*]," that is to say, of things which, "for though they happen frequently they do not happen always" (Cicero 1923, *On Divination* 2.5.13–14). Cicero retorts that physicians, pilots or military men continuously make predictions concerning future events, which are however based on science, experience, skill and wisdom. But in the absence of such professional knowledge, can there be

> any foreknowledge of things for whose happening no reason exists? For we do not apply the words "chance," "luck," "accident," or "casualty" except to an event which has so occurred or happened that it either might not have occurred at all, or might have occurred in any other way. How, then, is it possible to foresee and to predict an event that happens at random, as the result of blind accident, or of unstable chance? (*Ibid.*, 2.5.15).

Indeed—Cicero concludes—the very idea of foretelling what is random is self-contradictory! For this reason,

> it is not in the power even of God himself to know what event is going to happen accidentally and by chance [*casu et fortuito*]. For if He knows, then the event is certain to happen; but if it is certain to happen, chance [*fortuna*] does not exist. And yet chance does exist, therefore there is no foreknowledge of things that happen by chance (*ibid.*, II.7.18).

Cicero's reflections on divine foreknowledge provide us with a perfect bridge to the Christian Middle Ages, in which the divine predicates of omniscience and omnipotence forced the discussion about the status of chance, coincidence, fortune and luck in new directions, although the logical possibilities had already been defined by Greek and Latin philosophers. Irrespective of whether ancient philosophy was the cradle of Christianity or rather an obstacle to be overcome, we cannot understand medieval discussions without Greek philosophy.

It is evident that in a cosmos created and ruled over by an eternal, omniscient, and omnipotent God, mere chance can have no place. Whatever happens must have been known to God even before it happened; whether that implies that God also willed it, is a different and theologically difficult question. Is all "pro-vidence," in the sense of "fore-seeing," also "providence" in the sense of "benevolent guidance?" God must have foreseen the Fall of Adam and Eve; but it presumably was not an intended part of his plan.

However one may wish to settle this tricky issue, for Saint Augustine, the most influential of the Latin Church Fathers, it was obvious that there existed a personal type of divine providence, which implied that whatever happened, was—at least for God, the source of all providence—a rational event. This meant that no event was ultimately fortuitous and without reason: "those things that seem fortuitous come about by hidden forces" (Augustine 1841, *Quaestiones in Heptateuchum*, I. 91), just as generally, "the world is not governed by blind fate, but by the providence of a highest God, just as the Platonists also maintain" (Augustine 1998, 9.13.2). For most Christian authors, these two ideas were indeed linked, and necessarily so because of the divine predicates. On the one hand, there was God's omniscience, which left no room for mere chance in the sense of unpredictability—we have seen that Cicero had already pointed to this logical incompatibility even before the advent of Christianity. On the other hand, there was God's omnipotence, which implied that whatever happened, had to happen, and since God was benevolent, whatever happened, also had a positively providential aspect to it—an idea that Saint Augustine attributes to the Platonists.

2.5   Boethius

But if God is omnipotent and benevolent, and if everything is providential, how should we then explain the presence of evil in this world? This so-called problem of theodicy was addressed by another early Christian author, Boethius, whom we have encountered earlier in our essay, and who tried to correlate the three causal terms of "necessity," "free will" and "chance." Against the Stoics, Boethius insisted on the existence of a free will and argued against determinism; and against the Epicureans, he defended a plurality of causes, and rejected atomic monocausalism (Boethius 1891). While developing his solution to this problem, he drew a distinction between "divine providence" and "fate." These two terms, he explained, referred to the same thing, but did so from a different perspective:

> The mind of God has set up a plan for the multitude of events. When this plan is thought of as in the purity of God's understanding, it is called Providence, and when it is thought of with reference to all things, whose motions and order it controls, it is called by the name the ancients gave it, Fate. [...] Providence includes all things at the same time, however diverse or infinite, while Fate controls the motion of different individual things in different places and in different times. So this unfolding of the plan in time when brought together as a unified whole in the foresight of God's mind is Providence; and the same unified whole when dissolved and unfolded in the course of time is Fate.... (Boethius 2000, IV.6p).

Boethius applies a similar perspectival approach to the existence of chance. In a world governed by providence, there was of course no space for chance; still, there was a sense in which something could be said to "happen by chance":

> Whenever anything is done for one reason, but something other than what was intended happens on account of other reasons, it is called chance [casus]. [...] Therefore, we can define chance as an unexpected event brought about by a concurrence of causes which had other purposes in view. These causes come together because of that order which proceeds from inevitable connection of things, the order which flows from the source which is Providence and which disposes all things, each in its proper time and place (Boethius 2000, V.1p).

It is no coincidence that in the same chapter from which these quotes are drawn, Boethius refers to Aristotle's *Physics*. Indeed, Aristotle's *tuchê* and Boethius' *casus* have in common that they are the non-intended by-products of intended actions. We have earlier encountered Aristotle's example of the man who went to the market and there *happened* to encounter his debtor. Boethius' main example is also taken from Aristotle (*Metaphysics* V, 30), and is that of a man who goes to his field to plant a tree and *happens* to find a treasure. But what to us seems mere chance, is in reality only a concurrence (*concursus*) of causally accountable circumstances. After all, someone must have buried the gold in the field in the first place. For Boethius, the concept of *casus* is thus not only incompatible with providence, but also with the causal structure of the world. A real *casus* would not only be inexplicable, but would be uncaused, or, in Boethius' terms, *ex nihilo*. This identification of *casus* with *ex nihilo* events is not taken from Aristotle, but might be indicative of Boethius' debt to Stoic determinism.

2.6 Late Medieval Views on Chance

In fact, even when they read and used Aristotle on the issue of chance, Christian authors were generally driven by different concerns than their admired Greek preceptor. Several centuries after Boethius, for example, Peter Abelard defined "chance as an unexpected event" (*inopinatus eventus*). He insisted that what is to blame is not the event itself, but only our own lack of understanding: "the word 'chance' … denotes always ignorance" (Peter Abelard 1919, 426). The common medieval view that "chance" is always "in us, not in the things," agrees with Aristotle's view that chance does indeed denote ignorance, in the sense that it defies our scientific grasp of the underlying causal pattern, but it deviates from Aristotle in correlating this ignorance with the rare, irregular and indeed "casual" nature of a given event.

Apart from the obvious impact of a monotheistic conception of an all-powerful God running the universe on discussions regarding chance, fortune and accident, when one examines the later Middle Ages, one cannot but be impressed by the acuity with which these and related words were examined. Ever since the seventeenth century, the so-called scholastics have been derided because of the bookish nature of their knowledge claims and their delight in hair-splitting controversies. But precisely their trust in the authority of the authors of the books they commented on and their attention to even the most abstruse interpretative possibilities implied that they were good readers and careful observers of language. Having to examine and reconcile ideas from various traditions—Greek and Latin philosophy, Jewish and Christian theology—they were aware of the abundance of different terms that were used to express similar ideas. They noticed, for example, that *casus*, *contingentia* and *fortuna* described similar and often even identical events, although the meaning of some of the words was more general than that of others (e.g., John Buridan 1509, 36rb). They also noticed that *casus*, "chance," was applied to both causes and effects—an observation to which we have already drawn attention in our own introduction (e.g., Roger Bacon 1935, 116). Many of their considerations regarding contingency (notably in their analysis of the status of future contingents), non-essential predicates (*accidentia*), the concomitance of various "coinciding" causes, or the nature of "fortuitous events" (*eventus fortuiti*) were indeed ground-breaking.

Let us end our medieval section with Thomas Aquinas, who in his famous *Summa theologiae* examined the relation between chance, fate and divine providence. Invoking positions that we have encountered earlier in our chapter, Thomas refers to Aristotle's conception of chance, quotes Augustine's view that there is neither chance nor luck in the world as all events are foreseen, and cites Boethius, for whom the word "fate" referred to an inherent disposition of things by which divine Providence brings about the desired effects. Christianizing the Aristotelian distinction between the regularity encountered in the supralunary world and the disorder found in the sublunary world, Thomas explains that "what happens on earth accidentally, either in nature or in human affairs, is derived from a pre-ordaining cause, namely Divine Providence" (Thomas Aquinas 1964–1976, vol. 15, *Summa Theologiae, Part 1*, art. 116, qu. 1). Only when explained in terms of their proximate causes do things happen by luck or chance, but not when explained in terms of divine providence, whereby "nothing happens randomly in the world" (*ibid.*).

In the second book of the *Summa contra gentiles,* which deals with the creation of the world, Thomas devotes a chapter to the question of whether "the distinction of things," that is to say their separation into genera and species, is the result of chance. His answer, which relies on Aristotle's so-called hylemorphist doctrine, according to which all substances are constituted by matter and form, is that all individuals belonging to a species share the same form, and that it is matter that is responsible for individual differences. Given that "chance is found only in things that are possibly otherwise," Thomas argues that "the distinction of things in terms of species cannot be the result of chance," as the forms (which define the species) are by definition unchangeable. By contrast, differences between individuals belonging to the same species "can perhaps be the result of chance," because matter is "a reservoir of multiple possibilities" (Thomas Aquinas 1975a, II. 39). As Norman Kretzmann has explained, for Aquinas, the existence of, say, a particular pigeon is a chance state of affairs, not because it is uncaused, but because it is "the result of an unplanned convergence of two or more previously independent series of causes" (Kretzmann 1999, 208). Thomas seems to suggest that "the generating of individual members of species of plants or of non-human animals" may take place "apart from" (*praeter*), although "of course not contrary to, the intention of the creator/distinguisher" (*ibid.*, 209). Humans, "metaphysical hybrids" composed of a body and a soul, are the only individual beings whose coming-into-existence and life-course cannot take place without an divine intentional act (*ibid.*, 209).

In the third book of the *Summa contra gentiles,* Thomas invokes Aristotle's example of the casual encounter between a man and his debtor to show that providence does not exclude chance: "It would be contrary to the essential character of divine providence if all things occurred by necessity (…). Therefore, it would also be contrary to the character of divine providence if nothing were to be fortuitous and a matter of chance in things" (Thomas Aquinas 1975b, III. 74).

2.7   Chance, Necessity and Design in a Mechanistic Universe

From what little has been said, it must be clear that Thomas Aquinas involves God where he must, but for the rest tries to leave space for contingency. According to Anneliese Maier, this wiggling space was to disappear within a century after Thomas' death. Maier is convinced that a noteworthy development took place in the fourteenth century, which was going to shape the entire period up to the twentieth. Most scholastics had previously insisted, just like Thomas, that each and every natural event required a cause, but that not everything took place *ex necessitate*. In the fourteenth century, however, a more restrictive view came to prevail according to which contingency—understood as the contrary of necessity, as something that could be thus but also otherwise—could only be encountered in the realm of voluntary acts (Maier 1949, 241). Maier boldly suggests that from the fourteenth century to the advent of quantum mechanics in the twentieth, there existed an underlying consensus that excluded contingency from the natural world:

> [...] for the [divine] first cause, there exists no *Zufall* [chance/coincidence/randomness]. But this means: taken by itself, there exists no *Zufall* at all in the world, but only in a relative sense, *in respectu*, that is, only with respect to specific and particular causes and only for those who are not capable of surveying the *concursus causarum* [concourse of causes] (Maier 1949, 231, our translation).

As we will see below, Maier's bold thesis is probably mistaken for the nineteenth century, but it is quite convincing for the period that we tend to describe as the Scientific Revolution, and notably for the seventeenth century. That century witnessed the emergence of the idea of a physical world governed by laws of nature, which were universally valid and admitted no exception. In the mechanistic universe that became so fashionable in the second half of the century, nothing could happen at random, so that the word "chance" could at best designate events that provoked a subjective feeling of surprise while being inherently necessary.

It might at first sight appear paradoxical that the probability calculus originated precisely in that deterministically minded seventeenth century, in the hands of mathematicians like Blaise Pascal, Pierre de Fermat and Christiaan Huygens. Until the Renaissance, the adjective "probable" had been used to designate an opinion which was based not on a demonstration, but on a reliable authority (Byrne 1968). Only in the second half of the seventeenth century did a mathematical notion of probability emerge (Hacking 1975, 11). According to Ian Hacking, who in an unsurpassed historical analysis has reconstructed the history of probabilistic thinking, early-modern determinism, far from precluding any thought about randomness, in fact paved the way for the mathematical study of chance and probability (*ibid.*, 3). A similar point has been made by Lorain Daston, according to whom "determinism, far from stifling mathematical probability theory, actually promoted it" (Daston 1988, 37). To be sure, in his little *Liber de ludo aleae* ("Book on the game of dice") of 1520, Gerolamo Cardano had already tried to calculate the probability of various dice throws, but had still attributed the discrepancy between calculated and actual outcome to the intervention of *fortuna* (*ibid.*, 36). But once chance and *fortuna* had both been banned from the deterministic world of seventeenth-century natural philosophy, a new way of calculating probabilities had to emerge. As Hacking has pointed out, the early modern notion of probability is, however, "Janus-faced: on the one side it is statistical, concerning itself with stochastic laws of chance processes; on the other side it is epistemological, dedicated to assessing reasonable degrees of belief in propositions quite devoid of statistical background" (Hacking 1975, 12).

No one captures the substitution of the Lady of Chance, Fortuna, by a conception of chance as mathematical and epistemic probability better than the Scottish philosopher David Hume. In the chapter "Of Probability" of his *An Enquiry Concerning Human Understanding,* he introduced a crucial distinction between "Chance," written with a capital letter, and mere "chances":

> Though there be no such thing as Chance in the world (....) there is certainly a probability, which arises from a superiority of chances on any side; and according as this superiority increases, and surpasses the opposite chances, the probability receives a proportionable increase, and begets still a higher degree of belief or assent to that side, in which we discover the superiority. If a dye were marked with one figure or number of spots on four sides, and with another figure or number of spots on the two remaining sides, it would be more probable, that the former would turn up than the latter; though, if it had a thousand sides marked in the same manner, and only one side different, the probability would be much higher, and our belief or expectation of the event more steady and secure. This process of the thought or reasoning may seem trivial and obvious; but to those who consider it more narrowly, it may, perhaps, afford matter for curious speculation. (Hume 1748, Ch. 6).

Similarly, Hume's French contemporary, Voltaire, was convinced that "chance is nothing, and that we have invented this word to describe the known effect of un unknown cause." Voltaire expressed this view in *Le philosophe ignorant* (Voltaire 1766, Ch. 13) as well as in the entry "On atoms" of the *Philosophical Dictionary,* in which he explained that seventeenth-century mechanical philosophers

> distinguished what is good in Epicurus and Lucretius, from their chimeras, founded on imagination and ignorance (...). All have acknowledged that chance is a word without meaning. What we call chance can be no other than the unknown cause of a known effect. Whence comes it then, that philosophers are still accused of thinking that the stupendous and indescribable arrangement of the universe is a production of the fortuitous concurrence of atoms—an effect of chance? Neither Spinoza nor any one else has advanced this absurdity (Voltaire 1901, s.v.).

Although very critical of the Church and of revealed religion, Voltaire was convinced, as a Deist, that the existence of God could be inferred from the order of the natural world. In the entry on "God/Gods" of the *Philosophical Dictionary* he in fact claimed:

> Every work which shows us means and an end, announces a workman; then this universe, composed of springs, of means, each of which has its end, discovers a most mighty, a most intelligent workman. Here is a probability approaching the greatest certainty. (...) I am aware that various philosophers, and especially Lucretius, have denied final causes (...). To affirm that the eye is not made to see, nor the ear to hear, nor the stomach to digest—is not this the most enormous absurdity, the most revolting folly, that ever entered the human mind? (Voltaire 1901, s.v.).

Voltaire could not ignore the fact that Spinoza had launched a powerful attack against the doctrine of final causes. In the famous Appendix to the first book of his *Ethics,* Spinoza had argued that the idea that "God directs all things to a definite goal" was a widespread misconception, which hindered "the understanding of the concatenation of things." According to Spinoza, God is the only substance that exists and "acts solely by the necessity of his own nature." All things "are in God" and are "predetermined by God, not through his free will or absolute fiat, but from the very nature of God or infinite power." Being "ignorant both of things and their own nature," people wrongly "believe that there is an order in things" and that "God has created all things in order." This misconception, Spinoza maintained, is the product of a double fallacy. People mistakenly "think themselves free, inasmuch as they are conscious of their own volitions and desires" and from this they wrongly conclude that "all things in nature act as men themselves act, namely, with an end in view" (Spinoza 1883, 75–81; *Ethics*, Appendix to Part I).

In his *Philosophical Dictionary*, Voltaire tried to convince his readers that, contrary to Lucretius and other ancient philosophers, Spinoza could not "help admitting an intelligence acting in matter, and forming a whole with it." In Voltaire's eyes, Spinoza "did not understand himself":

> If this infinite, universal being thinks, must he not have design? If he has design, must he not have a will? Spinoza says, we are modes of that absolute, necessary, infinite being. I say to Spinoza, we will, and have design, we who are but modes; therefore, this infinite, necessary, absolute being cannot be deprived of them; therefore, he has will, design, power (Voltaire 1901, s.v. God/Gods).

That Spinoza would have rejected Voltaire's interpretation without further ado is clear from some letters he wrote to Hugo Boxel, a Dutch contemporary who tried to persuade him of the existence of ghosts. In a letter dated 21 September 1674, Boxel had claimed that "it appertains to the beauty and perfection of the universe" that "there are spirits of all sorts, but, perhaps, none of the female sex," adding that his reasoning would not convince those "who rashly believe that the world has been created by chance." In his answer, Spinoza could not avoid addressing the question, "whether the world was made by chance." He insisted that "chance and necessity are two contraries," so that

> he, who asserts the world to be a necessary effect of the divine nature, must utterly deny that the world has been made by chance; whereas, he who affirms, that God need not have made the world, confirms, though in different language, the doctrine that it has been made by chance; […]. I, myself, lest I should confound the divine nature with the human, do not assign to God human attributes, such as will, understanding, attention, hearing, &c. I therefore say, as I have said already, that *the world is a necessary effect of the divine nature, and that it has not been made by chance*" (Spinoza 1883, 381).

In his reply, which is unfortunately lost to us, Boxel must have objected that the opposite of "necessity" is not "chance," but "freedom." Spinoza reacted with astonishment:

I am […] at a loss for the reasons, with which you want to make me believe, that chance and necessity are not contraries. […] As soon as I affirm that heat is a necessary effect of fire, I deny that it is a chance effect. To say, that necessary and free are two contrary terms, seems to me no less absurd and repugnant to reason. For no one can deny, that God freely knows Himself and all else, yet all with one voice grant that God knows Himself necessarily (Spinoza 1883, 385).

Returning to Voltaire, we may now declare that although he did not do justice to Spinoza's philosophical views, he was yet quite right in stating that no early modern mechanical philosopher had regarded chance as an explanatory cause of physical phenomena. Other eminent examples confirm this opinion clearly. Pierre Gassendi, one of the founding fathers of early-modern atomism, explicitly claimed that "chance is nothing in itself (…), but the lack of foreknowledge and of the intention of an event" (Gassendi 1658, 2: 829a). He borrowed from ancient atomism the idea that all physical phenomena could be explained in terms of the motion of minute particles of matters, but criticized Epicurus for turning chance (*fortuna*) into a cause. In Gassendi's eyes, Epicurus' recourse to the swerve to explain the formation of the world and to account for human freedom was no convincing alternative to the determinism of Democritus and of the Stoics, because whatever happens "by a variety of motions, collisions, rebounds, swerves" still happens by necessity (*ibid.*, 2: 838). Margaret Osler has rightly pointed out that

in order to embrace the evident facts of both causal order and contingency within the bounds of his mechanical philosophy, Gassendi undertook a Christian reinterpretation of the concepts of fate, fortune, and chance […]. Fate is nothing more than God's decree, and fortune and chance are expressions of contingency in the world coupled with human ignorance of the causes of fortuitous events (Osler 1994, 92, with references to Sarasohn 1985).

Gassendi, who is usually dismissive of Aristotelian philosophy, here invoked, just as Boethius had done centuries earlier, Aristotle's example of the man who digs the ground to plant a tree and accidentally finds a treasure in order to explain that chance is the "concourse" of two independent causal chains (Gassendi 1658, 2: 828b). And, like Boethius, Gassendi also claimed that chance events are "part of divine providence," which "includes things which are foreseen as well as things which are unforeseen to humans" (*ibid.*, 2: 840b).

A position analogous to Gassendi's—accepting atomism, but rejecting Epicurus' random swerves as explanatory tool in physical and psychological matters—was endorsed by the chemist Robert Boyle, one of the central figures of the early Royal Society and the person to render "the mechanical philosophy" programmatic. In his *About the Excellency and Grounds of the Mechanical Hypothesis* (1674), Boyle wrote:

When I speak of the *Corpuscular* or *Mechanical* Philosophy, I am far from meaning with the Epicureans that Atoms, meeting together by chance in an infinite Vacuum, are able of themselves to produce the World, and all its Phaenomena (Boyle 2000a, 103; cf. Fig. 2).

In his treatise, Boyle took issue not only with Epicurus, but also with "some modern philosophers" who suggested that all God had to do in order "to make the

Fig. 2 The seventeenth century's difficulty of depicting "chance" and "randomness." It is literally no coincidence that of the 79 editions of Lucretius' *De rerum natura* printed between 1473 and 1725, only one contains a depiction of atoms, namely the third edition of Thomas Creech's English translation of 1683. But what an ambivalent image it is! We see Lucretius gesturing at dots descending—without any swerve!—from a celestial globe carrying the name "chance" (CASUS). How "chance" can generate these dot-atoms is left unexplained, and it also remains entirely unclear how a single type of dot-atoms can bring about the variety of life forms seen emerging out of mud at the bottom of this frontispiece. In case the dots descending in a diagonal shaft of light are intended as a reference to dust motes dancing in sunbeams, then the image is even more misleading, because Lucretius denies explicitly that the motes are atoms; they are merely similar to them (*rei simulacrum et imago*). (See Lüthy 2003, 122)

world" was to impart motion to "the whole mass of matter (...), the material parts being able by their own unguided motions to cast themselves into such a system." The type of mechanical philosophy that Boyle was defending was quite different, for it

> reaches but to things purely corporeal [and hence not to the soul], and distinguishing between the first *original of things*; and the subsequent *course of Nature*, teaches concerning the former, not only that God gave Motion to Matter, but that in the beginning He so guided the various motions of the parts of it, as to contrive them into the World (...) and establish'd those Rules of Motion, and that order amongst things corporeal, which we are wont to call the Laws of Nature. And having told this as to the former, it may be allowed as to the latter to teach, that the Universe being once fram'd by God, and the laws of motion being settled and all upheld by this incessant concourse and general Providence; the Phaenomena of the world this constituted, are Physically produced by the mechanical affections of the parts of matter, and what they operate upon one another according to Mechanical laws (Boyle 2000a, 103–104).

In this passage, Boyle addresses two of the most debated issues of early-modern philosophy, namely the nature of corporeal things and the relation between God and his creation. Boyle's first remark must be read as an answer to materialist philosophers like Thomas Hobbes, who believed that the soul, being corporeal, was subjected to the same immutable laws that governed the behaviour of physical bodies. The second remark is an expression of what John Henry called "an unmistakably voluntarist position" (Henry 2009, 94). Whereas intellectualists insisted that God had created the best of all possible worlds, of which there could only be one, so that God was limited in his choice, voluntarists regarded creation as an act of God's will and therefore as freely chosen. In the *Free Enquiry into the Vulgarly Received Notion of Nature* (1686), Boyle wrote:

> God is a most Free Agent, and Created the World, not out of necessity, but voluntarily, having fram'd It, as he pleas'd and thought fit, at the beginning of Things, when there was no Substance but Himself, and consequently no Creature, to which He could be oblig'd, or by which he could be limited (Boyle 2000b, 566).

In a similar vein, in a manuscript note redacted around 1672, Isaac Newton wrote: "The world might have been otherwise than it is (because there may be worlds otherwise framed than this). It was therefore no necessary but a voluntary & free determination that it should be thus" (Newton, MS Yahuda 21, fol. 2r, spelling adjusted, quoted from Henry 2009, 96).

The most famous clash between voluntarism and intellectualism is the controversy between Gottfried Wilhelm Leibniz and the Newtonian theologian Samuel Clarke. In his letters, Leibniz repeatedly stressed that nothing in nature "happens without a reason why it should be so, rather than otherwise" (Leibniz, *Second Letter,* § 1; in Alexander 1956, 16). If there really did exist an absolute space, as Newton believed, which was ontologically independent from the bodies contained in it, then the act of creation would have included an element of arbitrariness, for God would have had no reason to order objects "after one certain particular manner rather than otherwise" (Leibniz, *Third Letter,* § 5; *ibid.,* 26). Clarke agreed with Leibniz that "nothing is, without a sufficient reason, why it is, and why it is thus rather than otherwise," but in his eyes, "this sufficient reason is oft-times no other, than the will of God." For Clarke, to deny to God the power of determining "why this particular system of matter, should be created in one particular place, and that in another particular place," meant nothing less than "to take away all power of choosing, and to introduce fatality" (Clarke, *Second Letter,* § 1; *ibid.,* 20). On his account, then, intellectualism implied fatalism. Leibniz, on the other hand, accused Newton and Clarke of reintroducing chance into the world: "A will without reason, would be the chance of the Epicureans. A God, who should act by such a will, would be a God only in name" (Leibniz, *Fourth Letter,* § 18; *ibid.,* 39). Clarke rejected this objection. In his eyes, "the Epicurean chance is not a choice of will, but a blind necessity of fate" (Clarke, *Fourth Letter,* § 18; *ibid.,* 50). But Leibniz, conceiving of the relation between chance, choice, necessity and fate differently, retorted: "Epicurus' chance is not a necessity, but something indifferent. Epicurus brought it in on purpose to avoid necessity. 'T is true, chance is blind; but a will without motive would be no less blind, and no less owing to real chance" (Leibniz, *Fifth Letter,* § 39; *ibid.,* 79).

This opposition sheds much light on our issue. For Leibniz, the word "chance" designates the absence of a determining cause, and it can hence be applied to whatever happens without a reason. For Clarke, by contrast, "chance" implies "involuntariness," so that no free agent can be said to operate by chance:

> comparing the will of God, when it chooses one out of many equally good ways of acting, to Epicurus' chance, who allowed no will, no intelligence, no active principle at all in the formation of the universe; is comparing together two things, than which no two things can possibly be more different (Clarke, *Fifth Letter,* § 70; *ibid.,* 107–108).

Leibniz considered his own metaphysics, which was based on the idea that an omnipotent God cannot fail to choose the best, to be the only viable alternative to the determinism of Spinoza, according to whom everything that exists flows necessarily from the essence of God. In Clarke's eyes, however, Leibniz' worldview was as necessitarian as Spinoza's: to claim that "whatever God can do, he cannot but do (…) is making him a mere necessary agent, that is, indeed no agent at all, but mere fate and nature and necessity" (Clarke, *Fourth Letter,* § 22–23; *ibid.,* 50).

2.8   Hume's Critique of the Argument from Design

Precisely because the relation between chance, will, reason, and necessity can be thought of in such radically different ways, David Hume was to insist on the importance of agreeing over the definition of these terms. In his *Treatise of Human Nature*, Hume explains that no "freedom of indifference" can exist, if it is defined as "that which means a negation of necessity and causes":

> According to my definitions, necessity makes an essential part of causation; and consequently liberty, by removing necessity, removes also causes, and is the very same thing with chance. As chance is commonly thought to imply a contradiction, and is at least directly contrary to experience, there are always the same arguments against liberty or free-will. If any one alters the definitions, I cannot pretend to argue with him, until I know the meaning he assigns to these terms (Hume 2007, pt. 3, s. 1).

In other words, that voluntary actions are caused by the agent's will does not make them any less necessary than the behaviour of material objects: "the chance or indifference lies only in our judgment on account of our imperfect knowledge, not in the things themselves, which are in every case equally necessary, though to appearance not equally constant or certain" (*ibid.*).

The concept of "chance" plays an important role also in Hume's famous *Dialogues Concerning Natural Religion.* Cleanthes, one of the literary interlocutors, argues that the world exhibits too much order and harmony to be a mere product of chance:

> Throw several pieces of steel together, without shape or form; they will never arrange themselves so as to compose a watch. Stone, and mortar, and wood, without an architect, never erect a house. (…) The adjustment of means to ends is alike in the universe, as in a machine of human contrivance. The causes, therefore, must be resembling (Hume 1779, 56).

Hume's spokesman, Philo, suggests that Cleanthes' reasoning rests on a weak analogy. The dissimilitude between a house and the universe "is so striking, that the utmost you can here pretend to is a guess, a conjecture, a presumption concerning a similar cause." Moreover, one should not suppose that the attributes of God "have any analogy or likeness to the perfections of a human creature." We ascribe to God "Wisdom, Thought, Design, Knowledge (…) because these words are honourable among men," forgetting that "He is infinitely superior to our limited view and comprehension" (Hume 1779, 46).

Now, whereas Cleanthes argues that what cannot be the outcome of chance must be the result of design, Philo adds a third term to the disjunction, namely necessity. He illustrates his point by means of an interesting piece of mathematical reasoning:

> It is observed by arithmeticians, that the products of 9, compose always either 9, or some lesser product of 9, if you add together all the characters of which any of the former products is composed. Thus, of 18, 27, 36, which are products of 9, you make 9 by adding 1 to 8, 2 to 7, 3 to 6. Thus, 369 is a product also of 9; and if you add 3, 6, and 9, you make 18, a lesser product of 9. To a superficial observer, so wonderful a regularity may be admired as the effect either of chance or design: but a skilful algebraist immediately concludes it to be the work of necessity, and demonstrates, that it must for ever result from the nature of these numbers. Is it not probable, I ask, that the whole economy of the universe is conducted by a like necessity, though no human algebra can furnish a key which solves the difficulty? And instead of admiring the order of natural beings, may it not happen, that, could we penetrate into the intimate nature of bodies, we should clearly see why it was absolutely impossible they could ever admit of any other disposition? So dangerous is it to introduce this idea of necessity into the present question! and so naturally does it afford an inference directly opposite to the religious hypothesis! (Hume 1779, 168).

However ingenious Hume's triptych of possibilities, which is composed of design, chance and necessity may have been, and however modern Hume was in many other respects, with respect to biology, things turned out differently. What emerged in the late eighteenth century and culminated in the mid-nineteenth was an evolutionary account of life forms in which neither design, nor necessity, but chance would in fact provide the required *explanans*.

2.9   From Natural History to Darwinism

In the domain of natural history—what would later become biology and geology—
the eighteenth century ushered in a more chaotic world-view. God receded from his
previous role as the designing creator as well as the guarantor of an all-pervasive
necessity, as our world gradually turned out to have a tempestuous past made of ice
ages, inundations, volcanic eruptions, extinct species and ultimately of forms of life
that diversified in unpredictable ways in reaction to these circumstances.

   Indeed, an impressive and ever increasing battery of eminent authors emerged
who would deny the distinction, which Anneliese Maier ascribes to this time period,
between a contingent realm of human action and a deterministic realm of nature.
One may observe, beginning in the eighteenth century, an increasing insistence on
the accidental nature of all forms of life, including man. Julien Offray de La Mettrie,
in his famous *L'Homme machine*, provocatively stated that human existence had
been thrown upon the Earth *au hazard*, "just like mushrooms," mushrooms being at
the time in many quarters still seen as imperfect beings that were generated
spontaneously (La Mettrie 1764, 46). And as biologists began to get an inkling of
the changing morphology of species, they arrived at the concomitant idea of "innumerable multitude of individuals" produced by "chance" (*hazard*) and of "fortuitous combinations of the productions of nature," of which the species living
today are only "a small part of what blind fate [*un destin aveugle*] has produced"
(Maupertuis 1752).

   The epitome of that trend is of course Charles Darwin's *On the Origin of Species*
of 1859, which introduces the notion of a blind natural selection, which relies on a
very simple combination of factors: there is a random type of variation of traits
found among siblings (a longer or shorter neck, thicker or thinner fur, greater or
lesser need of water, etc.); a deadly struggle for survival due to the presence of
predators, a perennial excess of offspring and the resulting scarcity of food and
resources; and the resulting selection of those randomly generated traits that happen
to give their owners an advantage in the struggle for survival. These traits, selected
again and again across numerous generations, would eventually lead to such
modifications in a population that a new species or even genus could come about
(Darwin 1859). Importantly, there existed, for Darwin, no underlying evolutionary
direction or logic. The environmental factors were as accidental as the traits they
selected among the randomly generated variants. Whether a thicker fur happened to
be an advantage or a disadvantage for survival depended on changing weather
patterns, diseases, the presence of predators and many other unpredictable conditions. C. S. Lewis mocked this vision of nature in the first lines of his satirical
Evolutionary Hymn:

   Lead us, Evolution, lead us
   Up the future's endless stair;
   Chop us, change us, prod us, weed us.

For stagnation is despair:
Groping, guessing, yet progressing,
Lead us nobody knows where (Lewis 1964, 55).

Lewis parodies here a famous hymn by James Edmeston (1821), which to this day is found in all Anglican and Episcopalian hymnals and whose first verses sound as follows:

Lead us, heavenly Father, lead us
o'er the world's tempestuous sea;
guard us, guide us, keep us, feed us,
for we have no help but thee;
yet possessing every blessing,
if our God our Father be.

The opposition between the invocation of divine providence, in the original hymn, and Lewis' ironical description of the total absence thereof in an evolutionary process that chops and weeds aimlessly and without purpose and direction is stark. But the lack of providentialism is clearly expressed in Darwin's model:

In such case, every slight modification, which in the course of ages *chanced to arise*, and which in any way favoured the individuals of any of the species, by better adapting them to their altered conditions, would tend to be preserved; and natural selection would thus have free scope for the work of improvement. (Darwin 1859, Ch. 4).

Darwin honestly admitted that he had no idea about the forces that were responsible for the variability of traits found in offspring. After all, Mendel, genetics, and the discovery of DNA were later episodes in the history of biology. Nevertheless, his basic model has remained fairly intact, as has the role of chance in it. For example, modern biology speaks of the role of mutations in the evolution of species in terms of spontaneous mutations (such as molecular decay) or mutations due to errors occurring in the replication of DNA. The default process is faithful copying, but errors take place a bit like the sudden swerve or *klinamen* of atoms, unexplained deviations from the usual direction.[8]

In the eyes of the American philosopher, logician, chemist and mathematician Charles Sanders Peirce, Darwin's evolutionary theory in fact constituted strong evidence against a deterministic world-view. Quite generally, Peirce combated the idea that the universe was governed by strict laws, preferring to see mathematical laws of nature as nothing more than statistical approximations to general patterns or "habits," as he called them, which natural bodies tended to exhibit. In fact, taking recourse to *tuché*, the Greek word with which we begun our historical section, Peirce in 1892 coined the neologism "tychism" as the name of the view that the universe was characterized by "absolute chance," not by a deterministic type of "necessity." Peirce dismissed the idea "that every single fact in the universe is precisely determined by law" (Peirce 1892, 321). That mistaken idea had been around since the days of Democritus and the Stoics, but had in the meantime been

---

[8]See on this Han Brunner's chapter in this book.

clad in new scientific clothes, looking thus: "Given the state of the universe in the original nebula, and given the laws of mechanics, a sufficiently powerful mind could deduce from these data the precise form of every curlicue of every letter I am now writing" (*ibid.*, 323).

But—Peirce retorted—the so-called "laws of mechanics," like all laws of nature, were mere approximations. The more exact one's experimental measurements, the greater the deviations of the data from the mathematical ideal. In the essay's concluding dialogue between an imaginary determinist and Peirce, which starts with a discussion over whether the apparently random fall of a die is determined or not, the real force of tychism is finally introduced. In an evolving cosmos, which displayed ever-increasing complexity over time, all apparent mechanical regularity could at best be provisional. In other words, one had to admit "pure spontaneity or life as a character of the universe, acting always and everywhere though restrained within narrow bounds by law, producing infinitesimal departures from law continually, and great ones with infinite infrequency" (*ibid.*, 333–334).

2.10   Laplace's Determinism, Statistical Regularity and the New Physical Randomness

A similar recovery of chance and randomness took place in the domain of physics. This occurred, paradoxically enough, after these concepts had been quite thoroughly expelled from the exact sciences. We have seen earlier that when medieval philosophers such as Abelard claimed that "the word 'chance' … always denotes ignorance," they did so because they compared the low level of human comprehension with the omniscience of God, for whom nothing happened unexpectedly and for whom there existed no chance. But we also recall that in the seventeenth century, the idea emerged that if one managed to find all laws of nature, these would ultimately explain everything within a deterministic framework. In the latter paradigm, "chance" was no longer opposed to "divine providence," nor was it any longer the expression of the innate limits of human understanding. If humans attributed an event to "chance," this term had to do either with their personal ignorance of the physical laws causing the event in question, or else with the mathematical difficulty of deriving that particular event from the multiplicity of underlying causes and the respective laws governing them. This position was forcefully expressed by the French mathematician Pierre-Simon Laplace, whose name is associated with the development of statistical methods for calculating probabilities (*Théorie analytique des probabilités* 1812), with a scientific form of determinism, and with the concomitant elimination of divine causality from cosmology as well as physics quite generally. As the apocryphal story goes, he explained to Napoleon that he did not need God as a hypothesis ("je n'avais pas besoin de cette hypothèse-là"). His scientific determinism, in turn, expressed itself most famously in the notion of a "demon"—a kind of perfect intelligence—that could derive all current and future states of the world from a complete understanding of previous states:

> We may regard the present state of the universe as the effect of its past and the cause of its future. An intellect which at a certain moment would know all forces that set nature in motion, and all positions of all items of which nature is composed, if this intellect were also vast enough to submit these data to analysis, it would embrace in a single formula the movements of the greatest bodies of the universe and those of the tiniest atom; for such an intellect nothing would be uncertain and the future just like the past would be present before its eyes (Laplace 1902, 4).

This is of course the very theory to which Peirce alluded when he said that the modern version of determinism pretended that "every curlicue of every letter I am now writing" had been predetermined by the state of the first stellar nebulae. Now, "Laplace's demon" is expressly not a god, and certainly not a creator, but rather a calculating device of the type that we would nowadays identify with a supercomputer. Nor is he, or it, omniscient and in fact need not even be conscious. All it does is to deduce, on the basis a complete set of natural laws and an equally complete data set on all bodies in the world, mechanically, and with absolute certainty, the present and future behaviour of the world and all that is in it.

But then, the century that started with Laplace ended with Peirce's rejection of the possibility the former's omniscient demon. Quite generally, it witnessed what Ian Hacking has described as a veritable "erosion of determinism" (Hacking 1983, 445). This erosion took place not only in philosophy and biology, but also in the domain of physics.[9]

The probabilistic revolution in physics in fact clearly predated the advent of quantum theory. It all started in the mid-nineteenth century with what we now call "statistical mechanics" (Brush 1976, Ch. 4). This theory arose in the wake of thermodynamics and made the point that one fundamental assumption of classical physics, namely a complete specification of the state of a system as input for accurate and certain predictions (in keeping with the spirit of determinism), was hardly satisfied if the system in question consisted of 100,000,000,000,000,000,000,000 particles, as is typically the case for a gas in a container.

While at first sight, this problem might still seem solvable by the hypothetical Laplacian demon, the everyday phenomenon of irreversibility in macroscopic systems turned out to be inexplicable on the assumption that probability was merely a matter of ignorance. Incidentally, this issue remains unresolved until the present day (Sklar 2009; Uffink 2007). In the late nineteenth century, at any rate, its recognition served as a first admonition with regard to a possibly fundamental (or "irreducible") role for probability in physics (Uffink 2014).

Another challenge to classical physics that fed probabilistic reasoning and attitudes consisted in the observation of discrete and discontinuous phenomena

---

[9]The following six paragraphs have been written by Klaas Landsman; we have imported only minor modifications.

(especially at an atomic scale). Two of Einstein's four path-breaking articles in his *annus mirabilis* 1905 were concerned with such phenomena (Stachel 1998). The first article tackled the issue of Brownian motion—the motion of particles suspended in a fluid—and was based on the use of what is called "random walks," the latter term referring to a mathematical formalisation of a path such as that of a molecule in a liquid, which consists of a succession of random steps. Einstein's second article provided the first empirical confirmation of the quantum nature of light, that is to say, the fact that light manifests itself only in multiples of a basic unit. It was mainly the latter issue of discreteness and discontinuity, first discovered by Max Planck, Albert Einstein, and Niels Bohr, that eventually led to quantum mechanics (Jammer 1966).

After a period of confusion and crisis lasting from 1900 to 1925, which ended with the complementary work of Heisenberg on matrix mechanics in 1925 and of Schrödinger on wave mechanics in 1926, quantum mechanics was more or less finalized during the subsequent five years, apart from Werner Heisenberg and Erwin Schrödinger also through the remarkable contributions of Paul Dirac, Max Born, Pascual Jordan, Wolfgang Pauli, and John von Neumann. Quantum mechanics thereby replaced Newton's formalism of classical mechanics by a totally different mathematical scheme, whose physical interpretation has remained a matter of controversy to the present day (Jammer 1974).

Quantum mechanics has many strange features, all of which appear to be related (including non-locality, another holistic property called entanglement, as well as the phenomenon described as Schrödinger's Cat). What counts for our purpose is that its predictions are a priori probabilistic: instead of specifying one particular outcome of some physical process with certainty, as classical physics does (at least under ideal circumstances and for an ideal calculator such as Laplace's demon), quantum mechanics merely states a range of possible outcomes, even though each probability can be precisely predetermined. Indeed, quantum mechanics allows for the possibility of an absolutely random coin flip, realized, for example, by a single photon (that is, the basic quantum of light), which may or may not be transmitted by a polarizer, or by a spin measurement on an electron. Such quantum-mechanical coin flip devices are even commercially available from the Swiss company *ID Quantique*, which "commercializes a quantum random number generator, which is the reference in the gaming and lottery industries" (*ID Quantique* 2015). With this "random number generator," we return to several earlier themes, the *casus*, hazard, and the Wheel of Fortune, but now at the most basic level of matter, at which Laplace's demon expected to find nothing but predictable order.

The probabilistic interpretation of quantum mechanics had been in the air almost from the beginning, and notably, *à contre-coeur*, in the work of Einstein himself, but it was first explicitly proposed (and declared to be fundamental) in a paper by Born in 1926 on collision theory. This paper also provided a formula for the probabilities of the various outcomes, which is now known as "Born's Rule" and which forms the basis of practically all quantitative—and extremely successful—predictions of quantum theory. What is crucial in the present context is that Born's probabilistic interpretation of quantum physics was construed by him and his colleagues in terms of a turn to indeterminism. The latter idea was reinforced by Heisenberg's famous paper of 1927, which proposed the uncertainty relations now named after him. Heisenberg suggested that quantum mechanics was not only indeterministic in its inability to predict the outcome of a single experiment, but also in its failure to specify initial conditions with arbitrary accuracy (see Mehra and Rechenberg 2000 and 2001 for a historical overview of this episode). In the wake of Born's and Heisenberg's epoch-making papers, Niels Bohr (backed by most if not all of the other leading players except Einstein and Schrödinger) soon stepped forward as the champion of indeterminism, a position which, with the assistance of Heisenberg and Pauli, he successfully defended against Einstein's penetrating and relentless criticism during their famous debate from 1927 to 1949 (Bohr 1949).

The general perception among physicists is that Bohr emerged victorious from this debate with Einstein, and that determinism and hence the epistemic view of probability is a thing of the past, at least in fundamental physics, forever replaced by the indeterminism of quantum mechanics. Whether this view is really correct remains to be seen, however. It certainly cannot be *proved* mathematically that quantum mechanics, even if it should be the correct and ultimate theory of nature, implies indeterminism; acceptable models to the contrary exist (such as Bohmian mechanics). Furthermore, one should be open-minded to possible modifications of quantum mechanics, including underlying theories that would restore determinism, whilst reproducing its probabilistic predictions by averaging over so-called hidden variables. The nature of contemporary discussions is to put constraints on deterministic interpretations of quantum mechanics and on possible refinements thereof, as first attempted by von Neumann in 1932 and more successfully in John Stuart Bell's path-breaking work of 1964. Such constraints typically make such alternative theories unattractive, but not impossible. However, the discussion is ongoing, and the last word clearly hasn't been spoken yet.

With this mention of contemporary discussions in fundamental physics, our selective survey of almost 2500 years of philosophical and scientific reflections on chance, coincidence, fortune, randomness, luck and related concepts comes to a close. Does it tell us anything helpful? Maybe above all this, that in specific domains such as statistics, evolutionary biology, or quantum physics, the last three hundred years have generated specific technical sub-meanings of several of these terms. At the same time, it is also striking that none of the discussions seems to have come to a close. We have seen, for example, how the deterministic world-view of the ancient atomists was supplanted by Aristotelianism, which identified the atomists' "blind necessity" as "mere chance," rejecting it; how Christian philosophy tried to find degrees of freedom within a cosmos otherwise defined by an omnipotent, providential God, who was however not to be held responsible for everything, including evil, that occurred in it; how in the Newtonian age a scientific determinism returned to prominence, which was sometimes accompanied by an overt rejection of any divine agency; how this deterministic worldview was shattered by a quantum physics that seemed to locate indeterminacy, probabilistic and random behaviour at the lowest material and energetic levels and in the very laws that describe them; and finally how, from Einstein's protests until today, the hope of finding an intellectually more satisfactory, that is, deterministic model has never entirely vanished.

3   A Conclusion *ex negativo*

Our first, etymological, approach has taught us something about the common element of several of the words in our cluster. As our eyes are usually directed ahead of us, a falling object will tend to surprise us. We stop, in shock or pleasantly surprised, to contemplate the unexpected arrival. Such a situational and emotional description of our cluster of concepts—inspired as it was by the verb "falling" that underlies *casus*, "coincidence," "accident" as well as a number of words related to the falling of dice—will, however, only take us so far. In specific situations, such as when we receive a random assignment or try to calculate our chances of winning in the lottery, the archetypal situation of the falling object will seem quite remote. Moreover, we have seen that languages don't divide the words along similar lines. No English or French word is as broad as the German *Zufall*, and what has a neutral meaning in one language, such as the French *hazard* or the Latin *accidens*, has negative connotations in another.

Our second, main approach has been *begriffsgeschichtlich*. We don't need to repeat the conclusions of that section once more. Suffice it to say here that Hegel would be dismayed: we have not been able to detect any dialectical progress from the ancient Greeks up to today's physicists in the way in which scientists and philosophers resolved the perennial tension between the predictable and the unpredictable; between the necessary and the contingent; between necessity and chance; or to dismiss fate, fortune, the accidental and the random. Given the developments in evolutionary biology and quantum physics of the past 150 years, it seems rather as if "chance," "randomness," and "coincidence" had been restored to a place of respectability that they had previously lost. Indeed, whether our personal surprise at a given event is merely a sign of personal ignorance or is instead a necessary feature of this universe has once again been elevated to the status of unresolved question.

One thing is certain. Time and again, throughout our pages, it has become evident that any of the words with which we have been engaged could only be understood if we also understood the type of explanation that it attempted to exclude. And precisely because the alternative did not have a stable identity, it was obvious that its anti-pole also had to change meaning. In the course of our chapter, we have found the word "chance" opposed to "fate," "purpose," "providence," "natural laws," "determinism," or simply to "the knowledge of causes." Given this heterogeneous list, it is evident that the common opposite, "chance," was doomed to be a slippery concept.

The helpfulness of understanding our words *ex negativo*, that is, from their respective contraries, should be evident. It helps understand, for example, the conceptual clash between the ancient atomists and Aristotle. As we recall, Democritus and Leucippus had proposed that the world had come about by necessity, through a blind and mechanical process of atomic combination, because "nothing exists at random." But what they regarded as "necessity" was in Aristotle's terminology mere "chance," because the atomists' cosmogony took place without any plan or purpose. Or take the disagreements over whether divine providence allowed for any fortuitous events. If "chance" is taken to mean that something happened without divine foreknowledge, as Cicero postulated, providentialism is indeed incompatible with it. If "chance" means, by contrast, that "something other than was intended happens on account of other reason," as Boethius argued, then it is compatible with providentialism in the precise sense that the intentions in question are ours, not God's. Or, again, take the conflict between Leibniz and Clarke over whether "a will without reason" amounts to "the chance of the Epicureans" (Leibniz), or whether instead an act of will per definition excludes the "blind necessity of fate" (Clarke). Similarly, when Hugo Boxel objected that necessity was the contrary of freedom, not of chance, as Spinoza had assumed, the latter remarked on "the difficulties experienced by two people following different principles, and trying to agree on a matter." Finally, take the redefinition of "chance" in the early modern period, in which someone like David Hume could claim that there is "no such thing as Chance in the world" (Chance written with a capital 'C'), adding that in a probabilistic sense, one was justified in speaking of "chances," written with a lower cap.

It is often mockingly asserted that philosophy is that academic discipline that deals with questions that have no answers, or, more maliciously, that the reason why philosophers can still engage with two thousand year-old texts is because there has been no philosophical progress in all those centuries. If there should be any truth to this view, it must with equal right be applied to the philosophical aspects of all modern sciences (those grown-up daughters of what up to the seventeenth century was "natural philosophy"). After all, we have seen, maybe with surprise, how in each moment of scientific reflection on the relationship between natural causality, determinacy, and chance, the ancient Greek vocabulary tends to re-emerge. What has been overly evident in C. S. Peirce's decision to re-introduce a Greek term (namely *tuchê*) for a philosophy based on "chance" is true more generally. The "fortune" of our cluster of words has indeed followed the logic of the Wheel of Fortune: *tuchê*, "chance," or "randomness," temporarily deposed and "without kingdom," have returned to the top of the wheel, to rule.

In a book dealing with ancient Greek concepts of nature, the famous physicist Erwin Schrödinger once wrote:

> By the laws of physics we are forced in each moment to do whatever we do. What is the point then in considering whether it is right or wrong? Where is there any room for a moral law, if the omnipotent law of nature does not provide it with a chance to speak? Today, the antinomy is as unresolved as it was twenty-three centuries ago (Schrödinger 1956, 18).

Open Access This chapter is distributed under the terms of the Creative Commons Attribution-Noncommercial 2.5 License (http://creativecommons.org/licenses/by-nc/2.5/) which permits any noncommercial use, distribution, and reproduction in any medium, provided the original author(s) and source are credited. The images or other third party material in this chapter are included in the work's Creative Commons license, unless indicated otherwise in the credit line; if such material is not included in the work's Creative Commons license and the respective action is not permitted by statutory regulation, users will need to obtain permission from the license holder to duplicate, adapt or reproduce the material.

References

Alexander, H. G. (Ed.). (1956). *The Leibniz-Clarke correspondence. Together with extracts from Newton's Principia and Opticks*. Manchester: Manchester University Press.

Aristotle (1933). *Metaphysics* (H. Tredennick, Trans.) (Loeb Edition, 2 Vols.). London and Cambridge, MA: Harvard University Press.

Aristotle (1935). *Athenian constitution. Eudemian ethics. Virtues and vices* (H. Rackham, Trans.) (Loeb Edition). London and Cambridge, MA: Harvard University Press.

Aristotle (1937). *Parts of animals. Movement of animals. Progression of animals* (A. L. Peck, Trans.) (Loeb Edition). London and Cambridge, MA: Harvard University Press.

Aristotle (1957). *Physics* (P. H. Wicksteed & F. M. Cornford, Trans.) (Loeb Edition, 2 Vols.). London and Cambridge, MA: Harvard University Press.

Augustine (1841). *Quaestionum in Heptateuchum libri VII*. In *Sancti Augustini ... Opera omnia…* Vol. 3, pt. 1 (= *Patrologia Latina*, Ed. J. P. Migne, Vol. 34, pp. 547–824). Paris: Migne.

Augustine (1998). *The city of God against the Pagans* (R. W. Dyson, Ed., Trans.) (Cambridge Texts in the History of Political Thought). Cambridge: Cambridge University Press.

Bobzien, S. (1998). *Determinism and freedom in Stoic philosophy*. Oxford: Clarendon Press.

Bobzien, S. (2000). Did Epicurus discover the free-will problem? *Oxford Studies in Ancient Philosophy, 19*, 287–337.

Boethius, A. (1897). *The consolation of philosophy* (H. R. James, Trans.). Oxford: Oxford University Press.

Boethius, A. (2000). *The consolation of philosophy* (V. Watts, Trans.). London: Hamish Hamilton.

Boethius, A. (1891). *In librum De interpretatione editio secunda*. In J. P. Migne (Ed.), *Patrologia Latina* (Vol. 64). Paris: Migne.

Bohr, N. (1949). Discussion with Einstein on epistemological problems in atomic physics. In P. A. Schilpp (Ed.), *Albert Einstein: Philosopher-scientist* (pp. 201–241). La Salle: Open Court.

Boyle, R. (2000a). Considerations about the excellency and grounds of the mechanical hypothesis [1674]. In M. Hunter & E. B. Davis (Eds.), *The works of Robert Boyle* (Vol. 8, pp. 104–105). London: Pickering & Chatto.

Boyle, R. (2000b). Free enquiry into the vulgarly received notion of nature [1686]. In M. Hunter & E. B. Davis (Eds.), *The works of Robert Boyle* (Vol. 10, pp. 99–116). London: Pickering & Chatto.

Brush, S. (1976). *The kind of motion that we call heat*. Amsterdam: North-Holland.

Byrne, E. F. (1968). *Probability and opinion*. The Hague: Cadell.

Carmina burana: die Lieder der Benediktbeurer Handschrift (1979) (B. Bischoff, H. Hilka, O. Schumann, C. Fischer, H. Kuhn, G. Bernt, Ed., Trans.). Munich: Deutscher Taschenbuch Verlag.

Cherniss, H. F. (1935). *Aristotle's criticism of Presocratic philosophy*. Baltimore: Johns Hopkins Press.

Cicero (1923). *On old age. On friendship. On divination* (W. A. Falconer, Trans.) (Loeb Classical Library). London and Cambridge, MA: Harvard University Press.

Cicero (1941). *On the orator: Book 3. On fate. Stoic paradoxes. Divisions of oratory* (H. Rackham, Trans.) (Loeb Classical Library). London and Cambridge, MA: Harvard University Press.

Darwin, Ch. (1859). *On the origin of species by means of natural selection; or the preservation of favoured races in the struggle for life.* London: John Murray.

Daston, L. (1988). *Classical Probability in the Enlightenment.* Princeton: Princeton University Press.

Duden. Deutsches Universalwörterbuch (2007) (6th ed.). Mannheim, Leipzig, Vienna, Zurich: Dudenverlag.

Dudley, J. (2012). *Aristotle's concept of chance: Accidents, cause, necessity, and determinism.* Albany: SUNY Press.

Edmunds, L. (1972). Necessity, chance, and freedom in the early atomists. *The Phoenix, 26,* 342–357.

Epicurus (1926). *The extant remains* (C. B. Bailey, Ed., Trans.). Oxford: Clarendon Press.

Epicurus (1931). *Letter to Herodotus, Letter to Pythocles, Letter to Menoeceus, Principal sayings* = vol. 2 of *Diogenes Laertius: Lives of Eminent Philosophers* (2 Vols.) (R. D. Hicks, Ed.) (Loeb Edition). London and Cambridge, MA: Havard University Press.

Fowler, D. (1983). Lucretius on the Clinamen and 'Free Will' (II 251–93). *SYZÊTÊSIS: Studi sull'epicureismo greco e romano offerti a Marcello Gigante* (pp. 329–352.). Naples: Centro internazionale per lo studio dei papiri ercolanesi.

Furley, D. J. (1967). *Two studies in the Greek atomists.* Princeton: Princeton University Press.

Gassendi, P. (1658). *Opera omnia* (6 Vols.). Lyon: Laurent Anisson et Jean-Baptiste Devenet.

Hacking, I. (1975). *The emergence of probability: A philosophical study of early ideas about probability, induction and statistical inference.* Cambridge: Cambridge University Press.

Hacking, I. (1983). Nineteenth century cracks in the concept of determinism. *Journal of the History of Ideas, 44,* 455–75.

Henry, J. (2009). Voluntarist theology at the origins of modern science: A response to Peter Harrison. *History of Science, 47,* 79–113.

Historisches Wörterbuch der Philosophie (1971–2007) (13 Vols.) (J. Ritter, K. Gründer & G. Gabriel, Eds.). Basel: Schwabe AG.

Hume, D. (1748). *An enquiry concerning human understanding.* London: A. Millar.

Hume, D. (1779). *Dialogues concerning natural religions* (2nd ed.). London: s.n.

Hume, D. (2007). *A treatise of human nature* (Clarendon Edition). Oxford: Oxford University Press.

ID Quantique (2015). Retrieved on June 12, 2015 http://www.idquantique.com.

Jammer, M. (1966). *The conceptual development of quantum mechanics.* New York: McGraw-Hill.

Jammer, M. (1974). *The philosophy of quantum mechanics: The interpretations of quantum mechanics in historical perspective.* New York: McGraw-Hill.

John Buridan (1509). *Subtilissimae questiones super octo libros Aristotelis.* Paris: Denis Roce.

Kirk, G. S., Raven, J. E., & Schofield, M. (1983). *The Presocratic philosophers. A critical history with a selection of texts* (2nd ed.). Cambridge: Cambridge University Press.

Kretzmann, N. (1999). *The metaphysics of creation: Aquinas' natural theology in 'Summa contra gentiles' II.* Oxford: Clarendon Press.

Laertius, D. (1925). *Lives of eminent philosophers* (2 Vols.) (R. D. Hicks, Trans.) (Loeb Edition). London and Cambridge, MA: Harvard University Press.

La Mettrie, J.O. de (1764). *L'homme machine.* In *Oeuvres philosophiques* (Vol. 1). Amsterdam: s.n.

Laplace, P.-S. (1902). *A philosophical essay on probabilities* (6th ed.) (F. W. Truscott & F. L. Emory, Trans.). New York: Wiley.

Laplace, P.-S. (1812). *Théorie analytique des probabilités.* Paris: Mme. Ve. Courcier.

Lawrence, S. (2013). Fate and Chance. In H. M. Roisman, (Ed.), *The encyclopedia of Greek tragedy* (pp. 502–506). London: Wiley.

Lewis, C. S. (1964). *Poems* (W. Hooper, Ed.). London: Harcourt.

Lewis, C. T., & Short, Ch. (1879). *A Latin dictionary.* New York: Wiley.

Long, A. A. (1977). Chance and natural law in Epicureanism. *Phronesis, 22*, 63–88.
Lucretius (1924). *On the nature of things.* (W. H. D. Rouse, Trans.revised by M. F. Smith) (Loeb Edition). London and Cambridge, MA: Harvard University Press.
Lucretius (1947). *Titi Lucreti Cari De rerum natura libri sex* (3 Vols.) (C. B. Bailey, Ed. Trans.). Oxford: Clarendon Press.
Lüthy, C. H. (2003). The invention of atomist iconography. In W. Lefèvre, J. Renn & U. Schoepflin (Eds.), *The power of images in early modern science* (pp. 117–138). Basel/Boston/Berlin: Birkhäuser.
Maier, Anneliese. (1949). Notwendigkeit, Kontingenz und Zufall. In Maier, *Die Vorläufer Galileis im 14. Jahrhundert* (pp. 219–250). Rome: Edizioni di Storia e Letteratura.
Maupertuis, P. L. M. de (1752). *Essai de cosmologie.* In *Oeuvres* (4 Vols.) (Vol. 1, pp. 1–58). Lyon: J.M. Bruyset.
Mehra, J., & Rechenberg, H. (2000, 2001). *The Historical Development of Quantum Theory.* Vol. 6: *The Completion of Quantum Mechanics 1926–1941.* Part 1: *The Probabilistic Interpretation and the Empirical and Mathematical Foundation of Quantum Mechanics, 1926–1936.* Part 2: *The Conceptual Completion of Quantum Mechanics.* New York: Springer.
O'Keefe, T. (2005). *Epicurus on freedom.* Cambridge: Cambridge University Press.
Online Etymological Dictionary. (2015). Retrieved on May 15–20, 2015 http://www.etymonline. com.
Osler, M. J. (1994). *Divine will and the mechanical philosophy: Gassendi and Descartes on contingency and necessity in the created world.* Cambridge: Cambridge University Press.
Oxford English Dictionary (1989). (2nd ed.). Oxford: Oxford University Press.
Pacuvius (1897). *Scaenicae Romanorum poesis fragmenta* (O. Ribbeck, Ed.) (3rd ed.). Leipzig: Teubner.
Peirce, C. S. (1892). The doctrine of necessity examined. *The Monist, 2*, 321–37.
Peter Abelard (1919). In B. Geyer (Ed.), *Peter Abaelards philosophische Schriften. I. Die Logica 'Ingredientibus'* (pp. 1–109). Münster: Verlag der Aschendorffschen Verlagsbuchhandlung.
Plutarch (1976). *Moralia.* Vol. 13, pt. 2: *Stoic Essays* (H. Cherniss, Trans.) (Loeb Classical Library). London and Cambridge, MA: Harvard University Press.
Purinton, J. S. (1999). Epicurus on 'free volition' and the atomic swerve. *Phronesis, 44*, 253–99.
Roger Bacon (1935). *Quaestiones super libros VIII physicorum Aristotelis* [ca. 1250] (R. Steele, Ed.), in Bacon, *Opera hactenus inedita* (vol. 13). Oxford: Oxford University Press.
Sarasohn, L. T. (1985). Motion and morality: Pierre Gassendi, Thomas Hobbes and the mechanical world-view. *Journal of the History of Ideas, 46*, 363–79.
Sedley, D. (1988). Epicurean anti-reductionism. In J. Barnes & M. Mignucci (Ed.), *Matter and metaphysics* (pp. 295–327). Naples: Bibliopolis.
Sklar, L. (2009). Philosophy of statistical mechanics. *The Stanford encyclopedia of philosophy* (Summer 2009 Edition) (E. N. Zalta, Ed.). Retrieved on May 15, 2015 http://plato.stanford. edu/archives/sum2009/entries/statphys-statmech/.
Spinoza, B. de (1883). *The chief works of Benedict de Spinoza. On the improvement of the understanding. The ethics. Corresponedence* (R.H.M. Elwes, Trans. & Intro.). London: Bell.
Stachel, J. (Ed.) (1998). *Einstein's miraculous year. Five papers that changed the face of physics.* Princeton: Princeton University Press.
Thomas Aquinas (1964–1976). *Summa theologiae* (60 Vols.). Cambridge: Blackfriars.
Thomas Aquinas (1975a). *Summa contra gentiles* (J. F. Anderson, Ed., Trans.). Book 2: *Creation.* Notre Dame: University of Notre Dame Press.
Thomas Aquinas (1975b). *Summa contra gentiles* (V. J. Bourke, Ed., Trans.). Book 3: *Providence.* Notre Dame: University of Notre Dame Press.
Uffink, J. (2007). Compendium to the foundations of classical statistical mechanics, in Butterfield, J. & Earman, J. (Eds.), *Handbook for the philosophy of physics* (pp. 924–1074). Amsterdam: North Holland/Elzevier.

Uffink, J. (2014). Boltzmann's work in statistical physics, in E. N. Zalta (Ed.), *The Stanford Encyclopedia of Philosophy* (Fall 2014 Edition). Retrieved on May 15, 2015 http://plato.stanford.edu/archives/fall2014/entries/statphys-Boltzmann/.

Vogt, P. (2011). *Kontingenz und Zufall. Eine Ideen- und Begriffsgeschichte.* Berlin: Akademie Verlag.

Vollmer, M. (2009). Fortuna Diagrammatica. *Das Rad der Fortuna als bildhafte Verschlüsselung der Schrift* De Consolatione Philosophiae *des Boethius.* Frankfurt: Peter Lang.

Voltaire (1766). *Le Philosophe ignorant.* S.l., s.n.

Voltaire (1901). *The philosophical dictionary* (W. F. Fleming, Trans.). 5 vols. = *The Works of Voltaire. A Contemporary Version* (J. Morley, T. Smollett, W. F. Fleming, Eds. & Trans.) (Vols. 3–7). New York: E. R. DuMont.

The Mathematical Foundations of Randomness

Sebastiaan A. Terwijn

Abstract We give a nontechnical account of the mathematical theory of randomness. The theory of randomness is founded on computability theory, and it is nowadays often referred to as algorithmic randomness. It comes in two varieties: A theory of finite objects, that emerged in the 1960s through the work of Solomonoff, Kolmogorov, Chaitin and others, and a theory of infinite objects (starting with von Mises in the early 20th century, culminating in the notions introduced by Martin-Löf and Schnorr in the 1960s and 1970s) and there are many deep and beautiful connections between the two. Research in algorithmic randomness connects computability and complexity theory with mathematical logic, proof theory, probability and measure theory, analysis, computer science, and philosophy. It also has surprising applications in a variety of fields, including biology, physics, and linguistics. Founded on the theory of computation, the study of randomness has itself profoundly influenced computability theory in recent years.

1 Introduction

In this chapter we aim to give a nontechnical account of the mathematical theory of randomness. This theory can be seen as an extension of classical probability theory that allows us to talk about individual random objects. Besides answering the philosophical question what it *means* to be random, the theory of randomness has applications ranging from biology, computer science, physics, and linguistics, to mathematics itself.

The theory comes in two flavors: A theory of randomness for finite objects (for which the textbook by Li and Vitányi 2008 is the standard reference) and a theory for infinite ones. The latter theory, as well as the relation between the two theories of randomness, is surveyed in the paper (Downey et al. 2006), and developed more in

S.A. Terwijn (✉)
Department of Mathematics, Radboud University, P.O. Box 9010,
6500 Nijmegen, GL, The Netherlands
e-mail: terwijn@math.ru.nl

© The Author(s) 2016
K. Landsman and E. van Wolde (eds.), *The Challenge of Chance*,
The Frontiers Collection, DOI 10.1007/978-3-319-26300-7_3

full in the recent textbooks by Downey and Hirschfeldt (2010) and Nies (2009). Built
on the theory of computation, the theory of randomness has itself deeply influenced
computability theory in recent years.

We warn the reader who is afraid of mathematics that there will be formulas and
mathematical notation, but we promise that they will be *explained* at a nontechnical
level. Some more background information about the concepts involved is given in
footnotes and in two appendices. It is fair to say, however, that to come to a better
understanding of the subject, there is of course no way around the formulas, and we
quote Euclid, who supposedly told King Ptolemy I, when the latter asked about an
easier way of learning geometry than Euclid's Elements, that "there is no royal road
to geometry".[1]

2   What Is Randomness?

Classical probability theory talks about random objects, for example by saying that if
you randomly select four cards from a standard deck, the probability of getting four
aces is very small. However, every configuration of four cards has the *same* small
probability of appearing, so there is no qualitative difference between individual
configurations in this setting. Similarly, if we flip a fair coin one hundred times,
and we get a sequence of one hundred tails in succession, we may feel that this
outcome is very special, but how do we justify our excitement over this outcome? Is
the probability for this outcome not exactly the same as that of any other sequence
of one hundred heads and tails?

Probability theory has been, and continues to be, a highly successful theory, with
applications in almost every branch of mathematics. It was put on a sound mathematical foundation in (1933) by Kolmogorov, and in its modern formulation it is
part of the branch of mathematics called *measure theory*. (See Appendix A.) In this
form it allows us to also talk not only about randomness in discrete domains (such
as cards and coin flips), but also in continuous domains such as numbers on the real
line. However, it is important to realize that even in this general setting, probability
theory is a theory about *sets* of objects, not of individual objects. In particular, it does
not answer the question what an *individual* random object is, or how we could call
a sequence of fifty zero's less random than any other sequence of the same length.
Consider the following two sequences of coin flips, where 0 stands for heads and 1
for tails:

000000000000000000000000000000000000000000000000000

00001110011111011110011110010010101111001111010111

---

[1] As with many anecdotes of this kind, it is highly questionable if these words were really spoken,
but the message they convey is nevertheless true.

The first sequence consists of fifty 0's, and the second was obtained by flipping a coin fifty times.[2] Is there any way in which we can make our feeling that the first sequence is special, and that the second is less so, mathematically precise?

3 Can Randomness Be Defined?

A common misconception about the notion of randomness is that it cannot be formally defined, by applying a tautological reasoning of the form: As soon as something can be precisely defined, it ceases to be random. The following quotation by the Dutch topologist Freudenthal (1969) (taken from van Lambalgen 1987) may serve to illustrate this point:

> It may be taken for granted that any attempt at defining disorder in a formal way will lead to a contradiction. This does not mean that the notion of disorder is contradictory. It is so, however, as soon as I try to formalize it.

A recent discussion of randomness and definability, and what can happen if we equate "random" with "not definable", is in Doyle (2011).[3] The problem is not that the notion of definability is inherently vague (because it is not), but that no *absolute* notion of randomness can exist, and that in order to properly define the notion, one has to specify *with respect to what* the supposed random objects should be random. This is precisely what happens in the modern theory of randomness: A random object is defined as an object that is random with respect to a given type of definition, or class of sets. As the class may vary, this yields a *scale* of notions of randomness, which may be adapted to the specific context in which the notion is to be applied.

The first person to attempt to give a mathematical definition of randomness was von Mises (1919), and his proposed definition met with a great deal of opposition of the kind indicated above. Von Mises formalized the intuition that a random sequence should be *unpredictable*. Without giving technical details, his definition can be described as follows. Suppose that X is an infinite binary sequence, that is, a sequence

$$X(0), X(1), X(2), X(3), \ldots$$

---

[2]The author actually took the trouble of doing this. We could have tried to write down a random sequence ourselves, but it is known that humans are notoriously bad at producing random sequences, and such sequences can usually be recognized by the fact that most people avoid long subsequences of zero's, feeling that after three or four zero's it is really time for a one. Indeed, depending on one's temperament, some people may feel that the first four zero's in the above sequence look suspicious.

[3]The notion of mathematical definability is *itself* definable in set theory, see Kunen (1980, Chap. V). If "random" is equated with "not definable", then the following problem arises: By a result of Gödel (1940) *it is consistent with the axioms of set theory that all sets are definable*, and hence the notion of randomness becomes empty. The solution to this problem is to be more modest in defining randomness, by only considering more restricted classes of sets, as is explained in what follows.

where for each positive integer n, $X(n)$ is either 0 or 1. Suppose further that the values of X are unknown to us. We now play a game: At every stage of the game we point to a new location n in the sequence, and then the value of $X(n)$ is revealed to us. Now, according to von Mises, for X to be called random, we should not be able to predict in this way the values of X with probability better than $\frac{1}{2}$, no matter how we select the locations in X. A strategy to select locations in X is formalized by a *selection function*, and hence this notion says that no selection function should be able to give us an edge in predicting values from X. However, as in the above discussion on absolute randomness, in this full generality, *this notion is vacuous!* To counter this, von Mises proposed to restrict attention to "acceptable" selection rules, without further specifying which these should be. He called the sequences satisfying his requirement for randomness *Kollektiv*'s.[4]

Later Wald (1936, 1937) showed that von Mises' notion of Kollektiv is nonempty if we restrict to any *countable* set of selection functions.[5] Wald did not specify a canonical choice for such a set, but later Church (1940) suggested that the (countable) set of *computable* selection rules would be such a canonical choice. We thus arrive at the notion of *Mises–Wald–Church randomness*, defined as the set of Kollektiv's based on computable selection rules. This notion of random sequence already contains several of the key ingredients of the modern theory of randomness, namely:

- the insight that randomness is a *relative* notion, not an absolute one, in that it depends on the choice of the set of selection rules;
- it is founded on the theory of computation, by restricting attention to the *computable* selection functions (cf. Sect. 4).

Ville (1939) later showed that von Mises' notion of Kollektiv is flawed in the sense that there are basic statistical laws that are not satisfied by them. Nevertheless, the notion of Mises–Wald–Church randomness has been decisive for the subsequent developments in the theory of randomness.[6]

The Mises–Wald–Church notion formalized the intuition that a random sequence should be *unpredictable*. This was taken further by Ville using the notion of martingale. We discuss this approach in Sect. 7. The approach using Kolmogorov complexity formalizes the intuition that a random sequence, since it is lacking in recognizable structure, is hard to *describe*. We discuss this approach in Sect. 5. Finally, the notion randomness proposed by Martin-Löf formalizes the intuitions underlying classical probability and measure theory. This is discussed in Sect. 6. It is a highly remarkable fact that these approaches are intimately related, and ultimately turn out to be essentially equivalent. As the theory of computation is an essential ingredient in all of this, we have to briefly discuss it before we can proceed.

---

[4]For a more elaborate discussion of the notion of Kollektiv see van Lambalgen (1987).

[5]A set is called *countable* if its elements can be indexed by the natural numbers 0, 1, 2, 3, . . . These sets represent the smallest kind of infinity in the hierarchy of infinite sets.

[6]In the light of the defects in the definition of Mises–Wald–Church random sequences, these sequences are nowadays called *stochastic* rather than random.

4   Computability Theory

The theory of computation arose in the 1930s out of concerns about what is provable in mathematics and what is not. Gödel's famous incompleteness theorem from 1931 states, informally speaking, that in any formal system strong enough to reason about arithmetic, *there always exist true statements that are not provable in the system.* This shows that there can never be a definitive formal system encompassing all of mathematics. Although it is a statement about mathematical provability, the proof of the incompleteness theorem shows that it is in essence a result about *computability.* The recursive functions used by Gödel in his proof of the incompleteness theorem were later shown by Turing (1936) to define the same class of functions computable by a Turing machine. Subsequently, many equivalent definitions of the same class of computable functions were found, leading to a robust foundation for a general theory of computation, called *recursion theory*, referring to the recursive functions in Gödel's proof. Nowadays the area is mostly called *computability theory*, to emphasize that it is about what is computable and what is not, rather than about recursion.

Turing machines serve as a very basic model of computation, which are nevertheless able to perform any type of algorithmic computation.[7] The fortunate circumstance that there are so many equivalent definitions of the same class of computable functions allows us to treat this notion very informally, without giving a precise definition of what a Turing machine is. Thus, a *computable function* is a function for which there is an algorithm, i.e. a finite step-by-step procedure, that computes it. It is an empirical fact that any reasonable formalization of this concept leads to the same class of functions.[8]

Having a precise mathematical definition of the notion of computability allows us to prove that certain functions or problems are *not* computable. One of the most famous examples is Turing's Halting Problem:

Definition 4.1   The *Halting Problem* is the problem, given a Turing machine M and an input x, to decide whether M produces an output on x in a finite number of steps (as opposed to continuing indefinitely).

Turing (1936) showed that the Halting Problem is *undecidable*, that is, that there is no algorithm deciding it. (Note the self-referential flavor of this statement: There is no algorithm deciding the behavior of algorithms.) Not only does this point to a fundamental obstacle in computer science (which did not yet exist in at the time that

---

[7]It is interesting to note that the Turing machine model has been a blueprint for all modern electronic computers. In particular, instead of performing specific algorithms, Turing machines are *universally programmable*, i.e. any algorithmic procedure can be implemented on them. Thus, the theory of computation *preceded* the actual building of electronic computers, and the fact that the first computers were universally programmable was directly influenced by it (cf. Copeland 2008). This situation is currently being repeated in the area of *quantum computing*, were the theory is being developed before any actual quantum computers have been built (see e.g. Arora and Barak 2009).

[8]The statement that the informal and the formal notions of computability coincide is the content of the so-called *Church-Turing thesis*, cf. Odifreddi (1989) for a discussion.

Turing proved this result), but it also entails the undecidability of a host of other problems.[9] Its importance for the theory of randomness will become clear in what follows.

5 Kolmogorov Complexity

An old and venerable philosophical principle, called *Occam's razor*, says that when given the choice between several hypotheses or explanations, one should always select the simplest one. The problem in applying this principle has always been to determine which is the simplest explanation: that which is simple in one context may be complicated in another, and there does not seem to be a canonical choice for a frame of reference.

A similar problem arises when we consider the two sequences on page 3: We would like to say that the first one, consisting of only 0's, is simpler than the second, because it has a shorter *description*. But what are we to choose as our description mechanism? When we require, as seems reasonable, that an object can be effectively reconstructed from its description, the notion of Turing machine comes to mind. For simplicity we will for the moment only consider finite binary strings. (This is not a severe restriction, since many objects such as numbers and graphs can be *represented* as binary strings in a natural way.) Thus, given a Turing machine M, we define a string y to be a description of a string x if $M(y) = x$, i.e. M produces x when given y as input. Now we can take the *length* of the string y as a measure of the complexity of x. However, this definition still depends on the choice of M. Kolmogorov observed that a canonical choice for M would be a *universal* Turing machine, that is, a machine that is able to simulate all other Turing machines. It is an elementary fact of computability theory that such universal machines exist. We thus arrive at the following definition:

Definition 5.1 Fix a universal Turing machine U. The *Kolmogorov complexity* of of a finite binary string x is the smallest length of a string y such that

$$U(y) = x.$$

We denote the Kolmogorov complexity of the string x by $C(x)$.

Hence, to say that $C(x) = n$ means that there is a string y of length n such that $U(y) = x$, and that there is no such y of length smaller than n. Note that the definition of $C(x)$ *still* depends on the choice of U. However, and this is the essential point, *the*

---

[9]In 1936, Turing used the undecidability of the Halting Problem to show the undecidability of the *Entscheidungsproblem*, that says (in modern terminology) that first-order predicate logic is undecidable.

theory of Kolmogorov complexity is independent of the choice of U in the sense that when we choose a different universal Turing machine U' as our frame of reference, the whole theory only shifts by a fixed constant.[10] For this reason, the reference to U is suppressed from this point onwards, and we will simply speak about *the* Kolmogorov complexity of a string.

Armed with this definition of descriptive complexity, we can now define what it means for a finite string to be random. The idea is that a string is random if it has no description that is shorter than the string itself, that is, if there is no way to describe the string more efficiently than by listing it completely.

Definition 5.2 A finite string x is *Kolmogorov random* if $C(x)$ is at least the length of x itself.

For example, a sequence of 1000 zero's is far from random, since its shortest description is much shorter than the string itself: The string itself has length 1000, but we have just described it using only a few words.[11] More generally, if a string contains a regular pattern that can be used to efficiently describe it, then it is not random. Thus this notion of randomness is related to the *compression* of strings: If $U(y) = x$, and y is shorter than x, we may think of y as a *compressed* version of x, and random strings are those that cannot be compressed.

A major hindrance in using Kolmogorov complexity is the fact that *the complexity function C is noncomputable*. A precise proof of this fact is given in Appendix B (see Corollary B.2), but it is also intuitively plausible, since to compute the complexity of y we have to see for which inputs x the universal machine U produces y as output. But as we have seen in Sect. 4, this is in general impossible to do by the undecidability of the Halting Problem! This leaves us with a definition that may be wonderful for theoretical purposes, but that one would not expect to be of much practical relevance. One of the miracles of Kolmogorov complexity is that the subject *does* indeed have genuine applications, many of which are discussed in the book by Li and Vitányi (2008). We will briefly discuss applications in Sect. 11.

We will not go into the delicate subject of the history of Kolmogorov complexity, other than saying that it was invented by Solomonoff, Kolmogorov, and Chaitin (in that order), and we refer to Li and Vitányi (2008) and Downey and Hirschfeldt (2010) for further information.

---

[10]This is not difficult to see: Since both U and U' are universal, they can simulate each other, and any description of x relative to U can be translated into a description relative to U' using only a fixed constant number of extra steps, where this constant is independent of x.

[11]Notice that the definition requires the description to be a string of 0's and 1's, but we can easily convert a description in natural language into such a string by using a suitable coding, that only changes the length of descriptions by a small constant factor. Indeed, the theory described in this chapter applies to anything that can be represented or coded by binary strings, which includes many familiar mathematical objects such as numbers, sets, and graphs, but also objects such as DNA strings or texts in any language.

6 Martin-Löf Randomness

The notion of Martin-Löf randomness, introduced by Martin-Löf in (1966), is based on classical probability theory, which in its modern formulation is phrased in terms of *measure theory*. In Appendix A the notion of a measure space is explained in some detail, but for now we keep the discussion as light as possible.

The unit interval [0, 1] consists of all the numbers on the real line between 0 and 1. We wish to discuss probabilities in this setting by assigning to subsets A of the unit interval, called *events*, a probability, which informally should be the probability that when we "randomly" pick a real from [0, 1] that we end up in A. The *uniform* or *Lebesgue measure* on [0, 1] assigns the measure $b - a$ to every interval $[a, b]$, i.e. the measure of an interval is simply its length. For example, the interval $[0, \frac{1}{2}]$ has measure $\frac{1}{2}$, the interval $[\frac{3}{4}, 1]$ has measure $\frac{1}{4}$. Note that [0, 1] itself has measure 1.

Given this, we can also define the measure of more complicated sets by considering combinations of intervals. For example, we give the combined event consisting of the union of the intervals $[0, \frac{1}{2}]$ and $[\frac{3}{4}, 1]$ the measure $\frac{1}{2} + \frac{1}{4} = \frac{3}{4}$. Since the measures of the subsets of [0, 1] defined in this way satisfy the laws of probability (cf. Appendix A), we can think of them as *probabilities*.

A series of intervals is called a *cover* for an event A if A is contained in the union of all the intervals in the series. Now an event A is defined to have measure 0 if it is possible to cover A with intervals in such a way that the total sum of the lengths of all the intervals can be chosen arbitrarily small.

For example, for every real x in [0, 1], the event A consisting only of the real x has measure 0, since for every n, x is contained in the interval $[x - \frac{1}{n}, x + \frac{1}{n}]$, and the length of the latter interval is $2\frac{1}{n}$, which tends to 0 if n tends to infinity.

These definitions suffice to do probability theory on [0, 1], and to speak informally about picking reals "at random", but we now wish to define what it means for a *single* real x to be random. We can view any event of measure 0 as a "test for randomness", where the elements not included in the event pass the test, and those in it fail. All the usual statistical laws, such as the law of large numbers, correspond to such tests. Now we would like to define x to be random if x passes all statistical tests, i.e. x is not in any set of measure 0. But, as we have just seen in the example above, every single real x has measure 0, hence in its full generality this definition is vacuous. (The reader may compare this to the situation we already encountered above in Sect. 3 when we discussed Kollektiv's.)

However, as Martin-Löf observed, we obtain a viable definition if we restrict ourselves to a countable collection of measure 0 sets. More precisely, let us say that an event A has *effective measure 0* if there is a *computable* series of covers of A, with the measure of the covers in the series tending to 0. Phrased more informally: A has effective measure 0 if there is an *algorithm* witnessing that A has measure 0, by producing an appropriate series of covers for A. Now we can finally define:

Definition 6.1 A real x is *Martin-Löf random* if x is not contained in any event of effective measure 0.

It can be shown that with this modification random reals exist.[12] Moreover, *almost every real in* [0, 1] *is random*, in the sense that the set of nonrandom reals is of effective measure 0.

Note the analogy between Definition 6.1 and the way that Church modified von Mises definition of Kollektiv, as described in Sect. 3: There we restricted to the computable selection functions, here we restrict to the effective measure 0 events.

Identifying a real number x with its decimal expansion,[13] we have thus obtained a definition of randomness for infinite sequences. The question now immediately presents itself what the relation, if any, of this definition is with the definition of randomness of *finite* sequences from Sect. 5. A first guess could be that an infinite sequence is random in the sense of Martin-Löf if and only if all of its finite initial segments are random in the sense of Kolmogorov, but this turns out to be false. A technical modification to Definition 5.1 is needed to make this work.

A string y is called a *prefix* of a string y' if y is an initial segment of y'. For example, the string 001 is a prefix of the string 001101. Let us now impose the following restriction on descriptions: If $U(y) = x$, i.e. y is a description of x, and $U(y') = x'$, then we require that y is not a prefix of y'. This restriction may seem arbitrary, but we can motivate it as follows. Suppose that we identify persons by their phone numbers. It is then a natural restriction that no phone number is a prefix of another, since if the phone number y of x were a prefix of a phone number y' of x', then when trying to call x' we would end up talking to x. Indeed, in practice phone numbers are not prefixes of one another. We say that the set of phone numbers is *prefix-free*. We now require that the set of descriptions y used as inputs for the universal machine U in Definition 5.1 is prefix-free. Of course, this changes the definition of the complexity function $C(x)$: Since there are fewer descriptions available, in general the descriptive complexity of strings will be higher. The complexity of strings under this new definition is called the *prefix-free complexity*. The underlying idea of the prefix-free complexity is the same as that of Kolmogorov complexity, but technically the theory of it differs from Kolmogorov complexity in several important ways. For us, at this point of the discussion, the most important feature of it is the following landmark result. It was proven in 1973 by Claus-Peter Schnorr, one of the pioneers of the subject.

Theorem 6.2 (Schnorr 1973) *An infinite sequence X is Martin-Löf random if and only if there is a constant c such that every initial segment of X of length n has prefix-free complexity at least $n - c$.*

The reader should take a moment to let the full meaning and beauty of this theorem sink in. It offers no less than an equivalence between two seemingly unrelated

---

[12]The proof runs as follows: There are only countably many algorithms, hence there are only countably many events of effective measure 0, and in measure theory a countable collection of measure 0 sets is again of measure 0.

[13]We ignore here that decimal expansions in general are not unique, for example $0, 999\ldots =$ $1, 000\ldots$, but this is immaterial.

theories. One is the theory of randomness for finite sequences, based on descriptive complexity, and the other is the theory of infinite sequences, based on measure theory. The fact that there is a relation between these theories at all is truly remarkable.

7 Martingales

Thus far we have seen three different formalizations of intuitions underlying randomness:

  (i) Mises–Wald–Church randomness, formalizing unpredictability using selection functions,
 (ii) Kolmogorov complexity, based on descriptive complexity,
(iii) Martin-Löf randomness, based on measure theory.

Theorem 6.2 provided the link between (ii) and (iii), and (i) was discussed in Sect. 3. We already mentioned Ville, who showed that the notion in (i) was flawed in a certain sense. Ville also showed an alternative way to formalize the notion of unpredictability of an infinite sequence, using the notion of a *martingale*, which we now discuss.[14] Continuing our game-theoretic discussion of Sect. 3, we imagine that we are playing against an unknown infinite binary sequence X. At each stage of the game, we are shown a finite initial part

$$X(0), X(1), X(2), \ldots, X(n-1)$$

of the sequence X, and we are asked to bet on the next value $X(n)$. Suppose that at this stage of the game, we have a capital of d dollar. Now we may split the amount d into parts b_0 and b_1, and bet the amount b_0 that $X(n)$ is 0, and the amount b_1 that $X(n)$ is 1. After placing our bets, we receive a payoff $d_0 = 2b_0$ if $X(n) = 0$, and a payoff $d_1 = 2b_1$ if $X(n) = 1$. Hence the payoffs satisfy the equation

$$\frac{d_0 + d_1}{2} = d. \tag{1}$$

After placing our bets, we receive a payoff d_0 if $X(n) = 0$, and a payoff d_1 if $X(n) = 1$.

For example, we may let $b_0 = b_1 = \frac{1}{2}d$, in which case our payoff will be d, no matter what $X(n)$ is. So this is the same as not betting at all, and leaving our capital intact. But we can also set $b_0 = d$ and $b_1 = 0$. In this case, if $X(n) = 0$ we receive a payoff of $2d$, and we have doubled our capital. However, if it turns out that $X(n) = 1$, we receive 0, and we have lost everything. Hence this placement of the

---

[14]The word "martingale" comes from gambling theory, where it refers to the very dangerous strategy of doubling the stakes in every round of gambling, until a win occurs. With the stakes growing exponentially, if the win does not occur quickly enough, this may result in an astronomical loss for the gambler. In modern probability theory, the word "martingale" refers to a betting strategy in general.

bets should be made only when we are quite sure that $X(n) = 0$. Any other placement of bets between these two extremes can be made, reflecting our willingness to bet on $X(n) = 0$ or $X(n) = 1$.

After betting on $X(n)$, the value $X(n)$ is revealed, we receive our payoff for this round, and the game continues with betting on $X(n + 1)$.

Now the idea of Ville's definition is that we should not be able to win an infinite amount of money by betting on a random sequence. For a given binary string σ, let $\sigma^\frown 0$ denote the string σ extended by a 0, and $\sigma^\frown 1$ the string σ extended by a 1. Formally, a *martingale* is a function d such that for every finite string σ the martingale equality

$$\frac{d(\sigma^\frown 0) + d(\sigma^\frown 1)}{2} = d(\sigma) \tag{2}$$

holds. The meaning of this equation is that when we are seeing the initial segment σ, and we have a capital $d(\sigma)$, we can bet the amount $\frac{1}{2}d(\sigma^\frown 0)$ that the next value will be a zero, and $\frac{1}{2}d(\sigma^\frown 1)$ that the next value will be a one, just as above in Eq. (1). Thus the martingale d represents a particular *betting strategy*. Now for a random sequence X, the amounts of capital

$$d\big(X(0), \ldots, X(n-1)\big)$$

that we win when betting on X should not tend to infinity.[15]

As in the case of Mises–Wald–Church randomness and the case of Martin-Löf randomness, this definition only makes sense when we restrict ourselves to a countable class of martingales.[16] A natural choice would be to consider the *computable* martingales. The resulting notion of randomness was studied in Schnorr (1971), and it turns out to be *weaker* than Martin-Löf randomness. However, *there exists another natural class of martingales, the so-called c.e.-martingales,*[17] *such that the resulting notion of randomness is equivalent to Martin-Löf randomness.*

Thus Ville's approach to formalizing the notion of unpredictability using martingales gives yet a third equivalent way to define the same notion of randomness.

8 Randomness and Provability

By Gödel's incompleteness theorem (see Sect. 4), in any reasonable formal system of arithmetic, there exist formulas that are true yet unprovable. A consequence of this result is that there is no algorithm to decide the truth of arithmetical formulas.

---

[15]Ville showed that martingales provide an alternative, game-theoretic, formulation of measure theory: The sets of measure 0 are precisely the sets on which a martingale can win an infinite amount of money.

[16]Note that for every sequence X there is a martingale that wins an infinite amount of capital on X: just set $d(X(0) \ldots X(n-1)^\frown i) = 2d(X(0) \ldots X(n-1))$, where $i = X(n)$. However, in order to play this strategy, one has to have full knowledge of X.

[17]C.e. is an abbreviation of "computably enumerable". This notion is further explained in Sect. 8.

It follows from the undecidability of the Halting Problem (see Definition 4.1) that the set of formulas that are *provable* is also undecidable.[18] However, the set of provable formulas is *computably enumerable*, meaning that there is an algorithm that lists all the provable statements. Computably enumerable, or c.e., sets, play an important role in computability theory. For example, the set H representing the Halting Problem is an example of a c.e. set, because we can in principle make an infinite list of all the halting computations.[19] The complement $\overline{H}$ of the set H, consisting of all nonconvergent computations, is *not* c.e. For if it were, we could decide membership in H as follows: Given a pair M and x, effectively list both H and its complement $\overline{H}$ until the pair appears in one of them, thus answering the question whether the computation $M(x)$ converges. Since H is not computable, it follows that $\overline{H}$ cannot be c.e. Because the set of all provable statements is c.e., it also follows that not all statements of the form

$$\text{``}M(x)\text{ does not halt''}$$

are provable. Hence there exist computations that do not halt, but for which this fact is not provable! Thus we obtain a specific example of a true, but unprovable statement. The same kind of reasoning applies if we replace H by any other noncomputable c.e. set.

Now consider the set R of all strings that are Kolmogorov random, and let non-R be the set of all strings that are not Kolmogorov random. We have the following facts:

(i) non-R is c.e. This is easily seen as follows: If x is not random, there is a description y shorter than x such that $U(y) = x$. Since the set of halting computations is c.e., it follows that non-R is also c.e.
(ii) R is not c.e. This is proved in Theorem B.1 in Appendix B.

By applying the same reasoning as for H above, we conclude from this that there are statements of the form

$$\text{``}x\text{ is random''}$$

that are true, but not provable. This is Chaitin's version of the incompleteness theorem, cf. Chaitin (1974).[20]

---

[18] This follows by the method of arithmetization: Statements about Turing machines can be translated into arithmetic by coding. If the set of provable formulas were decidable, it would follow that the Halting Problem is also decidable.

[19] We can do this by considering all possible pairs of Turing machines M and inputs x, and running all of them in parallel. Every time we see a computation $M(x)$ converge, we add it to the list. Note, however, that we cannot list the converging computations in order, since there is no way to predict the running time of a converging computation. Indeed, if we could list the converging computations in order, the Halting Problem would be decidable.

[20] As for Gödel's incompleteness theorem, the statement holds for any reasonable formal system that is able to express elementary arithmetic. In fact, it follows from Theorem B.1 that any such system can prove the randomness of at most *finitely* many strings.

Chaitin also drew a number of dubious philosophical conclusions from his version of the incompleteness theorem, that were adequately refuted by van Lambalgen (1989), and later in more

9 Other Notions of Randomness

Mises–Wald–Church random sequences were defined using computable selection functions, and Martin-Löf random sequences with computable covers, which in Ville's approach correspond to c.e.-martingales. As Wald already pointed out in the case of Kollektiv's, all of these notions can be defined relative to any count-able collection of selection functions, respectively covers and martingales. Choosing computable covers in the case of Martin-Löf randomness gave the fundamental and appealing connection with Kolmogorov randomness (Theorem 6.2), but there are situations in which this is either too weak, or too strong. Viewing the level of com-putability of covers and martingales as a parameter that we can vary allows us to introduce notions of randomness that are either weaker or stronger than the ones we have discussed so far.

In his groundbreaking book (1971), Schnorr discussed alternatives to the notion of Martin-Löf randomness, thus challenging the status of this notion (not claimed by Martin-Löf himself) as the "true" notion of randomness.[21]

In studying the randomness notions corresponding to various levels of computabil-ity, rather than yielding a single "true" notion of randomness, a picture has emerged in which every notion has a corresponding context in which it fruitfully can be applied. This ranges from low levels of complexity in computational complexity theory (see e.g. the survey paper by Lutz 1997), to the levels of computability (computable and c.e.) that we have been discussing in the previous sections, to higher levels of computability, all the way up to the higher levels of set theory. In studying notions of randomness across these levels, randomness has also served as a unifying theme between various areas of mathematical logic.

The general theory also serves as a background for the study of specific cases. Consider the example of π. Since π is a computable real number, its decimal expan-sion is perfectly predictable, and hence π it is not random in any of the senses discussed above. However, the distribution of the digits $0, \ldots, 9$ in π appears to be "random". Real numbers with a decimal expansion in which every digit occurs with frequency $\frac{1}{10}$, and more general, every block of digits of length n occurs with frequency $\frac{1}{10^n}$, are called *normal* to base 10. Normality can be seen as a very weak notion of randomness, where we consider just one type of statistical test, instead of infinitely many as in the case of Martin-Löf randomness. It is in fact not known if

---

(Footnote 20 continued)
detail by Ratikaainen, Franzen, Porter, and others. Unfortunately, this has not prevented Chaitin's claims from being widely publicized.

[21] After Martin-Löf's paper (1966), the notion of Martin-Löf randomness became known as a notion of "computable randomness". As Schnorr observed, this was not quite correct, and for example the characterization with c.e.-martingales pointed out that is was more apt to think of it as "c.e.-randomness". To obtain a notion of "computable randomness", extra computational restrictions have to be imposed. Schnorr did this by basing his notion on Brouwer's notion of constructive measure zero set. The resulting notion of randomness, nowadays called Schnorr randomness, has become one of the standard notions in randomness theory, see Downey and Hirschfeldt (2010).

π is normal to base 10, but it is conjectured that π is indeed "random" in this weak sense. For a recent discussion of the notion of normality, see Becher and Slaman (2014).

10 Pseudorandom Number Generators and Complexity Theory

In many contexts, it is desirable to have a good source of random numbers, for example when one wants to take an unbiased random sample, in the simulation of economic or atmospheric models, or when using statistical methods to estimate things that are difficult to compute directly (the so-called Monte Carlo method). In such a case, one may turn to physical devices (which begs the question about randomness of physical sources), or one may try to generate random strings using a computer. However, the outcome of a deterministic procedure on a computer cannot be random in any of the senses discussed above. (By Theorem B.1 in Appendix B, there is no purely algorithmic way of effectively generating infinitely many random strings, and it is easy to see that a Martin-Löf random set cannot be computable.) Hence the best an algorithm can do is to produce an outcome that is *pseudorandom*, that is, "random enough", where the precise meaning of "random enough" depends on the context. In practice this usually means that the outcome should pass a number of standard statistical tests. Such procedures are called *pseudorandom number generators*. That the outcomes of a pseudorandom number generator should not be taken as truly random was pointed out by the great mathematician and physicist John von Neumann, when he remarked that

> Anyone who considers arithmetical methods of producing random digits is, of course, in a state of sin.[22]

Randomized algorithms are algorithms that employ randomness during computations, and that allow for a small probability of error in their answers. For example, the first feasible[23] algorithms to determine whether a number is prime were randomized algorithms.[24] An important theme in computational complexity theory is the extent to which it is possible to *derandomize* randomized algorithms, i.e. to convert them to deterministic algorithms. This is connected to fundamental open problems about the relation between deterministic algorithms, nondeterministic algorithms,

---

[22]The Monte Carlo method was first used extensively in the work of Ulam and von Neumann on the hydrogen bomb.

[23]In computational complexity theory, an algorithms is considered feasible if it works in *polynomial time*, that is, if on an input of length n it takes n^k computation steps for some fixed constant k.

[24]Since 2001 there also exist *deterministic* feasible algorithms to determine primality (Agrawal et al. 2004), but the randomized algorithms are still faster, and since their probability of error can be made arbitrary small, in practice they are still the preferred method.

and randomized computation.[25] Besides being of theoretical interest, this matter is of great practical importance, for example in the security of cryptographic schemes that are currently widely used. For an overview of current research we refer the reader to Arora and Barak (2009). It is also interesting to note that randomness plays an important part in many of the proofs of results about deterministic algorithms, that do not otherwise mention randomness.

11 Applications

As pointed out in Sect. 5 and Corollary B.2, due to the undecidability of the Halting Problem, the notion of Kolmogorov complexity is inherently noncomputable. This means that there is no algorithm that, given a finite sequence, can compute its complexity, or decide whether it is random or not. Can such a concept, apart from mathematical and philosophical applications, have any *practical* applications? Perhaps surprisingly, the answer is "yes". A large number of applications, ranging from philosophy to physics and biology, is discussed in the monograph by Li and Vitányi (2008). Instead of attempting to give an overview of all applications, for which we do not have the space, we give an example of one striking application, namely the notion of information distance. Information distance is a notion built on Kolmogorov complexity that was introduced by Bennett et al. (1998). It satisfies the properties of a metric (up to constants), and it gives a well-defined notion of distance between arbitrary pairs of binary strings. The computational status of information distance (and its normalized version) was unclear for a while, but as the notion of Kolmogorov complexity itself it turned out to be noncomputable (Terwijn et al. 2011). However, it is possible to *approximate* the ideal notion using existing, computable, compressors. This gives a computable approximation of information distance, that can in principle be applied to any pair of binary strings, be it musical files, the genetic code of mammals, or texts in any language. By computing the information distance between various files from a given domain, one can use the notion to classify anything that can be coded as a binary string. The results obtained in this way are startling. E.g. the method is able to correctly classify pieces of music by their composers, animals by their genetic code, or languages by their common roots, purely on the basis of similarity of their binary encodings, and without any expert knowledge. Apart from these applications, the notion of information distance is an example of a *provably* intractable notion, which nevertheless has important practical consequences. This provides a strong case for the study of such theoretical notions.

---

[25]The question about derandomization is embodied in the relation between the complexity classes P and BPP, see Arora and Barak (2009). This is a probabilistic version of the notorious P versus NP problem, which is about determinism versus nondeterminism. The latter is one of the most famous open problems in mathematics.

Appendix A. Measure and Probability

A *measure space* is a set X together with a function μ that assigns positive real values $\mu(A)$ to subsets A of X, such that the following axioms are satisfied:

(i) The empty set $\emptyset$ has measure 0.
(ii) If $A \cap B = \emptyset$, then $\mu(A \cup B) = \mu(A) + \mu(B)$. That is, if A and B are disjoint sets then the measure of their union is the sum of their measures.[26]

If also $\mu(X) = 1$ we can think of the values of μ as *probabilities*, and we call X a *probability space*, and μ a *probability measure*. If A is a subset of X, we think of $\mu(A)$ as the probability that a randomly chosen element of X will be in the set A. Subsets of X are also called *events*. In this setting the axioms (i) and (ii) are called the *Kolmogorov axioms* of probability. The axioms entail for example that if $A \subseteq B$, i.e. the event A is contained in B, that then $\mu(A) \leqslant \mu(B)$.

An important example of a probability space consists of the unit interval $[0, 1]$ of the real line. The *uniform* or *Lebesgue* measure on $[0, 1]$ is defined by assigning to every interval $[a, b]$ the measure $b - a$, i.e. the *length* of the interval. The measure of more complicated sets can be defined by considering combinations of intervals.[27]

Appendix B. The Noncomputability of the Complexity Function

In Zvonkin and Levin (1970) the following results are attributed to Kolmogorov.

Theorem B.1 *The set R of Kolmogorov random strings does not contain any infinite c.e. set.[28] In particular, R itself is not c.e.*

Proof Suppose that A is an infinite c.e. subset of R. Consider the following procedure. Given a number n, find the first string a enumerated in A of length greater than n. Note that such a string a exists since A is infinite. Since a is effectively obtained from n, n serves as a description of a, and hence the Kolmogorov complexity $C(a)$ is bounded by the length of n, which in binary notation is roughly $\log n$ (plus a fixed constant c independent of n, needed to describe the above procedure), where $\log$ denotes the binary logarithm. So we have that $C(a)$ is at most $\log n$. But since a is random (because it is an element of A, which is a subset of R), we also have that

---

[26]It is in fact usually required that this property also holds for countably infinite collections.

[27]The definition of a probability measure on the unit interval $[0, 1]$ that assigns a probability to *all* subsets of it is fraught with technical difficulties that we will not discuss here. This problem, the so-called *measure problem*, properly belongs to the field of set theory, and has led to deep insights into the nature of sets and their role in the foundation of mathematics (cf. Jech 2003).

[28]R itself is infinite, but by the theorem there is no way to effectively generate infinitely many elements from it. Such sets are called *immune*.

$C(a)$ is at least the length of a, which we chose to be greater than n. In summary, we have $n \leqslant C(a) \leqslant \log n + c$, which is a contradiction for sufficiently large n. $\square$

Corollary B.2 *The complexity function C is not computable.*

Proof If C were computable, we could generate an infinite set of random strings, contradicting Theorem B.1. $\square$

Open Access This chapter is distributed under the terms of the Creative Commons Attribution-Noncommercial 2.5 License (http://creativecommons.org/licenses/by-nc/2.5/) which permits any noncommercial use, distribution, and reproduction in any medium, provided the original author(s) and source are credited. The images or other third party material in this chapter are included in the work's Creative Commons license, unless indicated otherwise in the credit line; if such material is not included in the work's Creative Commons license and the respective action is not permitted by statutory regulation, users will need to obtain permission from the license holder to duplicate, adapt or reproduce the material.

References

Agrawal, M., Kayal, N., & Saxena, N. (2004). PRIMES is in P. *Annals of Mathematics, 160*(2), 781–793.

Arora, S., & Barak, B. (2009). *Computational complexity: A modern approach.* Cambridge University Press.

Becher, V., & Slaman, T. A. (2014). On the normality of numbers in different bases. *Journal of the London Mathematical Society, 90*(2), 472–494.

Bennett, C. H., Gács, P., Li, M., Vitányi, P. M. B., & Zurek, W. (1998). Information distance. *IEEE Transactions on Information Theory, 44*(4), 1407–1423.

Chaitin, G. J. (1974). Information-theoretic limitations of formal systems. *Journal of the ACM, 21,* 403–424.

Church, A. (1940). On the concept of a random sequence. *Bulletin of the American Mathematical Society, 46,* 130–135.

Copeland, B. J. (Fall 2008 ed.). *The modern history of computing.* The Stanford Encyclopedia of Philosophy.

Downey, R. G., & Hirschfeldt, D. R. (2010). *Algorithmic randomness and complexity.* Springer.

Downey, R. G., Hirschfeldt, D. R., Nies, A., & Terwijn, S. A. (2006). Calibrating randomness. *Bulletin of Symbolic Logic, 12*(3), 411–491.

Doyle, P. G. (2011). *Maybe there's no such thing as a random sequence,* manuscript.

Freudenthal, H. (1969). Realistic models in probability. In I. Lakatos (Ed.), *Problems in inductive logic.* North-Holland.

Gödel, K. (1940). *The consistency of the continuum-hypothesis.* Princeton University Press.

Jech, T. (2003). *Set theory* (3rd millennium ed.). Springer.

Kolmogorov, A. N. (1933). *Grundbegriffe der Wahrscheinlichkeitsrechnung.* Springer.

Kunen, K. (1980). *Set theory: An introduction to independence proofs.* North-Holland.

Li, M., & Vitányi, P. (2008). *An introduction to Kolmogorov complexity and its applications* (3rd ed.). Springer.

Lutz, J. H. (1997). The quantitative structure of exponential time. In L. A. Hemaspaandra & A. L. Selman (Eds.), *Complexity theory retrospective II* (pp. 225–254). Springer.

Martin-Löf, P. (1966). The definition of random sequences. *Information and Control, 9,* 602–619.

Nies, A. (2009). *Computability and randomness.* Oxford University Press.

Odifreddi, P. G. (1989). Classical recursion theory (Vol. 1). In *Studies in logic and the foundations of mathematics* (Vol. 125). North-Holland.

Schnorr, C. P. (1971). Zufälligkeit und Wahrscheinlichkeit. In *Lecture Notes in Mathematics* (Vol. 218). Springer.

Schnorr, C. P. (1973). Process complexity and effective random tests. *Journal of Computer and System Sciences, 7,* 376–388.

Terwijn, S. A., Torenvliet, L., & Vitányi, P. M. B. (2011). Nonapproximability of the normalized information distance. *Journal of Computer and System Sciences, 77,* 738–742.

Turing, A. M. (1936). On computable numbers with an application to the Entscheidungsproblem. In *Proceedings of the London Mathematical Society* (Vol. 42, pp. 230–265). Correction in *Proceedings of the London Mathematical Society* (Vol. 43, pp. 544–546) (1937).

van Lambalgen, M. (1987). *Random Sequences*. Ph.D. thesis, University of Amsterdam.

van Lambalgen, M. (1989). Algorithmic information theory. *Journal of Symbolic Logic, 54*(4), 1389–1400.

Ville, J. (1939). *Étude critique de la notion de collectif*, Monographies des Probabilités, Calcul des Probabilités et ses Applications, Gauthier-Villars.

von Mises, R. (1919). Grundlagen der Wahrscheinlichkeitsrechnung. *Mathematische Zeitschrift, 5,* 52–99.

Wald, A. (1936). Sur la notion de collectif dans la calcul des probabilités. *Comptes Rendus des Seances de l'Académie des Sciences, 202,* 180–183.

Wald, A. (1937). Die Widerspruchsfreiheit des Kollektivbegriffes der Wahrscheinlichkeitsrechnung. *Ergebnisse eines Mathematischen Kolloquiums, 8,* 38–72.

Zvonkin, A. K., & Levin, L. A. (1970). The complexity of finite objects and the development of the concepts of information and randomness by means of the theory of algorithms. *Russian Mathematical Surveys, 25*(6), 83–124.

Randomness and the Madness of Crowds

Utz Weitzel and Stephanie Rosenkranz

Abstract Human interaction often appears to be random and at times even chaotic. We use game theory, the mathematical study of interactive decision making, to explain the role of rationality and randomness in strategic behavior. In many of these situations, humans deliberately create randomness as a best response and equilibrium strategy. Moreover, once out of equilibrium, individual beliefs about the real intentions of others introduce significant randomness into otherwise quite simple and deterministic situations of interaction. In a second step we discuss the role of randomness on financial markets, which are prototypical institutions for the aggregation of individual behavior. As in certain simple games, financial markets can produce outcomes that are close to perfect randomness. In fact, random walks in financial returns are considered by most scholars to be efficient and desirable. Finally, we apply game theoretical insights to behavior on financial markets and show how strategic speculation on 'greater fools' can create a 'madness of crowds' that often ends in chaotic swings, bubbles and crashes.

1 Introduction

In 1720, Sir Isaac Newton was heavily invested in the South Sea bubble. When the stock bubble burst he lost a fortune of about £2.4 million (in present day terms) and was quoted as stating: "I can calculate the movement of the stars, but not the madness of crowds".

U. Weitzel (✉)
Department of Economics, IMR, Nijmegen School of Management,
Radboud University, Thomas van Aquinostraat 5.1.26, 6525 GD Nijmegen,
The Netherlands
e-mail: u.weitzel@fm.ru.nl; u.weitzel@uu.nl

U. Weitzel · S. Rosenkranz
Utrecht University School of Economics, Kriekenpitplein 21-22,
3584EC Utrecht, The Netherlands
e-mail: s.rosenkranz@uu.nl

© The Author(s) 2016
K. Landsman and E. van Wolde (eds.), *The Challenge of Chance*,
The Frontiers Collection, DOI 10.1007/978-3-319-26300-7_4

67

The interaction between humans does indeed often appear like madness, governed by error and randomness. There is, however, a scientific field that attempts to logically explain human interaction. Game theory is the mathematical study of interactive decision making and it has revolutionized the way we see and understand economics, politics, financial markets, and many other aspects of human society. Game theory also applies to other species than humans and has made important contributions in, for example, biology.

This chapter will introduce simple game theoretical concepts and financial market applications to explain how we interact in certain situations and what role randomness plays in our behavior. The central question is how people deal with strategic uncertainty, which is the uncertainty about other people's expectations and actions that we face in human interaction. We then apply this approach to financial markets and discuss how heterogeneous beliefs and errors in updating can create feedback cycles and the 'madness of crowds' Newton referred to.

2   Super-Humans Against Nature and the Rationality Assumption

2.1   A Single Random Event

Imagine a very simple game against nature.

> Coin toss: First, human bets on one side of the coin, heads or tails. Then, in the coin toss, nature shows one side of the coin.[1]

Many people see the throw of a dice or a coin toss as a prime example for natural randomness. For at least 5000 years, our ancestors used randomization devices.[2] But is a coin toss really random? This goes back to an age old discussion culminating in the question whether the world is predictable or unpredictable; whether everything is predetermined, or whether nature is inherently stochastic. During the Age of Enlightenment and the Industrial Revolution, Isaac Newton's advances in mechanics suggested that the universe is predictably governed by simple physical laws. This lead to the lofty notion that, one day, humans might be able to take full control over their fate with a world formula. In 1814, the French astronomer and mathematician

---

[1]Another example of such a situation is a farmer who decides at the beginning of a year whether to plant crops or not (human places a bet). There is an equal chance that the weather this year is good or bad for crops (toss of a coin). It is up to nature to determine the outcome.

[2]The oldest known dice were part of an 5000-year-old backgammon set, excavated at the Burnt City in southeastern Iran. In ancient times the outcome of a throw of a dice was seen as the decision of God. Consequently, dice were frequently used in important decisions.

Pierre-Simon Laplace famously described the idea of scientific determinism as a perfect intelligence for which there exists no uncertainty (Laplace 1814).[3]

Laplace's Demon, as his notion became known, comes close to the definition of rationality in game theory. A completely rational agent is a super-human, an artificial construct that comes in handy when economists and game theorists need to build models. Like Laplace's Demon, this super-human knows everything ('perfect knowledge') and can compute even the most complex problems with lightning speed. Another feature of this super-human is that she always strives to maximize her own utility.[4]

In Laplace's scientific determinism, a coin toss is a quite boring affair. So would be Roulette or wheels of fortune. A super-human would simply know what side of the coin nature would show and bet accordingly. Scientific determinism remained the official dogma throughout the 19th century. This drastically changed with the 'probabilistic revolution in physics' initiated by statistical mechanics in the mid nineteenth century and continued by quantum mechanics in the early twentieth century (see Lüthy and Palmerino in this book).

But even without assuming unpredictable quantum states in quantum systems we may not be able to forecast with certainty, even in Laplace's deterministic world. Early works, for example, by Henry Poincaré have shown that, in deterministic systems, infinitesimally small changes in starting conditions can produce unpredictable outcomes.[5] This insight is the foundation of deterministic chaos theory and it took nearly a 100 years for the 'chaos revolution' to fully unfold. In the late 1960s, the MIT meteorologist Edward Lorenz discovered what is commonly referred to as the 'butterfly effect'.[6] In the 1970s, several mathematicians proved that simple nonlinear dynamic systems can produce irregular long run behavior and chaotic behavior without external random disturbance (Ruelle and Takens 1971; Li and Yorke 1975).[7] In nonlinear dynamic systems, predictions about the future become progressively worse when we do not have absolutely perfect knowledge of the initial

---

[3]See the contribution of Lüthy and Palmerino in this book for a more detailed discussion.

[4]Utility maximization is a tricky concept, which many mix up with ruthless money-making and egoism. First, utility is more than simply making money. Feeling happy, receiving love or any other positive sensation can also be a utility that people strive to maximize. This all depends on personal preferences. Given a choice between money and friendship, one person might prefer the former and another the latter. Second, a human can gain satisfaction (utility) from helping others. Did Mother Teresa only help others or also herself? Hence, being 'altruistic' can be perfectly in line with the definition of own utility maximization and rationality.

[5]In 1887, king Oscar II of Sweden promised a prize for the best answer to the question 'Is our solar system stable?' Poincaré showed that the motion in a simple three-body system—such as sun, earth and moon—that interact through gravitational attraction, can be sensitively dependent on initial conditions and become highly irregular and unpredictable.

[6]Lorenz and his team were running weather simulations on a computer and suddenly realized that rounding errors in the third decimal of just one measurement in one corner of their map (a 'flap of a butterfly') were able to change predictions in another area from clear skies to thunderstorms.

[7]A well-known application is logistic population growth in biology (May 1976).

state. Thus, even after the discovery of quantum physics, chaos theory re-introduced indeterminism 'through the back door' and at a surprisingly fundamental level.[8]

We will come back to deterministic chaos in complex systems in Sect. 5.3. For the time being, it is important to note that, according to quantum physics, but also to chaos theory, even a perfectly rational Laplacian super-human—without any restriction in knowledge and cognition—would approach a simple coin toss against nature in the same way as normal humans would: as a game of pure chance. This is in line with game theory where a perfectly rational agent is still exposed to randomness. When facing a coin flip, a rational decision maker, even when equipped with perfect knowledge, will not know whether the outcome will be head or tails.

2.2 Repeated Random Events

Fortunately, once faced with many independent coin tosses, our perfectly rational super-human can forecast the future very well.

> Repeated coin toss: We start with no money and every minute nature offers us a coin toss where we can either win one dollar (heads) or lose one dollar (tails). Our lifetime wealth then develops according to what is known as a 'random walk': we start at zero and might win a dollar, then another dollar (two dollars of wealth), then we may lose five dollars in a row (minus three dollars wealth), but then we win some money again, and so on.

What is our average lifetime wealth? According to the law of large numbers and the central limit theorem we can be almost certain to have earned an average of zero. Why? We have an equal chance to win or lose one dollar, on average, zero dollar. With millions of coin tosses, the gains and losses almost perfectly cancel each other out. On average, we expect to gain or lose nothing. We therefore also say that the expected value of such a coin toss is zero.

There is a catch, however. An expected value of zero dollar does not mean that we actually receive zero dollar. The expected value of a single coin toss is zero, but we still know for sure that the outcome will not be zero. Equally, just because we know that the average wealth over our life time is going to be very close to zero, our final wealth at the end of our life-time will most probably not be zero. In fact, our final wealth will probably be substantially above or below zero. Our final wealth is not an average but a single realization and it is impossible to predict this exact point. Hence, even if we are confident in predicting averages, we are not very good at exact point predictions.

Figure 1 shows this intuitively with a Galton board, named after the English scientist Sir Francis Galton. The horizontal position of the red ball dropped into the Galton board represents the wealth level and the pegs represent the coin tosses. Every time the red ball hits a peg there is an equal chance to fall to the left hand side (loss of one dollar) or the right hand side (gain of one dollar). Each red ball follows

---

[8]We thank Klaas Landsman for valuable contributions to this and the previous paragraph.

Fig. 1 Galton board

a random walk and many of these random walks (red balls) produce a binomial distribution of final wealth levels, as approximated by the distribution of red balls at the bottom of the Galton board. As binomial distributions are symmetric, the expected value of random walks, the average, is zero (the middle slot at the bottom). The large majority of the red balls, however, does not land in the middle slot. Therefore, although we can be quite sure to expect an average of zero wealth, individual final wealth levels are most probably not zero and the exact final wealth level (final slot) of one single ball is unpredictable.

2.3 Risk Preferences

How much would we bet on a single coin toss against nature in which we can win or lose one dollar? This depends on our risk preferences. The expected value is zero, so if we are risk-neutral we should offer the expected value, which is zero. This makes us indifferent between playing the game or not. But we might be risk-seeking. As the final wealth level of a single coin toss is certainly not zero, we might want to bet on the positive outcome of the coin toss and pay anything from 1 to 99 cents for playing the game. How much we are willing to pay for playing the coin toss is an indication of our risk-seekingness. Conversely, we might have a preference to prevent losses and to—at least partially—safeguard our current wealth level. In this case we are risk-averse and we require nature to pay us some amount from 1 to 99 cents to take the risk (play the game). The more risk averse we are, the more attractive nature must make the game for us to accept it. So, how much we are willing to pay/accept to play the game depends solely on our personal risk preferences. This also applies to fully rational super-humans. We assume that personal preferences are given and stable, but heterogeneous across individuals. Rational players are not necessarily risk-neutral. They can have any risk preference and maximize their payoff conditional on their preference. Moreover, we can have different types of preferences, not only for risk, but also for altruism or equality or with regard to other economic and social dimensions.

3   Super-Humans Against Super-Humans

The crucial characteristic of fully rational super-humans is that they have perfect knowledge about the rules of the game and know that this also applies to all other players, including the knowledge that they are also fully rational. The latter is called the 'common rationality assumption'. Given this definition of rationality, let's see what happens if two super-humans play the following game.

> Centipede game: Two super-humans, Superboy and Supergirl, play ball with each other. Nature randomly gives Supergirl the ball. She can decide to throw the ball to Superboy, or not. If she passes the ball, Superboy can decide to throw it back, or not. The game is finished either after 100 passes or if one of the two players decides not to pass the ball anymore. Nature also puts a number on the ball and increases it by 10 with every pass. When Supergirl gets the ball from nature, the number on the ball is 10. After the first pass, Superboy catches a ball displaying 20 on it. With the next pass the number changes to 30, and so on. If a player decides not to pass the ball, s/he gets the number on the ball paid out in dollar and the other player gets the same number divided by 10. Hence, the holder of the ball receives $10 + n \times 10$ dollar and the other player $(10 + n \times 10)/10 = 1 + n$ dollar after n passes.

Assuming that both players prefer to earn some money over nothing at all, how many passes do we observe between the two players? In game theory, analyses typically start at the end and then move backwards to the beginning. This is what we call 'backward induction'. After 100 successfully completed passes, Supergirl will get the ball back and receive 1010 dollar. But Superboy can see this coming and therefore does not pass the last ball back. Then Superboy gets 1000 dollar and Supergirl 100. Knowing this, Supergirl would not even pass the second-to-last ball to Superboy. Knowing this, Superboy would not make the pass before that one, and so on. Hence, when Supergirl receives the ball from nature, she does not even do the first pass and takes the 10 dollar. Superboy receives one dollar.

Supergirl's behavior is an equilibrium strategy. Under the common rationality assumption, Supergirl knows the equilibrium strategy of Superboy (keeping the ball) and she cannot benefit from changing her chosen strategy, while Superboy keeps his strategy unchanged. This applies to both players as none of the two players would pass the ball if randomly chosen by nature as first receiver. The current set of strategies and the corresponding payoffs constitute a Nash equilibrium, named after the mathematician and John Nash.[9]

Backward induction is often not very intuitive, which is one of the reasons why we have to assume super-humans. In many games only super-humans are actually able to 'see' the end of the game, keep it in mind, rationally backward induct, find the game-theoretical equilibrium strategy and finally play the corresponding equilibrium behavior flawlessly right from the beginning. Also, under the common rationality assumption we assume that everybody in the game is a super-human and everybody knows this. This takes all randomness out of the centipede game. Does this mean that randomness never plays a role for super-humans and always leads to determinism unless a mechanistic randomization device is introduced? Not quite. The point is that Supergirl may know everything about Superboy's reasoning, preferences and incentives, but this does not mean that Superboy's actions are always predictable. In fact, there are games where fully rational players want to be as unpredictable as possible.

> Rock-Paper-Scissors: Supergirl and Superboy simultaneously choose either Rock, Paper or Scissors. Rock beats Scissors, Paper beats Rock, Scissors beats Paper.

Each strategy has a $1/3$ chance to win, $1/3$ chance to draw and $1/3$ chance to lose. If Supergirl thinks that Superboy always plays Rock she could beat him with always choosing Paper. But this is not a Nash equilibrium as Superboy could improve on this strategy set by always choosing Scissors. This, again, would lead Supergirl to always choose Rock, and so it goes round and round. The only equilibrium strategy in this situation is to mix the three options as randomly as possible in order to win a least in $1/3$ of all tries, draw in $1/3$, and lose in $1/3$. So, the solution to this game is to play sequences that are perfectly random and unpredictable, just like a three-sided dice would be.

This is harder than we think. Humans are not very good at simulating random patterns. For example, in 'randomizing' we often underestimate clustering. This is the so-called gambler's fallacy, which describes the phenomenon that humans tend

---

[9]It does not make a difference if the two players communicate with each other. Whatever Superboy promises, he cannot commit to it. Therefore his answer is cheap talk. In fact, given his monetary preferences he has a clear commitment to keep the ball, because this maximizes his payoff. Knowing this, Supergirl will keep the ball even if Superboy promises to pass it back.

to expect a coin toss to show tails with a higher probability after a sequence of heads (Tversky and Kahneman 1971, 1974). In other situations we may fall prey to the hot-hand-effect (Gilovich et al. 1985). Here, we tend to believe that a series of heads indicates a higher likelihood of heads in future coin tosses.[10]

Of course, Supergirl and Superboy can randomize perfectly so that both win, draw and lose with equal probability over the long run. But as a thought experiment, let's take the Laplacian view to the extreme and see what would happen if fully rational super-humans would really know *everything* with absolute certainty. What would happen if the brains of two players are two completely transparent randomization devices (we basically see all neurons fire) and both players are able to perfectly anticipate—as a point prediction—what the other side will choose in the next round? In this situation, the only equilibrium strategy for both players would be to always play the same as the other so that every game ends in a draw.[11] But what happens if a draw is not an option?

> Matching pennies: Supergirl and Superboy each choose either heads or tails simultaneously. So, they both toss a virtual coin. Supergirl wins if the two coins match (heads and heads or tails and tails). Superboy wins in all other cases (coins do not match).

As in Rock-Paper-Scissors the Nash equilibrium is a mixed equilibrium strategy where both players have to perfectly randomize in order to win/lose half the time. Draw, however, is not an outcome. Thus, if super-humans could perfectly look into each other's brains, both players would constantly point-predict the opponent's intention for the next move, update, change their own intentions and point-predict again, only to realize that the opponent's intention has changed accordingly, and so on. In this setting, both players are frozen in an infinite optimization without the ability to act. This may be where free will or emotions are ultimately needed as 'circuit-breaker'. It may be that "to make a decision, emotion is the necessary trigger (and) without emotion, one would be reduced to the state of an idiot savant who goes on endlessly calculating without the ability to make a choice" (Olsen 1998).

---

[10]This phenomenon is found in sports, where people falsely attribute skill to a random series of wins and therefore believe that the team will win again. The same also applies to the believe that random successes in the past in investment performance will continue in the future. The hot-hand-effect applies less to situations where people have to randomize themselves, but more to situations where people have to correctly 'read' or identify random patterns produced by others.

[11]In terms of payoff it would not even matter whether two super-humans always play draw or perfectly randomize and win, draw and lose equally often. All that matters is that both know with certainty which of the two meta-strategies they will play: a perfect point-prediction of each other's next draw or a perfect randomization across the three options rock, paper, and scissors.

4 Humans Against Humans

4.1 Bounded Rationality

Rationality requires extreme assumptions concerning players cognitive abilities: perfect knowledge about all factors that affect the decision to be taken—so basically about everything—and virtually infinite computing abilities to derive rational expectations forecasts and optimal decisions. Needless to say that we are no super-humans. And needless to say that no economist seriously believes that human behavior is always fully rational. Rationality is only a benchmark model, but a very powerful one. It allows us to analyze benchmark behavior, which, under evolutionary pressure and over time, is theoretically more successful in dealing with nature and its randomness than any other model. Nevertheless, it is far from present in every human, in all situations, or at all times. In the 1950s, Herbert Simon advocated the concept of bounded rationality, a more realistic description of human behavior where agents have limited computing capacities and information (Simon 1955). Instead of perfectly optimal decision rules, boundedly rational players use short-cuts, rules of thumb, or so-called heuristics to overcome 'uncomputable' problems. These heuristics are not necessarily optimal or perfect but in complex environments they may perform reasonably well (for a discussion see Gigerenzer and Selten 2002). By using heuristics we inevitably make mistakes, which may be random but can also be biased.

4.2 Beliefs

As we cannot know everything, we are uncertain about the actions and beliefs (and beliefs about the beliefs) of others. This is commonly referred to as strategic uncertainty. Let's assume that Superboy and Supergirl in the above ball game (centipede game) can actually make mistakes. In other words, they are not super-humans anymore but simply humans: Girl and Boy. Let's also assume, that Girl, who received the ball from nature first, actually passes the ball to Boy. Remember that this is a move that super-humans would never do because it is no Nash equilibrium. However, as we now look at humans, there is a possibility that Boy receives the ball and suddenly has to form a belief about Girl's motives for passing the ball. Here are some beliefs that Boy might hold about Girl:

1. Girl *violates rationality* and made a mistake. She passed the ball, because she simply did not understand the game properly. She did not backward induct and did not realize that passing the ball in the first place is not fully rational.
2. Girl has *other preferences* (other than purely monetary ones). Maybe she passed the ball because she is altruistic and actually wants Boy to get the profit from the game. So, Girl actually gets more utility out of giving Boy the profits than keeping the ball and the money to herself.

3. Girl aims for a *more efficient outcome*. As the pot is increasing for both with every pass, Girl might expect that Boy colludes with her against nature. After the last pass, Girl and Boy would have extracted the highest possible profit from nature. For this, however, Girls would have to believe that Boy passes the last ball back to her (or have altruistic preferences).

Of course, the dilemma of the situation is that Boy does not know what Girl's underlying motivation was when she passed the ball. Boy has to form a belief about Girl's intentions, but he cannot know for sure. To make matters worse, in a world of many players, there are many possible beliefs and weighted mixtures of beliefs about each other's underlying motivations.

With certain assumptions, game theory can deal with these situations. For example, let us assume that all deviations from the rational equilibrium are because of the first of the above reasons. If people make independent and unbiased mistakes and we know about this, then Boy can compute how probable it is that Girl makes another mistake.[12] If players believe in a sufficiently high error rate, they end up in a 'Quantal Response Equilibrium' (QRE) of passing the ball at least once (McKelvey and Palfrey 1995, 1998). In fact, experimental evidence shows that the vast majority of people pass the ball more than once. Repeated rounds of this game also show, however, that the experienced error rate in the population in early rounds feeds into people's behavior in later rounds, which can then be explained quite rationally in a QRE sense (McKelvey and Palfrey 1992).

The basic reasoning in the centipede game is not restricted to sequential moves but can also take place in a one shot decision as the following example shows.

> Guessing game: Every person in a larger group is asked to privately pick a number from 0 to 100 and write it on a piece of paper. An experimenter collects the numbers and computes the average. The person with the number that is closest to $2/3$ of the group average wins. These rules are known to everybody before they pick the number (Moulin 1986).

Let's assume that everybody in the room (except you) randomly picks a number. Then the group's average would be 50 and you would pick $2/3 * 50 = 33$. If everybody thinks that, everybody would pick 33 and you should pick $2/3 * 33 = 22$. Then again, if everybody does that you should pick $14.\bar{6}$, $9.\bar{7}$, 6.5 etc. until you reach 0. Depending on their number, players exhibit distinct, boundedly rational levels of cognitive reasoning (Nagel 1995). Players with no level of reasoning ('Level 0') pick a random number, 'Level 1' players pick 33, 'Level 2' players pick 22, and so forth. In experiments, most players reveal first- and second-order depth of reasoning (Nagel 1995; Camerer et al. 2004).

Under the common rationality assumption, there is no strategic uncertainty about the others. Hence, if all players have an infinite level of reasoning, all players

---

[12]Of course, it might also be that Girl did not make a mistake at all but instead assumed that Boy would make a mistake. She might have passed the ball in the expectation that Boy erroneously passes it back. Hence, if we assume mistakes, observing a 'mistake' might not actually be a real error, but rational speculation on the other side making one. See Osborne (2003) for a discussion on this.

choose the number 0, which is the Nash equilibrium of this game. Zero is the only value where everybody in the group can win.

In a QRE-world, however, where we believe that some of us makes mistakes, 0 would not be a best response or equilibrium. We would have to pick a positive number, but which one exactly solely depends on our belief about the error rate of the other people in the group. Thus, to win this game in the real world, rational players should not choose the theoretical Nash equilibrium but a positive number. Interestingly, when doing so, we cannot tell anymore from the outside whether the winner was extremely rational or made a mistake and was simply lucky.

There are several other models that try to explain the real-world deviations from the Nash equilibrium in both the centipede and the guessing game (a.k.a. beauty contest). Cognitive hierarchy models, for example, assume that each player has a finite depth of reasoning and believes that s/he is the most sophisticated player in the game. Thus, in the guessing game, a Level 2 player will assume that all others are Level 1 and therefore choose 22. A Level 3 player expects all others to be Level 2 and chooses 14.$\bar{6}$, and so on.[13] Another branch of game theory, referred to as 'global games', attempts to deal with the second of the above reasons (other preferences), by assuming various simultaneous payoff structures that each player may face with a certain probability (Carlsson and Damme 1993).

In essence, all models advance possible ways how certain beliefs about other players' actions and beliefs are formed. Depending on these beliefs, practically all out-of-equilibrium outcomes can be reached. However, as all models plausibly describe experimentally observed outcomes, we still lack a fundamental under-standing of belief formation processes. How are initial beliefs (priors) about others are formed under strategic uncertainty? How quickly do people learn and in which way?[14] A common assumption is that people form expectations and update their beliefs about the real state of the world according to some learning scheme (Sargent 1993). Many studies in neuroscience, particularly in the area of sensorimotor control, suggest that our brain is a Bayesian prediction machine.[15] We would not be able to catch a ball without continuous forecasting and updating of priors about its most likely trajectory (Doya 2007). When it comes to cognitive processes, however, other studies have shown that we are not very good at Bayesian updating. For

---

[13]In the centipede game, if Girl is Level 0 (non-strategic), she will compare the payoffs at each possible endpoint of the game. As the pot is increasing for both with every pass, she will note that her highest reward results from Boy passing the ball on the final round. Girl will thus choose to always pass the ball. If Girl is Level 1, she will note that this outcome is not feasible for Boy on the last round and choose not to pass the ball on her last round. If Girl is Level 2, she expects that Boy is Level 1 and that he will, therefore, anticipate her ending the game on her last round. She therefore chooses to end the game on the second to last round, and so on.

[14]For example, in the centipede game, assume that Boy believes Girl is rational, but then he suddenly gets the ball. How did Boy come to his initial belief in the first place, and how does he adapt his belief given that Girl did not behave accordingly?

[15]Also see the chapter of Bekkering, van Elk and Friston in this book.

example, in the assessment of probabilities, people have been shown to neglect base rates (Kahneman and Tversky 1973). In stock markets, investors seem to over- and under-react to different types of news (De Bondt and Thaler 1985). Alternative models, for example, reinforcement learning and adaptive learning of simple forecasting heuristics or anchor and adjustment processes, are cognitively less demanding and allow for more errors (Kahneman 2003; Tversky and Kahneman 1974; Hommes 2013). At the extreme end of the spectrum, some psychologists argue that beliefs come first and that the brain is nothing more than a chatterbox that rationalizes beliefs ex post. The brain looks for patterns in sensory data and infuses them with meaning, forming beliefs. Then, it primarily focuses on the selection of confirmatory evidence that reinforces those beliefs in a positive feedback loop.[16]

4.3   Speculation

With heterogeneous beliefs and different levels of reasoning, speculation comes into play. We focus on financial speculation, which aims at making a profit from price movements in a market, even if these price movements are completely unrelated to the fundamental value of the underlying asset or its proceeds (e.g., dividends or interest).[17] This can be seen in the following adaptation of the centipede game from (Moinas and Pouget 2013).

> Bubble game: An asset, commonly known to have no fundamental value, is traded in a sequential market with three traders. At each point in the sequence, an incoming trader has two choices. S/he can either accept a buy offer at a given price and offer it to the next trader in line at a higher price, or s/he can reject the buy offer, which leaves the current owner stuck with a worthless asset. The last trader in the sequence cannot sell the asset anymore. Thus, when traders buy the asset, they effectively speculate on not being last and on being able to sell it to the next trader at a higher price. Traders do not know their position in the market sequence. They do, however, receive a signal about their position. This signal is the price of the asset that has been offered to them. The higher the offered price the higher the probability of being last in the sequence.

Figure 2 shows a graphical representation of the game. All traders receive one dollar initial capital. Trader 1 is offered to buy the asset at a randomly drawn price P_1 by nature.[18] Trader 1 does not know whether the offer comes from nature or a

---

[16]A recent bestseller of psychologist and science historian Michael Shermer has popularized this view (Shermer 2012).

[17]Despite many disadvantages and public criticism, speculation also has positive functions, for example, to provide liquidity in financial markets, which makes it easier or even possible for others to offset risk.

[18]As the random price can be above 1 dollar, we assume that a financial partner (who is not part of the game) provides each player with sufficient capital to be able to buy the asset. When selling the asset the financial partner gets all the profits except for 10 dollar which the trader receives.

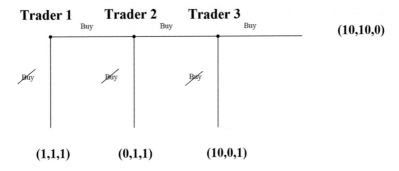

Fig. 2 Bubble game (extensive form)

previous trader (as s/he does not know her position in the sequence for sure).[19] When Trader 1 rejects the offer the game ends and all traders earn one dollar of initial capital. When Trader 1 accepts the offer, the asset is offered to Trader 2 at a price $P_2 > P_1$. When Trader 2 rejects, the game ends: Trader 1 earns nothing and Trader 2 and 3 each earn their initial capital (one dollar). When Trader 2 accepts, Trader 1 successfully sells the asset and earns 10 dollar. Trader 2 then offers the asset to Trader 3 at $P_3 > P_2$. When Trader 3 rejects, the game ends and Trader 2 is stuck with the worthless asset (Trader 1 gets 10 dollar, see above). As Trader 3 does not know for sure whether s/he is last in row she might buy the asset, but will be unable to resell. In this case Trader 3 gets nothing and Trader 1 and 2 each enjoy 10 dollar profit from successful reselling.

The Nash equilibrium of the bubble game is very similar to the centipede game: due to backward induction no trader should buy the asset. Thus, the first, randomly drawn price P_1 of the asset will not be accepted by Trader 1. Accordingly, the market value for the asset is equal to its fundamental value, namely 0. In their experiments, however, Moinas and Pouget (2013) find substantial trading of this worthless asset and the formation of significant price bubbles. Theoretically, the QRE povides the best explanation for this buying behavior (Moinas and Pouget 2013). Traders seem to believe that their fellow traders down the line will make mistakes. It is therefore rational for them to speculate on such mistakes and buy the asset as long as the probability to sell it to someone next in line is high enough. This result is very much in line with the famous 'greater fool theory' (Long et al. 1990), which suggests that rational traders buy overvalued assets in the expectation that a

---

[19]There are only two cases where traders can know their position for sure: when the offered price is the minimum or the maximum of the range of randomly drawn prices, which signal with certainty that they are at the first or last position in the sequence, respectively. For all other prices, however, traders can only infer a probability not to be last.

'greater fool' down the line will mistakenly buy the asset at an even higher price.[20] In fact, experimental tests show that individuals who speculate a lot in this game also produce stronger bubbles and crashes in more realistic and dynamic double auction trading environments (Janssen et al. 2015).

5 The Madness of Crowds

As explained in the previous section, speculators may try to ride a bubble in the belief that there are enough fools out there to buy them out. This can be a rational strategy and there are many scientific models that explain the existence of such rational bubbles in financial markets (see Stracca (2004) for an overview). It seems that there are potentially enough greater fools out there for more professional traders to speculate on. Heterogeneous agent models in finance assume that market participants are very different, not only with respect to preferences but also in terms of market experience, financial literacy and speculative sophistication (Hommes 2006). Empirical studies show that private traders, who are considered to be less sophisticated than professional traders, do not gain from their trading on average and actually underperform after deduction of transaction costs. Instead of (noise) trading, private investors could have made more money buy simply investing into a broadly diversified stock market index and do nothing (Barber and Odean 2000).

Speculating on greater fools, however, entails the risk to exit the market too late when not enough fools are left to buy the overpriced stocks. To complicate matters it is possible that speculators feed on each other, mistaking purchases of other speculators as noise. As in the guessing game it is often hard to tell whether a winning bid was really smart or simply lucky, particularly when there is a lot of noise. Warren Buffet, one of the richest and most successful investors of all time, once warned: "Nothing sedates rationality like large doses of effortless money. After a heady experience of that kind, normally sensible people drift into behavior akin to that of Cinderella at the ball. They know that overstaying the festivities— that is, continuing to speculate in companies that have gigantic valuations relative to the cash they are likely to generate in the future—will eventually bring on pumpkins and mice. But they nevertheless hate to miss a single minute of what is one helluva party. Therefore, the giddy participants all plan to leave just seconds before midnight. There's a problem, though: They are dancing in a room in which the clocks have no hands."[21]

---

[20]'Greater fools' are also often called 'noise traders', because they are seen to buy and sell assets in financial markets at random, like 'white noise'. Classical examples of noise traders are inexperienced individuals who inherit some money and decide to invest it in some random portfolio in the stock market.
[21]Warren Buffet, Letter to the Shareholders of Berkshire Hathaway Inc., 2000, p. 14.

5.1 Luck Versus Skill

This raises the question how speculators can be viewed as professional rational agents who exploit noise traders and, at the same time, as 'giddy Cinderellas' who miss the point of exit. The answer is that, although professional traders and sophisticated speculators may not be greater fools, even they cannot beat the market in the long run, which makes them fools, too; maybe lesser fools, but fools after all. This notion is a direct implication of the efficient market hypothesis (EMH), which states that nobody can systematically beat the market. The value of a financial asset is defined by its expected future cash flow, discounted to its present value. Through the market mechanism, all relevant forecasts of market participants are compounded in market prices. If financial markets are efficient, which means that all information about possible future states of nature and cash flows are impounded in market prices instantaneously, then the residual price movements must be triggered by genuine surprises, which nobody has seen coming and which are therefore, by definition, a random walk (Fama 1965).

For a graphic representation, let's extend the Galton board in Figure 0 to 1000 rows of pegs, run a couple of balls through it and track their paths. Figure 3 shows some of the random walks of these balls, turned by 90° so that they now 'fall' horizontally along the x-axis of 1000 pegs. Remember that this is equivalent to a 1000 coin tosses in which we can either lose or gain a dollar. Most random walks will deviate substantially and for longer periods from wealth levels of zero. Two thirds can deviate as far as ± 31.70 dollars, indicated by the two dotted lines, which are defined by $\sigma \times \sqrt{n}$: the standard deviation of the coin toss ($\sigma = 1$) and the number of tosses ($n = 1000$). One third of all random walks will deviate at some point to wealth levels above and below $\sigma \times \sqrt{n}$, as the two outliers show with wealth levels of ± 100 dollars.[22]

As the EMH predicts, the random walks in Fig. 3 have a high resemblance with stock price charts. In fact, some surveys indicate that stock market traders and other financial professionals cannot reliably tell the difference between random walks and real stock price developments (Siegel 2013). Many studies in financial economics show that the performance of the vast majority of financial professionals is due to (random) luck and not skill (Fama and French 2010; Malkiel 1995). Luck to be active in a certain period and in a certain class of investments. As a famous multi-annual experiment by the Wall Street Journal showed there is a very high likelihood that a dart-throwing monkey is an equally 'skilled' stock market forecaster as professional investment advisers (Porter 2005). If an investment manager

---

[22]Theoretically, if enough red balls fall through the Galton board, 1000 pegs or coin flips can produce a sequence of 1000 heads, leading to a final wealth of 1000 dollar. This is equivalent to Émile Borel's infinitely typewriting ape, published in 1913. At one point in time, by chance, this ape will have produced the Bible or Hamlet or any other finite text.

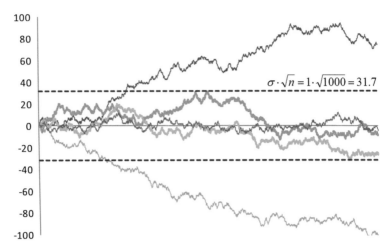

Fig. 3 Random walks

has an exceptional track record of past investments, there is a good chance that we have met the upper outlier random walk in Fig. 3 and not somebody who can consistently predict super-investments that others simply did not see. The catch with random walks is that the expected value of all future coin flips does not change and always remains zero, no matter at which point we currently are. This is what mathematicians and finance scholars call a 'martingale': at each point in a realized random sequence, the conditional expectation of the next value in the sequence is equal to the current value, irrespective of the preceding sequence. The martingale property of asset returns in efficient financial markets is the reason why governments warn clients that past investment performance provides no indication for the future. Unfortunately, too many investors believe that significant positive deviations from the x-axis are a signal of skill and not luck (Hoffmann and Post 2014).[23] In doing so, they fall prey to the self-attribution bias, which is the tendency to attribute success to one's own disposition and failure to external forces (Miller and Ross 1975; Feather and Simon 1971).

The prevalence of the EMH is the reason why traders say that there is 'no free lunch at Wall Street'. You cannot simply predict future stock prices from some charts (its preceding sequence) and make some easy money. Even news, when publicly available, cannot be used as forecasting and trading advantage as it is almost instantaneously compounded in the market price. In many financial markets computer algorithms are involved in more than half of all financial transactions. Algorithms trade in milliseconds, impounding new information in prices much

---

[23]For a vivid description of the pitfalls of randomness that financial traders falls prey to, also see Nassim Nicholas Taleb's bestseller 'Fooled by randomness' (Taleb 2005).

quicker than any human trader could, which has a positive effect on the informa-tiveness of prices (Chaboud et al. 2014).[24]

The bottom line is, that efficient financial markets are very good in 'producing' random walks. There is a broad consensus in the academic finance community—including many critics of the EMH—that, because of the efficiency of most financial markets, it is very hard, if not impossible, for traders to systematically beat the market (Stracca 2004).[25] In the end we are all greater or lesser fools in light of the self-produced randomness on financial markets.

5.2   *No Free Lunch ≠ the Price Is Right*

The EMH is probably the most powerful and, at the same time, most hotly debated principle in Finance. This was demonstrated in 2013, when the Nobel Prize in Economics was awarded to three eminent scholars: Eugene Fama, father of the EMH; Robert Shiller, an outspoken critic of the EMH, and Lars Peter Hansen, who offered an econometric compromise between the two. The EMH has two implica-tions: one is that we cannot beat the market (no free lunch); the other is that, because of this informational efficiency, the market price we observe is a correct estimate of a financial asset's future cash flows a.k.a. its fundamental or intrinsic value (the price is right). The former looks at price changes (returns), the latter at price levels. In the former we are in a world of arbitrage which exploits temporary differences between prices.[26] In the latter we are in a world of market timing, over-/undervaluation and mean reversion, which exploit differences to fundamental values. It is the latter of the two worlds in which we believe to observe 'madness' in markets: bubbles and crashes that—with hindsight—seem to be everything but 'the right price'.[27] As much as financial scholars agree on the former, that we cannot beat the market, they are critical about the latter, the claim that the price is always right (Stracca 2004).

---

[24]The implications of algorithmic trading for social welfare are less clear. The informational efficiency by speeding up price discovery with machines may not be socially efficient if traders overinvest in technology due to adverse selection (Biais et al. 2011).

[25]This insight has led to the phenomenal growth of index funds, which specialize in automatic and therefore very cost-effective investments in large, diversified index portfolios (the market return), without the pretense of being able to beat the market.

[26]A classic example is triangular arbitrage in currency markets. If we pay 2 euro for 1 dollar, 1 dollar for 1 pound, and 1.5 euro for 1 pound, then it makes sense to buy pounds with euros (1.5:1), sell pounds for dollars (1:1), and sell dollars for euros (1:2) until all three exchange rates are perfectly balanced.

[27]A prominent example is the 'tulipmania' in March 1637 in the United Provinces (now the Netherlands), where a single tulip bulb reached prices of more than 3000 guilders (florins), which was about 10 times the annual income of a skilled craftsman. Note that many of the peak prices were quoted in futures contracts which were later changed by decree into options contracts. Thus, despite extreme price quotes, it is questionable whether much money had changed hands between buyers and sellers (Thompson 2006).

To unravel this apparent contradiction we have to understand that the EMH rests on three, progressively weaker conditions, any one of which will lead to market efficiency: (i) full rationality, (ii) independent deviations from rationality, and (iii) unlimited arbitrage (Shleifer 2000). Proponents of the EMH argue that, even if conditions (i) and (ii) do not hold, which is widely accepted, any systematic pricing errors (biases) will be arbitraged away by more sophisticated traders. Critics of the EMH argue that the potential of arbitrageurs to reduce mispricing is limited: arbitrage is not riskless, in many situations there exist severe liquidity constraints to arbitrage against the market, and arbitrage requires substantial investments in ICT, real-time data, and human capital to succeed in a very competitive business (Shleifer and Vishny 1997). Hence, even if there is no free lunch, because the market does not offer any *feasible* arbitrage opportunities, this does not necessarily lead to a convergence of prices to fundamental values (Stracca 2004). This has been demonstrated by Robert Shiller, who is well-known for his early warnings of a housing price bubble in a comparatively inefficient market with very limited arbitrage possibilities.[28] A related criticism is that arbitrage is limited, because arbitrageurs themselves are boundedly rational. Then less rational traders (greater fools) are driven out of the market by more rational traders (lesser fools) so that nobody can beat the market anymore, but this does not exclude that assets are mispriced. Overall, "the existence of a pricing bias due to behavioral factors is indeed fully compatible with rational expectations and a random walk behavior of asset prices" (Stracca 2004 p. 395).

5.3   From Mispricing to Madness

An important difference between economics and natural sciences is that today's economic decisions and actions depend on today's beliefs and expectations about the future (which again can differ from tomorrow's belief about the future). The predictions, expectations or beliefs of agents about the future are part of a highly endogenous, dynamic and nonlinear feedback system which requires a *theory of expectations* (Hommes 2013). An early and mathematically very elegant theory of expectations was the rational expectations hypothesis (Muth 1961; Lucas Jr 1972):

---

[28]Accordingly, Shiller calls for more financial innovation that allow trading of risks that really matter: "Had there been a well-developed real estate market before the financial crisis of 2008, it would plausibly have reduced the severity of the crisis, because it would have allowed, even encouraged, people to hedge their real estate risks. The severity of that crisis was substantially due to the leveraged undiversified positions people were taking in the housing market, causing over 15 million US households to become underwater on their mortgages, and thus reducing their spending. There is no contradiction at all in saying that there are bubbles in the housing market and yet saying that we ought to create better and more liquid markets for housing" (Shiller 2014, p. 1511).

under assumptions of rationality this hypothesis provides a rational expectations equilibrium (REE), where expectations and realizations, on average, coincide. Theoretically, in an efficient market with risk neutral agents, prices correctly reflect all possible future states of an asset's cash flows (discounted at the risk free rate) and their true, physical (objective) probabilities. Hence, from efficient risk neutral market prices we can infer state price probabilities that coincide with objective probabilities.[29]

The REE refers to situations where we play Roulette with well-defined states, probability distributions and expected values. We refer to this kind of uncertainty as risk. Risk can be seen as a very special case of uncertainty, but it is not the norm. Most decisions in life are taken without knowing objective probabilities or all possible states, often referred to as ambiguity (Wakker 2010). Ambiguous situations provide a fertile breeding ground for very heterogenous beliefs and expectations (Stahl 2013) which agents have to learn about. As learning is not perfect, boundedly rational systems can be complex, nonlinear and dynamic (Hommes 2013). In such an environment, strategic uncertainty about the beliefs and behavior of others can easily create nonlinear feedback cycles. This would not be a problem if the system eventually converges to the REE.[30] There are many examples, however, where bounded rationality leads to deterministic chaos that makes predictions virtually impossible and forecasts become practically random. Econometric time series studies did not succeed in ruling out randomness in stock price data (or deterministic chaos) and there is strong evidence for nonlinear dependence (Hommes 2013). Hence, while fully informed rational expectations are self-fulfilling in the REE, less informed prophecies can also be self-fulfilling in boundedly rational systems under ambiguity.

A typical example of such a feedback cycle are situations where fundamental values themselves are affected by market evaluations. To illustrate this, take a look at the market price of Tesla Motors as shown in Fig. 4. In mid 2014, the electric car company is trading at a market value of more than half that of General Motors, Ford, and Honda. Each of those established companies had more than 50 times the annual revenues as Tesla. "Pure electric cars remain a niche market, making up <1 % of total U.S. car sales. And within that, Tesla is a niche product. Its Model S

---

[29]When markets reflect risk aversion, state price probabilities for undesirable (desirable) states are higher (lower) than objective probabilities (Bossaerts and Oedegaard 2000). The equivalent martingale measure (EMM) is a probability measure in mathematical finance that adjusts the observed state price probabilities of future outcomes such that they incorporate investors' risk preferences. The EMM is a central tool in arbitrage pricing. It reflects the probability distribution under which all possible bets are fair given complete markets and no-arbitrage conditions.

[30]Attempts by finance theorists to reconcile evidence of individual non-rational behavior with aggregate rationality at the market level through learning and evolutionary selection has proved difficult as they required a number of demanding conditions (see Sect. 5.2 and Stracca (2004) for a discussion).

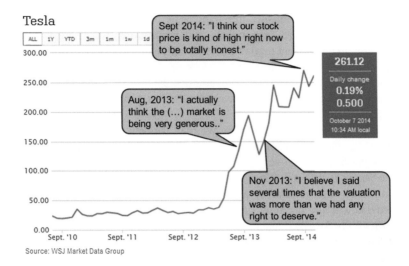

Fig. 4 Stock price of Tesla motors, 2010–2014

costs about $75,000, while prices for the Leaf start around $30,000 and the Volt around $35,000."[31] Moreover, in 2014 Tesla sold less e-cars than Nissan.[32]

Is Tesla a bubble? Interestingly, Tesla's CEO himself, Elon Musk, repeatedly remarked that he considered the stock to be overvalued (see quotes in Fig. 4). Indeed, there are indications that the price is partially driven by speculation.[33] It may therefore be rational, albeit risky, for investors to ride the bubble as long as others are still buying. In support of the latter, apparently many people believe that Tesla will lead a revolution in the car industry. In fact, the high share price, possibly also driven by pure financial speculation, provided enough funding for Tesla to make some very expensive investments in potentially game-changing projects.[34] Thus, if shareholders' beliefs have been over-optimistic originally, precisely this deviation from otherwise rational expectations, possibly reinforced by rational speculation, may have provided Tesla with the necessary capital to make their beliefs more realistic.

Even with hindsight it will be difficult to disentangle the underlying effects in Tesla's stock price development. "There is often a tendency (probably because

---

[31]According to marketwatch.com, Oct 3, 2014 2:59 p.m. ET.

[32]In September 2014, the most sold e-car was Nissan Leaf (2881 units), followed by Tesla's Model S (1650 units) and Chevrolet's Volt (1394) and BMW's 3i (1022).

[33]In a cryptic tweet in October 2014, Musk mentioned "D and something else". As a popular investor news site, MarketWatch.com, reported, "Musk's cryptic tweets last Thursday—and the rampant speculation they have fueled since—have pushed Tesla (...) shares about 9 % higher from their Wednesday close."

[34]Tesla announced that they invest 5 billion US$ in a lithium-ion battery Gigafactory with a planned production that exceeds the world capacity of 2013. Tesla also embarked on building an ambitious network of Supercharger stations along roads to facilitate longer distance journeys.

economists are themselves affected by hindsight bias) to regard a certain development caused by market developments as unavoidable (supporting the idea of exogenous rationality). But it can sometimes be the result of a self-fulfilling spiral in which the prime mover is indeed an 'endogenous' market whimsical move. (…) The issue of the feedback mechanism seems most relevant in this respect. Thus far, there has been no systematic attempt to address the issue of the feedback from market prices to fundamentals, and only some informal speculations have been provided (Shiller 2000a, b; Daniel et al. 2002)" (Stracca 2004, p. 397).

6   Conclusion

Interactions between people are rich in randomness, consciously produced or unintended. The fertilization of economics and finance with psychological ideas and evidence allows for new insights in dealing with randomness in human interactions, but it also adds to the risk of being less parsimonious (Tirole 2002). A useful feature of many game theoretical models and the classical REE is that they impose a strong discipline on the degrees of freedom in economic models. Boundedly rational models run the risk of incorporating too much randomness and freedom as if anything goes. "To avoid 'ad hoccery', a successful bounded rationality research program needs to discipline the class of expectations and decision rules" (Hommes 2013, p. 9). In doing so, and in order to understand 'madness' in markets, more investigation in social psychology rather than individual psychology is needed. We need to understand how randomness can be channeled at the aggregate level in social and economic systems, for example through the synchronization of expectations with improved market structures and communication (see, e.g., Shiller 2000a, b).

Open Access This chapter is distributed under the terms of the Creative Commons Attribution-Noncommercial 2.5 License (http://creativecommons.org/licenses/by-nc/2.5/) which permits any noncommercial use, distribution, and reproduction in any medium, provided the original author(s) and source are credited. The images or other third party material in this chapter are included in the work's Creative Commons license, unless indicated otherwise in the credit line; if such material is not included in the work's Creative Commons license and the respective action is not permitted by statutory regulation, users will need to obtain permission from the license holder to duplicate, adapt or reproduce the material.

References

Barber, B. M., & Odean, T. (2000). Trading is hazardous to your wealth: The common stock investment performance of individual investors. *The Journal of Finance, 55*(2), 773–806.
Biais, B., Foucault, T., & Moinas, S. (2011). *Equilibrium high frequency trading. Manuscript.* Retrieved August 17, 2012.

Bossaerts, P. L., & Oedegaard, B. A. (2000). *Lectures on Corporate Finance.* Singapore: River Edge, New Jersey: World Scientific Pub Co Inc.

Camerer, C. F., Ho, T.-H., & Chong, J.-K. (2004). A cognitive hierarchy model of games. *The Quarterly Journal of Economics, 119*(3), 861–898.

Carlsson, H., & Damme, Ev. (1993). Global games and equilibrium selection. *Econometrica, 61*(5), 989–1018.

Chaboud, A. P., Chiquoine, B., Hjalmarsson, E., & Vega, C. (2014). Rise of the machines: Algorithmic trading in the foreign exchange market: Rise of the machines. *The Journal of Finance, 69*(5), 2045–2084.

Daniel, K., Hirshleifer, D., & Teoh, S. H. (2002). Investor psychology in capital markets: Evidence and policy implications. *Journal of Monetary Economics, 49*(1), 139–209.

De Bondt, W. F. M., & Thaler, R. (1985). Does the stock market overreact? *The Journal of Finance, 40*(3), 793–805.

Doya, K. (2007). *Bayesian brain: Probabilistic approaches to neural coding.* Cambridge: MIT Press.

Fama, E. F. (1965). Random walks in stock-market prices. *Financial Analysts Journal, 21,* 55–59.

Fama, E. F., & French, K. R. (2010). Luck versus Skill in the cross-section of mutual fund returns. *The Journal of Finance, 65*(5), 1915–1947.

Feather, N. T., & Simon, J. G. (1971). Attribution of responsibility and valence of outcome in relation to initial confidence and success and failure of self and other. *Journal of Personality and Social Psychology, 18*(2), 173–188.

Gigerenzer, G., & Selten, R. (Eds.). (2002). *Bounded rationality: The adaptive toolbox* (reprint edition). Cambridge, MA: The MIT Press.

Gilovich, T., Vallone, R., & Tversky, A. (1985). The hot hand in basketball: On the misperception of random sequences. *Cognitive Psychology, 17*(3), 295–314.

Hoffmann, A. O. I., & Post, T. (2014). Self-attribution bias in consumer financial decision-making: How investment returns affect individuals' belief in skill. *Journal of Behavioral and Experimental Economics, 52,* 23–28.

Hommes, C. (2013). *Behavioral rationality and heterogeneous expectations in complex economic systems.* Cambridge: Cambridge University Press.

Hommes, C. H. (2006). Heterogeneous agent models in economics and finance, Chapter 23. In L. Tesfatsion & K. L. Judd (Eds.), *Handbook of computational economics* (Vol. 2, pp. 1109–1186). Elsevier.

Janssen, D.-J., Weitzel, U., & Füllbrunn, S. (2015). *Speculative bubbles—an introduction and application of the speculation elicitation task (SET).* SSRN Scholarly Paper ID 2577867, Social Science Research Network, Rochester, NY.

Kahneman, D. (2003). Maps of bounded rationality: Psychology for behavioral economics. *The American Economic Review, 93*(5), 1449–1475.

Kahneman, D., & Tversky, A. (1973). On the psychology of prediction. *Psychological Review, 80*(4), 237–251.

Laplace, P. (1814). *A philosophical essay on probabilities.*

Li, T.-Y., & Yorke, J. A. (1975). Period three implies chaos. *The American Mathematical Monthly, 82*(10), 985–992.

Long, J. B. D., Shleifer, A., Summers, L. H., & Waldmann, R. J. (1990). Noise trader risk in financial markets. *Journal of Political Economy, 98*(4), 703–738.

Lucas, R. E, Jr. (1972). Expectations and the neutrality of money. *Journal of Economic Theory, 4*(2), 103–124.

Malkiel, B. G. (1995). Returns from investing in equity mutual funds 1971 to 1991. *The Journal of Finance, 50*(2), 549–572.

May, R. M. (1976). Simple mathematical models with very complicated dynamics. *Nature, 261*(5560), 459–467.

McKelvey, R. D., & Palfrey, T. R. (1992). An experimental study of the centipede game. *Econometrica, 60*(4), 803–836.

McKelvey, R. D., & Palfrey, T. R. (1995). Quantal response equilibria for normal form games. *Games and Economic Behavior, 10*(1), 6–38.

Mckelvey, R. D., & Palfrey, T. R. (1998). Quantal response equilibria for extensive form games. *Experimental Economics, 1*(1), 9–41.

Miller, D. T., & Ross, M. (1975). Self-serving biases in the attribution of causality: Fact or fiction? *Psychological Bulletin, 82*(2), 213–225.

Moinas, S., & Pouget, S. (2013). The bubble game: An experimental study of speculation. *Econometrica, 81*(4), 1507–1539.

Moulin, H. (1986). *Game theory for the social sciences* (revised edition). New York: NYU Press.

Muth, J. F. (1961). Rational expectations and the theory of price movements. *Econometrica, 29*(3), 315–335.

Nagel, R. (1995). Unraveling in guessing games: An experimental study. *The American Economic Review, 85*(5), 1313–1326.

Olsen, R. A. (1998). Behavioral finance and its implications for stock-price volatility. *Financial Analysts Journal, 54*(2), 10–18.

Osborne, M. J. (2003). *An introduction to game theory* (1st edition). New York: Oxford University Press.

Porter, G. E. (2005). The long-term value of analysts' advice in the wall street journal's investment dartboard contest. *Journal of Applied Finance, 14*(2), Fall/Winter 2004.

Ruelle, D., & Takens, F. (1971). On the nature of turbulence. *Communications in Mathematical Physics, 20*(3), 167–192.

Sargent, T. J. (1993). *Bounded rationality in macroeconomics: The arne ryde memorial lectures.* OUP Catalogue: Oxford University Press.

Shermer, M. (2012). *The believing brain—how we construct beliefs and reinforce them as truths.* New York: St. Martin's Griffin.

Shiller, R. J. (2000a). *Irrational exuberance.* Princeton: Princeton University Press.

Shiller, R. J. (2000b). Measuring bubble expectations and investor confidence. *Journal of Psychology and Financial Markets, 1*(1), 49–60.

Shiller, R. J. (2014). Speculative asset prices. *American Economic Review, 104*(6), 1486–1517.

Shleifer, A. (2000). *Inefficient markets: An introduction to behavioral finance.* Oxford, U.S.A: Oxford University Press.

Shleifer, A., & Vishny, R. W. (1997). The limits of arbitrage. *The Journal of Finance, 52*(1), 35–55.

Siegel, J. J. (2013). *Stocks for the long run: The definitive guide to financial market returns & long-term investment strategies* (5th ed.). New York: McGraw-Hill.

Simon, H. A. (1955). A behavioral model of rational choice. *The Quarterly Journal of Economics, 69*(1), 99–118.

Stahl, D. O. (2013). Heterogeneity of ambiguity preferences. *Review of Economics and Statistics, 96*(4), 609–617.

Stracca, L. (2004). Behavioral finance and asset prices: Where do we stand? *Journal of Economic Psychology, 25*(3), 373–405.

Taleb, N. N. (2005). *Fooled by randomness: The hidden role of chance in life and in the markets* (2nd updated edition). New York: Random House Trade Paperbacks.

Thompson, E. A. (2006). The tulipmania: Fact or artifact? *Public Choice, 130*(1–2), 99–114.

Tirole, J. (2002). Rational irrationality: Some economics of self-management. *European Economic Review, 46*(4–5), 633–655.

Tversky, A., & Kahneman, D. (1971). Belief in the law of small numbers. *Psychological Bulletin, 76*(2), 105–110.

Tversky, A., & Kahneman, D. (1974). Judgment under uncertainty: Heuristics and biases. *Science, 185*(4157), 1124–1131.

Wakker, P. P. (2010). *Prospect theory: For risk and ambiguity.* Cambridge, NY: Cambridge University Press.

Randomness and the Games of Science

Jelle J. Goeman

Abstract Recently it has become clear that too many findings reported in the scientific literature are irreproducible. We study the causes of this phenomenon from a statistical perspective. Although a certain amount of irreproducible research is unavoidable due to the randomness inherent to scientific observation, two related phenomena conspire to increase the proportion of such findings: publication bias, i.e. the custom that negative findings are usually not published, and confirmation bias, i.e. the human inclination to interpret observations in a way that confirms prior beliefs. Both biases are poorly held in check in the current scientific publication model in which there is no explicit role for the views of a critic, i.e. a scientist with opposing theoretical views. We argue that if researchers are able to play the critic's role imaginatively, they will publish science of higher methodological quality that is not only more reproducible, but also more relevant for theory. To allow for this, we must promote a different view on statistical methodology, seeing statistics not as the gatekeeper of scientific evidence, but as a language scientists may use to discuss uncertainty when they talk about the implications of observations for theory.

1 Introduction

In 2009, a highly remarkable scientific experiment was performed by Bennett, Baird, Miller and Wolford, four American brain researchers. They used functional magnetic resonance imaging (fMRI), a brain imaging technique, to determine which brain areas respond to emotional stimuli in a test subject. The subject was shown

This text is based on my inaugural lecture "Toevalstreffers" (in Dutch), held on June 20, 2014 at Radboud University Nijmegen.

J.J. Goeman (✉)
Faculty of Medical Sciences, Radboud University, Nijmegen, The Netherlands
e-mail: jelle.goeman@radboudumc.nl

© The Author(s) 2016
K. Landsman and E. van Wolde (eds.), *The Challenge of Chance*,
The Frontiers Collection, DOI 10.1007/978-3-319-26300-7_5

several emotionally laden pictures and was asked to verbalize the emotion shown. The display of pictures was alternated with rest, and by comparing the brain readings between exposure and rest, the researchers were able to clearly identify a brain area that showed a response to the stimulus offered (Bennett et al. 2011).

What was so remarkable about this experiment? Certainly not the idea of measuring brain response to pictures using fMRI; this had been done countless times by other researchers in the past. Also not the statistical methods used to find the relevant brain regions by comparing exposure and rest states; the same techniques had been used in many influential publications in brain imaging before. The originality of the study lay in the choice of the test subject. This was not, as usual, a human, but an Atlantic salmon. Moreover, the salmon was stone dead, having been bought in the local supermarket on the very morning of the experiment.

The paper describing the experiment, when finally published, created quite a storm among brain imaging researchers, and was credited with the Ig Nobel prize in 2012.[1] Apparently, standard imaging techniques with standard analysis methods could produce clearly nonsensical results. In the future, the authors of the salmon experiment argued, more stringent statistical methods should be used in fMRI research that have a smaller risk of false positive results. As a result of this paper, methodological standards in brain imaging have increased substantially in the last few years. However, the salmon experiment not only had implications for future research, but also casts doubt on past results. How many published papers on brain imaging would have used the same methods as the salmon experiment to come to equally wrong conclusions? How reliable, then, is the brain imaging literature?

Other authors in other fields have also raised questions about the reliability of the scientific literature. Prominent among these is the epidemiologist John Ioannidis with his (2005) essay "Why most published research findings are false." Ioannidis argued quite generally from statistical arguments that a large proportion of the results presented in medical publications can be expected to be wrong. This proportion may differ between subfields of medicine, and depends on several factors, which we will come back to later. He comes to several surprising conclusions, among which one is that 'hot' scientific fields, in which many teams work on the same problems, and scientific breakthroughs are eagerly anticipated, are especially prone to produce unreliable findings. Consequently, results in high status journals, such as *Nature* and *Science*, would be especially unreliable.

Ioannidis' theoretical arguments have been confirmed by researchers that have actually tried to reproduce published scientific results. The results of such attempts have varied greatly. In psychology, where Ioannidis' arguments can be expected to hold as well, the journal *Social Psychology* published a special issue that reported replications of 13 recent studies (Klein et al. 2015). In 10 out of 13 cases, the effects reported in the original papers were found again, although often with a smaller magnitude. One study was on the borderline, replicating with a very small effect. The other 2 studies (14 %) failed to reach the same conclusions. More dramatic was

---

[1]The Ig Nobel Prizes honor scientific achievements that make people laugh, and then think.

the experience reported by Begley and Ellis (2012), scientists working at Amgen, a pharmaceutical company in California, who tried to replicate the results of many 'landmark' papers describing promising drug targets. They failed in no less than 47 out of 53 cases (89 %). Statisticians have tried to quantify the proportion of unreliable results in larger chunks of the scientific literature. Jager and Leek (2014) estimated the proportion of unreliable results in the whole medical literature at 14 %. Statisticians commenting on this effort almost invariably stressed that the percentage is very likely to be an underestimate, and possibly a severe one. The discussion of the reliability of scientific results has also reached the popular media, where regularly a bleak image is sketched of science in crisis. As to the cause and prevention of the unreliability scientific results, different opinions are voiced. Two competing explanations dominate the debate.

According to the first, scientists striving for fame and status deliberately engage in 'sloppy science'. They make their results look better than they are in order to publish them in higher ranking journals. Results are not fabricated, and 'sloppy science' is not the same as downright fraud, but 'sloppy' scientists are accused of wilfully neglecting proper checks and validations in order to publish more quickly. In variants of this argument, scientists are the victims rather than the perpetrators, as they are forced into their behavior by external institutional pressures. Because of savage competition between scientists and the demands from universities and funding agencies for ever longer lists of publications, scientists would have no choice but to engage in this type of dubious behavior.

A second explanation does not blame the scientists, but the methods they use. Since statistical methods are supposed to protect scientists against spurious findings, a high incidence of unreliable scientific results clearly indicates a design error in these methods. The type of statistical method most commonly denounced is the hypothesis test and the p-value, which, as critics point out, are frequently misunderstood and often used in a wrong way. Some authors argue that these methods should be banned altogether, a policy recently implemented by the journal *Basic and Applied Psychology* (Trafimow and Marks 2015). Some commentators advocate different statistical methods instead, e.g. Bayesian statistics. Others such as the editors of *Basic and Applied Psychology* simply advise against all advanced statistical methods, advocating simple descriptive statistics instead.

Interestingly, these two explanations suggest radically different solutions to the problem of unreliable results in science. If 'sloppy science' is the problem, scientists should be forced to adhere more strictly to proper statistical methodology. They should be kept in check by statisticians, who would then be cast into the role of policing various fields of science. Conversely, if statistics itself is the problem, the solution would be to free scientists from the influence of statisticians as much as possible. Scientists would then either convert to a completely different way of doing statistics, or just report their findings unencumbered by any need to demonstrate statistical significance.

More statistics or less? Which is better for the advancement of science? Which of the two explanations for the current flood of irreproducible research is the right one? Discussing the second explanation first, we will first review where

randomness and irreproducibility in science come from, and discuss the way statistical methods deal with this. We will explain that randomness is inherent to
scientific observation, and that statistics provides scientists with a way to discuss
the implications of this randomness on their experiments. Next, to shed light on the
first explanation, we discuss several models for the way scientists interact with each
other. We emphasize the important role of critics with different theoretical views in
scientific inquiry, arguing that statistical reasoning is an essential part of the dialogue between scientist and critic. Finally, we look at the current publication model
for reporting scientific results, and how it encourages a different, much more
mechanical view of statistics. In this view statistics is seen as an arbiter of truth
rather than as a language for discussing uncertainty. Rejecting both of the explanations given above, we will argue that it is primarily this distorted view of statistical methods that explains the current reproducibility crisis in science.

2 Randomness in Science

"Everything changes and nothing remains still; you cannot step twice into the same
stream" said the Greek philosopher Heraclitus, stressing the ever-changing nature of
reality. This truism applies very much to research, where no two experiments ever
return exactly the same result: different subjects respond differently to treatment,
and measurements are always variable. Randomness is inherent to scientific
observation.

Randomness, moreover, is bound to produce flukes. Since scientific observation
is subject to variability, seemingly meaningful patterns that the researcher observes
may well be one-time events rather than repeatable ones. For example, the patients
in a treated group may happen to recover very well, while the patients in the
untreated group do poorly, all because of their own particular reasons not related to
the treatment. To the researcher this may suggest a strong effect for a treatment that
is in reality not effective. When the experiment is subsequently replicated by the
same group of scientists or by a different one, the spurious patterns are very likely
not observed again. Irreproducible results, therefore, are a fundamental consequence of randomness in scientific observation, and are unavoidable even in the
most meticulous and honest scientific practice. We can, however, try to limit the
frequency of the occurrence of such result. This is what statistics tries to do.[2]

Statistical theory makes an explicit distinction between the *sample,* i.e. the
concrete observations the researcher has in hand, and the *population,* i.e. a larger
pool that these observations were drawn from. For example in a preelection poll, the

---

[2]This statistical view on (lack of) reproducibility is a limited one. There are of course many other
ways in which research can be irreproducible, for example because of systematic measurement
error, such as when the CERN-OPERA group in 2011 reported neutrino's that traveled faster than
light, or downright fraud, such as for example with the Dutch social psychologist Diederik Stapel.
See Baggerly and Coombes (2009) for an shocking account of how wrong things can go.

sample consists of the voters that have been interviewed by the pollsters, whereas the population is the much larger group of all voters. In many cases the 'population' is more abstract, such as in a lab experiment, where the sample might consist of a number of measurements the scientist has made, and the population we assume they have been drawn from is then the abstract collection of all possible measurement outcomes.[3]

The distinction between sample and population allows for an explicit definition of what replication of scientific experiments means. From a statistical perspective replication of an experiment means taking a new sample from the same population. Each sample is similar to the population it is drawn from, but deviates from the population in its own random way. Irreproducible findings then are statements that hold for a particular sample, but not for the underlying population, so that they do not typically occur again in other samples.

The central tenet of statistics is that we are not generally interested in the capricious sample, but only in the stable population behind it. Descriptive statistics describing the sample are therefore of limited use. We use the sample only as a means to learn about the population, a type of reverse engineering that we call *statistical inference*.[4] To do this in a quantitative way we must make an additional assumption on the manner in which the sample was obtained from the population, typically that it was drawn randomly. This assumption makes the powerful mathematical instrument of probability theory available that describes exactly how much the sample and the population are likely to differ, which in turn allows us to quantify the reliability of inferences about the population.

In particular, we can quantify the probability of drawing a wrong conclusion about the population from a sample. If we assume that a researcher has set out to find a certain relationship or pattern, i.e. to make a scientific discovery, then we can distinguish two possible erroneous conclusions. In the first place, the pattern can be visible in the sample, but not in the population. We call this a *false positive* or a *false discovery*. Secondly, the pattern can be present in the population, but obscured in the sample, called a *false negative*. While both types of errors are harmful, false discoveries are generally considered the more serious of the two. Where a false negative represents a waste of resources because a scientific experiment fails to produce a result, a false positive typically initiates an even greater waste of resources, as it will often be a trigger for misguided follow-up research. In terms of scientific progress, a false negative is a failure to take a step forward, but a false discovery is a step in the wrong direction.

With limited resources it is impossible to prevent both false positive and false negative results completely. A researcher could be very restrained, only publishing a result if there is ample evidence. Such a researcher will incur many false negative results while avoiding false positives. Conversely, an audacious researcher

---

[3]In such situations statistics is very explicitly platonic in its philosophy. It supposes that the unobservable abstract population really exists and is of more interest than the observable sample.

[4]As opposed to descriptive statistics, which describe the sample.

publishing results on precarious evidence can expect to have many false positives and few false negatives. Both researchers, however, risk both false positive and false negative results. The only way to avoid false positives completely is never to publish, and the only way to avoid false negatives completely is to always to publish, regardless of the evidence. The inherent randomness of scientific inquiry causes it to have elements of a game of chance. Even the best designed experiment may, by sheer bad luck, produce a sample that is different from the underlying population in crucial aspects and that therefore suggests a wrong conclusion.

False positive and false negative results are an inevitable consequence of the randomness of scientific data. They are not caused by statistical thinking, or inherent to any particular statistical method. Rather, by making the distinction between sample and population explicit statistics provides a language to discuss randomness of empirical data. Avoiding inferential statistics as *Applied Social Psychology* proposed, mostly ignores the problem. Switching to a different statistical framework, such as the Bayesian, merely rephrases it. Wrong conclusions will result from empirical research whatever methods we use, and this fact must be somehow taken into account.

3 The Likelihood of Irreproducible Research

The outcome of the experiment is never fully under the researcher's control, but the probability with which an adverse outcome occurs can be. One way to take randomness into account is to control the probability of an adverse outcome (a false positive or a false negative result). To avoid large differences between researches regarding the reliability of the evidence they present, in most scientific fields the acceptable risk of a false positive result is pre-specified for all researchers. It is conventionally set to 5 %, which implies that 19 out of 20 times that a researcher performs an experiment the result should not be a false positive, and should therefore be reproducible at least in the limited statistical sense.

This may seem to imply that 19 out of 20 published scientific results are reliable. Ioannidis, however, argued that this is not the case. This ratio of 19 out of 20 represents the perspective of the researcher, but is not immediately relevant from the perspective of the readers of the scientific literature. Even if 95 % of the time researchers produce results that are not false positives, this does not mean that 95 % of all scientific publications are not false positives. This is because negative results, being less newsworthy, are seldom published. Looking only at published results, the proportion of false positives is likely to be much higher than 5 %.

The argument follows from Bayes' rule. It is most conveniently illustrated with a table. Suppose that 200 experiments have been carried out by researchers in a certain field of science in a certain period of time. Sometimes the conjecture the researchers set out to prove was correct, sometimes it was not. For some experiments the researchers accumulated enough evidence to prove the conjecture; for others they were not. Based on these two dichotomies we can summarize these 200 experiments in a 2 × 2 contingency table. If we suppose that half of the conjectures that researchers try to prove are in fact true, then we have 100 experiments on true and false conjectures each. If 5 % false positive results are allowed, then 5 out of 100 experiments on false conjectures re- gardlessly accumulate enough evidence lead to a publication. Conversely, researchers typically accept a 20 % chance of false negative results, so that 80 out of the other 100 experiments lead to a publication. These numbers are summarized in Table 1. As readers of the scientific literature we only see the 85 published results, not the 115 experiments in which the researchers failed to demonstrate their point. The percentage of false positive results among the publications is 5/85 = 6 %, clearly more than 5 %, but not dramatically so.

This changes if we think of a field in which researchers try much more ambitious conjectures. Let us suppose that instead of 50 %, only 10 % of the conjectures that the researchers attempt are in fact true. In this case we can create a similar table, which will look like the one in Table 2. Now the researchers have to work a lot harder for their publications, and only 25 publications result from their 200 experiments. More importantly, the percentage of irreproducible findings soars to 9/25 = 36 %.

The percentage of irreproducible results can also be high if many of the experiments on true conjectures are underpowered, i.e. if researchers have a small probability of finding evidence for a conjecture even if it is true. If we would have 50 % true conjectures as in Table 1, but only for 30 out of 100 true conjectures enough evidence would be accumulated, then the proportion of false positive would be as high as 5/35 = 14 %, as we can see in Table 3. In general, even when the percentage of false positive results per experiment is at most 5 %, the percentage of false positive, i.e. irreproducible results will be large if most of the conjectures researchers set out to prove are false, or if the probability of accumulating enough evidence for publication of a true result is low.

It is interesting to note that in both Tables 2 and 3 we see that the percentage of experiments that leads to a publication is relatively low: 12.5 and 17.5 %,

Table 1 Illustration of Ioannidis' argument with 50 % true conjectures

	True conjecture	False conjecture	Total
Evidence for conjecture	80	5	85
No evidence for conjecture	20	95	115
Total	100	100	200

Table 2 Illustration of Ioannidis' argument with 10 % true conjectures

	True conjecture	False conjecture	Total
Evidence for conjecture	16	9	25
No evidence for conjecture	4	171	175
Total	20	180	200

Table 3 Illustration of Ioannidis' argument: underpowered studies

	True conjecture	False conjecture	Total
Evidence for conjecture	30	5	35
No evidence for conjecture	70	95	165
Total	100	100	200

respectively. One of the things that is crucial for judging the viability of scientific findings is therefore the success rate, i.e. the proportion of failed experiments for every successful one. This success rate is typically hidden from the view of the reader of the scientific literature, who only gets to see the successful experiments. The resulting selection bias, also known as *publication bias,* is inherent to the publication model that is currently dominant in science. Here, the initiative for performing experiments and publishing about them lies with the researchers. The experiment has clearly defined positive and negative outcomes, with positive outcomes being the only ones of real interest. The scientific readership has an exclusively passive role, only taking note of the experiment at a late stage after an apparent positive result has been obtained. Even the reviewers and editors who judge the manuscript are limited to retrospective checking of quality and plausibility. In this model no one except the researchers themselves can see the success rate. No one except the researchers themselves can therefore judge the probability that published results are false positives.

A third way, however, in which the proportion of false positive results in the literature may be high is when there is a large probability that evidence is seemingly found for a conjecture that is wrong. This probability is supposed to be at most 5 %, but it can be much larger because of the well-known psychological mechanism of *confirmation bias.* This is a natural tendency to look for evidence that supports our initial views, and to discard evidence that seems to counter those. Confirmation bias is a very strong force in human thinking, and one which is very difficult to counter. In research, confirmation bias works in rather the same way as publication bias, but at an earlier stage.

Confirmation bias in science may arise for example when there are multiple ways to perform an experiment, a number of statistical models and tests that can be used, or a number of ways to pre-process the data prior to that analysis. Some of these methods are better than others, but which ones those are is often not clear. If an experiment does not give the result that the researcher expected, this may therefore be due to several reasons. Of course the researcher's theory may be false, but it is also likely that something just went wrong in the experiment or that the right analysis method has not been chosen. It is perfectly reasonable, then, and scientifically sensible, to redo the experiment or the analysis. If a second experiment or a reanalysis now turns out to support the scientist's views, a natural explanation will be that there was an error in the first experiment or analysis, which has been corrected by the second.

In practice, researchers therefore do not usually perform one single analysis, but perform several, selecting relatively favorable ones by their confirmation bias. Even if every individual experiment yields a false positive result only once every 20 times, a series of experiments like this may easily have a much larger probability a false positive result, because a researcher trying to demonstrate something that is not true will make several attempts, each of which again has a probability of a seemingly favorable result. When the existence of confirmation bias is taken into account in Ioannidis' argument, it is easy to see that it will result in an even larger proportion of false positive results in the scientific literature.

Ioannidis' simple reasoning can be used to pinpoint areas in science in which we would expect false positive rates to be exceptionally high. These are for example areas with small studies that have low power, areas with exploratory studies where error control is lacking, areas in which statistical methods are not well standardized so that many will be tried out, areas with cheap but difficult experiments in which it is accepted that many experiments fail. However, these are especially those areas in which the scientific conjectures are a long shot, so that most of them are actually false. The resulting findings, paradoxically, are typically the most newsworthy ones which tend to get the attention of the high profile journals. As a rule of thumb, according to Ioannidis' analysis, the more excitement surrounding a scientific result, the greater the probability that it is a false positive. As an extreme case, Ioannidis also describes the existence of *null fields,* areas of research based on false prepositions, in which all researchers are working on research conjectures that are not true. From the reader's perspective, it is difficult to unmask such a field, because the failed experiments remain under the waterline, and a steady trickle of promising results will still be published, especially if many researchers are working in the area. Note that null fields are often sparked by an initial false positive result.

Confirmation and publication bias work together to increase the number of irreproducible results in the scientific literature. The argument we have given here is reminiscent of the 'sloppy science' argument for explaining irreproducible research described in the introduction, but subtly and importantly different. The 'sloppy science' argument implies wilful neglect of proper checks on scientific quality by scientists eager to publish, either because of their own ambition or because they are forced by external pressures. The argument implicitly assumes that if there would be no sloppy science (i.e. if scientists would adhere to statistical rules) there would not be many false positive results. Although it is true, of course, that 'sloppy science', when practiced, would increase confirmation biases and lead to irreproducible results, not all confirmation bias arises from 'sloppy science'. It is also clear from Ioannidis' arguments that large proportions of false positive findings would still arise if 'sloppy science' would cease to exist. Both publication and confirmation bias are inherent to the publication model used to disseminate results in science. We will discuss that model later in more detail, but first look at alternatives.

4 The Dialogue with the Critic

The discussion so far carried an implicit assumption about the way scientists communicate with each other. We take it for granted that they do so via scientific publications, which are well-prepared solitary efforts by a single research group, made public after extensive quality checking by editors and reviewers. This is the current dominant model for science, but it is not the only possible model. To see how other models might function, it is helpful to look back into the history of science. Current science has an amazing productivity in terms of sheer volume of knowledge, but early pre-20th century science has an even more surprising productivity if we take into account the relatively small number of scientists active at the time. In this period, when the foundations of many modern fields were laid out, how did science progress?

Let us illustrate this with an example. In the eighteenth century two Italian scientists were interested in electricity and its relationship to life. It was known that application of static electricity to the limbs of dead animals could cause them to jerk in movements similar to those a living creature would make. Surely, therefore, there was a relationship between electricity and life. This was at least the opinion of Luigi Galvani, a researcher from Bologna. He believed that electricity was an essential life force in animals. According to him, static electricity was sent to the muscles, where it was stored and used as energy for movement. By applying external electricity to the limbs, the researcher released the reservoir of 'animal electricity', thus causing the movement that was observed. Not everyone agreed with his views, however. Alessandro Volta, from Pavia, did not agree with Galvani's views. He did not believe in reservoirs of animal electricity, but held the opinion that it was the externally applied electricity alone that caused the movements.

Galvani and Volta corresponded extensively on this issue, each trying to convince the other. In 1781 Galvani performed what he thought was the definitive experiment. He hung a dead frog on an iron wire on which he had also attached a copper wire. When he touched the frog's leg with the copper, it jerked in the same way as when he applied static electricity to the frog's leg. The interpretation, to Galvani, was obvious. No outside static electricity had been applied, and still the frog's leg had moved. The electricity for the movement must have come from inside the frog. Volta replicated the experiment, getting exactly the same result. However, he remained unconvinced, while Galvani set out his grand theory of animal electricity in a large monograph entitled *De Viribus Electicitatis*.[5]

Volta still maintained that the electricity that caused the frog's movement must be external, but for a long time he stood alone in his opinion. Only many years later, in 1800, was he able to show that contact between two different metals, such as the copper and iron used by Galvani, may generate a minute electrical current, and that

---

[5]Although Galvani's theory turned out to be wrong, this is not irreproducible research in the statistical sense. All experiments the theory was based on were reproducible. Reproducibility is necessary but not sufficient for good theory.

this current was sufficient to cause the jerking of the frog's leg. The electricity was external after all. The exchange between Galvani and Volta has been of crucial importance both for physiology and for physics, as Volta's insights eventually led him to develop the first battery.

It is helpful to look more closely into the dialogue between these two scientists, which represented a type of scientific interaction quite typical for their time. We see a hefty competition that is fueled by irreconcilable theoretical views. Despite, or perhaps because of their differences the two researchers remain in frequent contact. Each tries to challenge his opponent by designing and performing an experiment of which he expects that the result will be in concordance with his own theory while at odds with his opponent's. In this 'duel', it is natural for each of the scientists to immediately try to replicate any crucial experiments in order to try to understand the results and to dismiss them should they turn out to be irreproducible. Volta never believed the results of Galvani's experiment until he had seen them with his own eyes. When he did see them, he still had his own explanation for the result, of course. Reproducibility of an experiment is not enough; in the end it is the implications of the experiment for theory that matter.

A competitive collaboration between scientists with diametrically opposed theoretical ideas can lead to research of high methodological quality, as we can see in the example of Galvani and Volta. For Galvani's experiments, Volta functions as a professional critic, always alert to false assumptions, wrongly designed experiments or hasty conclusions. Galvani could count on Volta immediately replicating every crucial experiment, attacking any weak spots in the design. Irreproducible research would be immediately exposed by him. Moreover, the competition with Volta gave focus to Galvani's experiments. It was not enough if his experiments lent support to his own theory, but they had to simultaneously discredit Volta's. Only experiments for which Volta and Galvani would expect a different result would be relevant to their argument.

The insight that collaboration between scientists with different views can be highly productive motivated the psychologist Willem Hofstee to advocate a 'wager model' for scientific research.[6] In this model, a scientist who wants to conduct an experiment first tries to find a scientist with different theoretical views and who, on the basis of these views, expects different findings from the experiment than the researcher him or herself. Let us call this scientist the critic. He will play a similar role as Volta in Galvani's experiments. If a critic cannot be found it is not necessary to perform the experiment, since no one would be surprised by the results. Such experiments apparently have no implications for theory. Once a critic is found, the researcher and the critic should sit together to discuss the details of the way the experiment will be performed, making sure that methodological biases do not favor the researcher or the critic. The experiment can proceed when both scientists agree on its validity, and it should possibly be executed in duplicate in both labs to

---

[6]'Weddenschapsmodel' in Dutch (Hofstee 1980). My translation.

prevent confirmation bias. An experiment set up in this way will have scientific merit whether the outcome is positive or negative for the researcher, and the researchers should commit themselves to publication whatever the outcome. From their competing theoretical views, it is likely that the two researchers will disagree on the final interpretation, with the 'losing' side trying to salvage their theory by alternative explanations.

The name of wager model has been appropriately chosen for two reasons. Firstly, because it suggests a clear investment of both parties into the experiment, with a commitment for each party to 'pay up' and proceed with the publication even in case of an adverse outcome. Secondly, because the word wager invokes the image of betting, suggesting that an element of chance plays a role. In fact, this is usually the case. As we have described above, the competing researchers will have to draw their conclusions on the basis of a sample, while their theoretical dispute is about the underlying population. Since the sample is variable, the risk is that the experiment favors the researcher although the critic's theory is right, or vice versa. This risk the contestants should be prepared to take.

Statistics can help to even the odds for both parties. In fact, the original framework of statistical hypothesis testing as proposed by Neyman and Pearson is highly suitable for the wager model. It uses a 'null hypothesis' representing the critic's view and an 'alternative hypothesis' representing the researchers view, and treats them symmetrically. The famous lemma of Neyman and Pearson tells us how to summarize the data most effectively in order to discriminate between these two hypotheses. The probabilities of a false conclusion favoring either the researcher or the critic can easily be calculated. Using this information a decision boundary can be set in such a way that the wager is a fair one, and the investment can be calculated that is needed to make the probability of both erroneous conclusions acceptably small. The statistician, therefore, has all the tools to stand as a natural arbiter between the researcher and the critic.

Like with the exchange between Galvani and Volta, close attention to methodology is naturally built into the wager model. The crucial element in both cases is the influential presence of a critic. The critic will insist on publication in those cases in which the researcher may not want to publish, thus countering publication bias. The critic will not share the confirmation bias of the researcher because of his competing theoretical views, and will thus be vigilant to counter it. The wager model thus avoids both confirmation and publication bias in a natural way. Since Ioannidis' causes for the large number of false positive results in the literature do not apply, we could expect far fewer irreproducible results if this model would be widely adopted. Sadly, this model is hardly ever used in practice, for various historical, psychological, practical and institutional reasons that we will not explore here.

The value of the wager model here is that provides a very useful ideal that can be used to study the current publication model of science, which we can see as an approximation to the wager model. This perspective will help to understand the methodology better, and also the extent to which this methodology is appropriate.

5   Publishing

Current research practice almost never involves an explicit critic. In contrast to the wager model we can refer to the dominant scientific model as a 'betting model'. It differs from the wager model mostly by the fact that the critic is abstracted and impersonal.

How does this work? Let us first review an example in which the model works very well.

A group of nutrition researchers from Amsterdam led by Martijn Katan wanted to demonstrate that the consumption of sugar through soft drinks makes children gain weight. This may seem obvious, but other researchers (and soft drink companies) maintained that children would automatically compensate for their sugar intake by being more active or eating less of other foods, negating the weight gain of the sugar intake. To prove their point, Katan's group enrolled 650 children in several schools and randomly allocated them into two groups. The first group was handed out a daily sugared soft drink. The second group received a daily sugar-free version. The two drinks tasted the same and the children and their parents were kept in the dark as to which child received which drink. After 1.5 year the researchers measured the weight gain of each of the children. They found that on average the children who drank the sugared drink gained one kilo more weight than the children who drank the sugar-free version. They submitted a description of the experiment and their conclusions to the *New England Journal of Medicine,* writing that consumption of sugar via soft drinks does indeed cause substantial weight gain in children. His manuscript was judged and commented on by an editor and two or more anonymous referees, and found acceptable for publication (De Ruyter et al. 2012b).

Before the study was started, the precise design of the study was laid down in a study protocol published separately (De Ruyter et al. 2012a).[7] This protocol stipulated exactly how the study would be executed, what measurements would be taken at what time, what statistical analyses would be performed and what would be done with the data (or the absence of data) of children who did not follow the study to the end. The protocol also motivates the number of participating children. This was chosen in such a way that if Katan's theory was right and children would indeed gain weight as a result of drinking soft drinks, Katan would have 80 % chance of demonstrating it with this trial.

If we compare the approach that Katan followed with the wager model of Hofstee, then we can easily see a number of parallels. Katan investigated an issue about which there was clear disagreement in the field. Katan did not explicitly involve a scientist of a different opinion on the matter at stake, but if we imagine that he would have, the design of the experiment would probably have been very similar. He built in many of the methodological checks that would have resulted from negotiation with a critic and which make the experiment impartial to either

---

[7]This is usual in clinical trials but not in nutrition research.

outcome, such as the blinding of the children and their parents for the type of drink received. In fact, Katan even put himself at a disadvantage when he accepted a 20 % chance of not being able to demonstrate his case even if he was right, against only 5 % for the absent critic. The protocol thus serves as a strong protection of the interest of the critic.

We call the model that Katan uses a betting model, since it is similar to the wager model, except that the researcher plays the game essentially against himself. Katan also played the role of the critic. Other critical scientists, namely Katan's peers, did come into play, but only at the peer review stage after the experiment had been conducted and reported. Like the critic in the wager model, they judged whether they were convinced by Katan's experiment. However, their role was in many respects very different from the role of the critic. They became involved only at a very late stage, and their power to influence the experiment was therefore extremely limited. Moreover, they had the power to influence whether the experiment would be published, a power that the critic in the wager model does not have. Reviewers do not themselves play the role of the critic, they can only judge whether Katan himself played that role convincingly.

The statistical framework that Katan used to analyze the outcome of his experiment, i.e. Fisher's approach to hypothesis testing, clearly reflects the characteristics of the betting model. In contrast to the symmetric framework of Neyman and Pearson that was suitable for the wager model, Fisher's approach is asymmetric. The null hypothesis, which represents the critic's opinion, becomes more formalized, and assumes a greater importance than the alternative hypothesis. Central to Fisher's approach is the concept of a p-value. This value between 0 and 1 is a measure of how extreme the outcome of the experiment would be from the critic's point of view. High values indicate outcomes that conform to the critic's theory. Low values indicate outcomes that are difficult to reconcile with it, but which would more easily fit the researcher's perspective. The p-value can therefore be seen as a quantitative measure that describes to what extent the absent critic is convinced by the outcome of the experiment. Numerically, the p-value is calibrated to take small values below 0.05 only 5 % of the time if the null hypothesis is true, i.e. the critic is right. Conventionally, this five percent is the threshold below which the critic will be convinced. With a p-value below this cut-off, the researcher may claim to have a convincing (in statistical parlance: 'significant') result.

We can see that the absent critic's role and opinions have been completely formalized in this approach. Katan found that children who drink a daily sugared beverage gained about kilo of weight in a year. He also maintained that these results were very difficult ($p = 0.001$) to reconcile with the theoretical view that it does not matter for children's weight whether or not they drink sugar. How convincing this latter statement is crucially depends on how well Katan represented this theoretical view that he did not himself support. We have seen that Katan built in all kinds of safeguards into his experimental design, such as the blinding and the protocol, to protect the experiment from his own biases. Essentially, these measures limit his own freedom in analyzing his results, evening out the odds between him and the critic, and by doing that making the outcome more convincing.

   Not all research is as well designed. Headlines in newspapers in 1995 announced that eating tomatoes would dramatically decrease the risk of prostate cancer. Surprisingly, the beneficial effect was not found in fresh tomatoes, but rather in tomato concentrate in the form of ketchup, pizza, tomato soup and even potato crisps with ketchup flavor. The source of the news was a publication by a group led by Edward Giovannucci from Harvard (Giovannucci et al. 1995). According to him, the substance lycopene, found abundantly in tomato concentrate, eliminated the free radicals which caused the cancer. Giovannucci's article has had a major impact, with over a 1000 citations in the scientific literature over the last twenty years. How did Giovannucci come to his conclusion? He asked a large group of health professionals to fill out food intake questionnaires, focusing on intake of 46 vegetables and fruits. Next, he followed his subjects in time to see who would develop prostate cancer, to check whether people who ate more or less of certain foodstuffs would on average develop prostate cance more frequently. In only 4 of the 46 food types he investigated was he able to find the relationship he was looking for, supported by p-values smaller than 0.05. Upon closer examination, those four were all related to industrially processed tomatoes. A plausible explanation was found in the lycopene theory, and this was the result that was highlighted in the publication.

   How convincing is the result? To answer this question it is helpful to imagine how the investigation would have turned out if Giovannucci would have involved a critic. We have to remember that Giovannucci did not yet have his theory about lycopene when he started his study, so that at the moment he contacted a critic, he would have only had a relatively vague theory that the risk of prostate cancer might be influenced by diet. We can therefore suppose that such a critic would be skeptical about this idea, maintaining that the risk of prostate cancer might depend on all manner of things, such as genetic and lifestyle factors, but that food intake did not matter. To settle this difference of opinion, it would be unethical and unpractical to use a clinical trial design such as the one that Katan followed, and Giovannucci and his critic would have quickly decided to study observational data. This is a methodological quagmire because it is difficult to distinguish the effects of different factors. For example, people who eat more vegetables typically also exercise more and are more highly educated. If we find that people who eat more vegetables have less prostate cancer, is that due to the vegetables or due to the exercise? Still, discussing these issues at length, it is conceivable that Giovannucci and an open-minded critic might have come to a wager. Would that wager have taken the form described as the evidence in the eventual paper?

   Giovannucci investigated 46 different foodstuffs separately, calculating a separate p-value for each of them. In 4 out of these 46 did he find a p-value smaller than 0.05. In terms of the betting model with which we can interpret the meaning of these p-values, this is equivalent to betting against the critic 46 times, of which he lost 42 times and won only 4. If a real critic would be present, it is likely that he or she would claim victory over Giovannucci rather than the other way around. If we remember that p-values are calculated in such a way that the critic will lose the bet about one out of twenty times even when the critic is right, we can expect

Giovannucci to win about 2.3 times out of 46 even when there is no relationship between diet and prostate cancer. Winning at least 4 times in this situation is not an unlikely event, with an occurrence almost 20 %.[8] Under a wager model, therefore, the conclusion of the study would most likely have been support of the critic's view that diet and prostate cancer are unrelated. If the four foodstuffs for which a relationship is suggested may be the product of chance, it is especially unlikely that a critic would be convinced by the mechanistic explanation about lycopene, made up only after the experiment. The critic may have wondered what explanations Giovannucci might have come up with had four other foodstuffs come out.

The difference between Giovannucci and Katan does not lie in the statistical methods they used. These are broadly the same. The difference is in the way they realized the meaning of the methods they used. Katan took great care to look at his own experiment from the perspective of a critic, taking that point of view into account in every aspect of the study. Giovannucci seems to have done this to a much lesser extent. He applies the rules of the statistical methods he uses, but he does not seem to realize that the results he presents are not as convincing as they have to be. Interestingly, also the reviewers who deemed his work suitable for publication did not notice this.

It is of course the reviewer's job to check a manuscript's quality before advising publication. We could expect that reviewer's take the same perspective as the critic, checking manuscripts meticulously for methodological errors, vigilantly aware of possible confirmation bias on the side of the researcher. In practice, sadly, this is not the rule. Since the reviewers come into play at a late stage, after the experiment has been carried out and reported, many the important problems resulting from confirmation bias remain invisible to them.[9] For example, they cannot see how many other analysis methods the researcher tried, or what the original hypothesis was that the experiment was designed for. Moreover, reviewers tend to focus much more on the conclusions of the papers than on the methods. This was demonstrated in 1998 by Fiona Godlee, editor of the British Medical Journal. She sent an article with 8 deliberate serious methodological errors to more than 200 regular reviewers of her journal. On average, each reviewer only observed 2 of the 8 errors. Of the reviewers, 33 % suggested to accept the article with only minor changes, while only 30 % advised to reject it (Godlee et al. 1998). Reviewers naturally bring their own confirmation bias. When they disagree with the conclusions they will study the methods much more critically than when they agree with them.

The betting model used for scientific publication can best be described as a watered-down version of the wager model. It calls for the scientist to win a bet against a critic of his or her own making, and it is completely up to him or her how

---

[8]Calculated under the assumption of independence. If—as is likely here—the p-values are dependent, this probability will typically be even larger.

[9]This is not the case for the paper of Giovannucci, who (to his credit) makes his confirmation bias very explicit in the description of the experiment and the analysis. The reviewers should have protested and demanded a proper multiple testing correction here.

formidable an opponent the critic is. Some researchers play the role of the critic very convincingly, others just set up a straw man. The statistical methods used are the same in both cases.

6   Speaking About Uncertainty

We now return to the original question about the role of statistics in creating or preventing irreproducible science. We have seen that the randomness of scientific observation makes it impossible to forestall irreproducible results completely, but that two types of bias may dramatically increase the proportion of such findings in the scientific literature: publication bias and confirmation bias. Both of these are tied closely to the current model we use for communication of scientific results via publications, a model I have called the betting model.

At first sight the role of statistics in this betting model seems a rather mechanical one, emphasizing the calculations that have to be done and the cut-offs that have to be exceeded 'to get the statistics right' and to achieve the necessary statistical proof needed for publication. This is often how statistics is taught, as a cookbook full of prescriptions that researchers have to follow in order to analyze their data in a correct way. This mechanical view underrates the role that statistics can play in scientific discourse. In the mechanical view, statistics is seen as an arbiter of truth. This is something it cannot be. Statistics is just a language researchers can use to speak about chance and uncertainty.

To be relevant for scientific progress, experiments must be designed and ana- lyzed in such a way that they make a difference, changing at least some people's opinions about theory. To be convincing requires to be empathic, studying the other side's arguments and taking them seriously. The betting model, as we have seen, only works well if the researcher is prepared to take a critical point of view throughout the design and analysis of his experiment, while maintaining focus on the theoretical issues at stake. A scientific experiment is only valuable if it furthers theoretical discussion in some way.

The scientific attitude necessary for this is under pressure in many countries due to the demands on scientists to publish and acquire grants. In this rat race publi- cations are often viewed as personal achievements of scientists, and as end products rather than as arguments in an ongoing scientific discussion. Regarding a publi- cation as a personal achievement emphasizes competition between scientists for honors, instead of their collaboration on furthering theory. It is based on the mis- conception that the essence of science is competition between individuals rather than between theories. Regarding publications as end products promotes the idea that the publication should present definite proof. This, in turn, encourages the mechanical perspective on methodology and statistics.

It may be clear that throwing inferential statistics out of the window represents a step back, leaving us with no language to even discuss the problem of irrepro- ducible research. However, having statisticians police scientists is equally pointless if these checks are only executed at the final stage when the experiment has already been performed. At this stage, much of the confirmation bias is not visible anymore, and should any clear mistakes be found, there is no way to mend them. In the words of the famous statistician Ronald Fisher 'To consult the statistician after an experiment is finished is often merely to ask him to conduct a post mortem examination. He can perhaps say what the experiment died of' (Fisher 1938). Moreover, involving statisticians in the role of arbiters only serves to emphasize the mechanical view of statistics. This will hamper the discussion between scientists about uncertainty more than it will stimulate it.

Reduction of the proportion of irreproducible research findings calls for a renewed interest in methodology. The mechanistic view of statistical analysis has made many scientists see methodology and statistics as a necessary evil. Better understanding of methodology might help scientists to think about statistics in terms of convincing rather than in terms of proof, and to see how statistical language is a necessary element of the dialogue between researchers with opposing views. The wager model, even if not practical, may help as a thought experiment for researchers setting up an experiment, and may help to create awareness of confirmation biases, and to design more imaginative experiments. To facilitate this thought experiment in the absence of a critic with opposing theoretical views, collaboration with a neutral methodologist may be a good alternative.

Open Access This chapter is distributed under the terms of the Creative Commons Attribution-Noncommercial 2.5 License (http://creativecommons.org/licenses/by-nc/2.5/) which permits any noncommercial use, distribution, and reproduction in any medium, provided the original author(s) and source are credited. The images or other third party material in this chapter are included in the work's Creative Commons license, unless indicated otherwise in the credit line; if such material is not included in the work's Creative Commons license and the respective action is not permitted by statutory regulation, users will need to obtain permission from the license holder to duplicate, adapt or reproduce the material.

References

Baggerly, K. A., & Coombes, K. R. (2009). Deriving chemosensitivity from cell lines: Forensic bioinformatics and reproducible research in high-throughput biology. *The Annals of Applied Statistics, 3*, 1309–1334.

Begley, C. G., & Ellis, L. M. (2012). Drug development: Raise standards for preclinical cancer research. *Nature, 483*(7391), 531–533.

Bennett, C. M., Baird, A. A., Miller, M. B., & Wolford, G. L. (2011). Neural correlates of interspecies perspective taking in the post-mortem Atlantic salmon: An argument for proper multiple comparisons correction. *Journal of Serendipitous and Unexpected Results, 1*, 1–5.

De Ruyter, J. C., Olthof, M. R., Kuijper, L. D. J., & Katan, M. B. (2012a). Effect of sugar-sweetened beverages on body weight in children: Design and baseline characteristics of the double-blind, randomized intervention study in kids. *Contemporary Clinical Trials, 33*(1), 247–257.

De Ruyter, J. C., Olthof, M. R., Seidell, J. C., & Katan, M. B. (2012b). A trial of sugar-free or sugar-sweetened beverages and body weight in children. *New England Journal of Medicine, 367*(15), 1397–1406.

Fisher, R. A. (1938). Presidential address. *Sankhya: The Indian Journal of Statistics, 4*, 14–17.

Giovannucci, E., Ascherio, A., Rimm, E. B., Stampfer, M. J., Colditz, G. A., & Willett, W. C. (1995). Intake of carotenoids and retino in relation to risk of prostate cancer. *Journal of the National Cancer Institute, 87*(23), 1767–1776.

Godlee, F., Gale, C. R., & Martyn, C. N. (1998). Effect on the quality of peer review of blinding reviewers and asking them to sign their reports: A randomized controlled trial. *JAMA, 280*(3), 237–240.

Hofstee, W. (1980). *De empirische discussie: Theorie van het sociaal-wetenschappelijk onderzoek*. Amsterdam: Boom Koninklijke Uitgevers.

Ioannidis, J. (2005). Why most published research findings are false. *PLoS Medicine, 2*(8), e124.

Jager, L. R., & Leek, J. T. (2014). An estimate of the science-wise false discovery rate and application to the top medical literature. *Biostatistics, 15*(1), 1–12.

Klein, R.A., Ratliff, K.A., Vianello, M., Adams Jr., R.B., Bahnlk, S., Bernstein, M.J., et al. (2015). Investigating variation in replicability. *Social Psychology*.

Trafimow, D., & Marks, M. (2015). Editorial. *Basic Applied Social Psychology, 37*, 1–2.

The Fine-Tuning Argument: Exploring the Improbability of Our Existence

Klaas Landsman

> A mild form of satire may be the appropriate antidote. Imagine, if you will, the wonderment of a species of mud worms who discover that if the constant of thermometric conductivity of mud were different by a small percentage they would not be able to survive. (Earman 1987, p. 314).

Abstract Our laws of nature and our cosmos appear to be delicately fine-tuned for life to emerge, in a way that seems hard to attribute to chance. In view of this, some have taken the opportunity to revive the scholastic Argument from Design, whereas others have felt the need to explain this apparent fine-tuning of the clockwork of the Universe by proposing the existence of a 'Multiverse'. We analyze this issue from a sober perspective. Having reviewed the literature and having added several observations of our own, we conclude that cosmic fine-tuning supports neither Design nor a Multiverse, since both of these fail at an explanatory level as well as in the more quantitative context of Bayesian confirmation theory (although there might be other reasons to believe in these ideas, to be found in religion and in inflation and/or string theory, respectively). In fact, fine-tuning and Design even seem to be at odds with each other, whereas the inference from fine-tuning to a Multiverse only works if the latter is underwritten by an additional metaphysical hypothesis we consider unwarranted. Instead, we suggest that fine-tuning requires no special explanation at all, since it is not the Universe that is fine-tuned for life, but life that has been fine-tuned to the Universe.

1 Introduction

Twentieth Century physics and cosmology have revealed an astonishing path towards our existence, which appears to be predicated on a delicate interplay between the three fundamental forces that govern the behavior of matter at very small distances and the long-range force of gravity. The former control chemistry

K. Landsman (✉)
Faculty of Science, Radboud University, Nijmegen, The Netherlands
e-mail: landsman@math.ru.nl

© The Author(s) 2016
K. Landsman and E. van Wolde (eds.), *The Challenge of Chance*,
The Frontiers Collection, DOI 10.1007/978-3-319-26300-7_6

and hence life as we know it, whereas the latter is responsible for the overall evolution and structure of the Universe.

- If the state of the hot dense matter immediately after the Big Bang had been ever so slightly different, then the Universe would either have rapidly recollapsed, or would have expanded far too quickly into a chilling, eternal void. Either way, there would have been no 'structure' in the Universe in the form of stars and galaxies.
- Even given the above fine-tuning, if any one of the three short-range forces had been just a tiny bit different in strength, or if the masses of some elementary particles had been a little unlike they are, there would have been no recognizable chemistry in either the inorganic or the organic domain. Thus there would have been no Earth, no carbon, et cetera, let alone the human brains to study those.

Broadly, five different responses to the impression of fine-tuning have been given:

1. *Design:* updating the scholastic *Fifth Way* of Aquinas (1485/1286), the Universe has been fine-tuned with the emergence of (human) life among its designated purposes.[1]
2. *Multiverse:* the idea that our Universe is just one among innumerably many, each of which is controlled by different parameters in the (otherwise fixed) laws of nature. This seemingly outrageous idea is actually endorsed by some of the most eminent scientists in the world, such as Martin Rees (1999) and Steven Weinberg (2007). The underlying idea was nicely explained by Rees in a talk in 2003, raising the analogy with 'an 'off the shelf' clothes shop: "if the shop has a

[1]"The Fifth Way is based on the directedness of things. We observe that some things which lack awareness, namely natural bodies, act for the sake of an end. This is clear because they always or commonly act in the same manner to achieve what is best, which shows that they reach their goal not by chance but because they tend towards it. Now things which lack awareness do not tend towards a goal unless directed by something with awareness and intelligence, like an arrow by an archer. Therefore there is some intelligent being by whom everything in nature is directed to a goal, and this we call 'God'." Translation in Kenny (1969, p. 96), to whom we also refer for a critical review of Aquinas's proofs of the existence of God. It is a moot point whether the Fifth Way is really an example of the medieval Argument of Design, which Aquinas expresses elsewhere as: "The arrangement of diverse things cannot be dictated by their own private and divergent natures; of themselves they are diverse and exhibit no tendency to form a pattern. It follows that the order of many among themselves is either a matter of chance or must be attributed to one first planner who has a purpose in mind." (Kenny 1969, p. 116). Everitt (2004) distinguishes between the Argument *to* Design and the Argument *from* Order, respectively, both of which may still be found in modern Christian apologists such as Swinburne (2004), Küng (2005), and Collins (2009), rebutted by e.g., Everitt (2004) and Philipse (2012). It is clear from his writings (such as the General Scholium in *Principia*) that Isaac Newton supported the Argument from Design, followed by Bentley (1692). Throughout early modern science, the gradual 'reading' of the 'Book of Nature', seen as a second 'book' God had left mankind next to the Bible, was implicitly or explicitly seen as a confirmation of Design (Jorink 2010). Paley (1802) introduced the famous watchmaker analogy obliterated by Dawkins (1986). See also Barrow and Tipler (1986) and Manson (2003) for overviews of the Argument from Design.

large stock, we're not surprised to find one suit that fits. Likewise, if our universe is selected from a multiverse, its seemingly designed or fine-tuned features wouldn't be surprising." (Mellor 2002).

3. *Blind Chance:* constants of Nature and initial conditions have arbitrary values, and it is just a matter of coincidence that their actual values turn out to enable life.[2]

4. *Blind Necessity:* the Universe could not have been made in a different way or order, yet producing life is not among its goals since it fails to have any (Spinoza 1677).[3]

5. *Misguided:* the fine-tuning problem should be resolved by some appropriate therapy.

We will argue that whatever reasons one may have for supporting the first or the second option, fine-tuning should not be among them. Contemporary physics makes it hard to choose between the third and the fourth option (both of which seem to have supporters among physicists and philosophers),[4] but in any case our own sympathy lies with the fifth.

First, however, we have to delineate the issue. The *Fine-Tuning Argument*, to be abbreviated by FTA in what follows, claims that the present Universe (including the laws that govern it and the initial conditions from which it has evolved) permits life only because these laws and conditions take a very special form, small changes in which would make life impossible. This claim is actually quite ambiguous, in (at least) two directions.

1. The FTA being counterfactual (or, in Humanities jargon, being 'what if' or 'alternate' history), it should be made clear what exactly is variable. Here the range lies between raw Existence itself at one end (Rundle 2004; Holt 2012; Leslie and Kuhn 2013) and fixed laws of nature and a Big Bang with merely a few variable parameters at the other (cf. Rees 1999; Hogan 2000; Aguirre 2001; Tegmark et al. 2006).
   Unless one is satisfied with pure philosophical speculation, specific technical results are only available at the latter end, to which we shall therefore restrict the argument.

2. It should be made clear what kind of 'life' the Universe is (allegedly) fine-tuned for, and also, to what extent the emergence of whatever kind of life is deemed merely possible (if only in principle, perhaps with very low probability), or

---

[2]In the area of biology, a classical book expressing this position is Monod (1971).

[3]The most prominent modern Spinozist was Albert Einstein: "there are no *arbitrary* constants of this kind; that is to say, nature is so constituted that it is possible logically to lay down such strongly determined laws that within these laws only rationally completely determined constants occur (not constants, therefore, whose numerical value could be changed without destroying the theory." (Einstein in Schilpp 1949, p. 63).

[4]The famous ending of *The First Three Minutes* by the physicist Weinberg (1977)—"The more the universe appears comprehensible, the more it also appears pointless."—could be bracketed under either.

likely, or absolutely certain. For example, should we fine-tune just for the possible existence of self-replicating structures like RNA and DNA,[5] or for "a planet where enough wheat or rice could be cultivated to feed several billion people" (Ward and Brownlee 2000, p. 20), or for one where morally (or indeed immorally) acting rational agents emerge (Swinburne 2004), perhaps even minds the like of Newton and Beethoven?

It seems uncontroversial that at the lowest end, the Universe should exhibit some kind of order and structure in order to at least enable life, whereas towards the upper end it has (perhaps unsurprisingly) been claimed that essentially a copy of our Sun and our Earth (with even the nearby presence of a big planet like Jupiter to keep out asteroids) is required, including oceans, plate tectonics and other seismic activity, and a magnetic field helping to stabilize the atmosphere (Ward and Brownlee 2000).[6]

For most of the discussion we go for circumstances favoring simple carbon-based life; the transition to complex forms of life will only play a role in discussing the fine-tuning of our solar system (which is crucial to some and just a detail to others).[7]

According to modern cosmology based on the (hot) Big Bang scenario,[8] this means that the Universe must be sufficiently old and structured so that at least galaxies and several generations of stars have formed; this already takes billions of years.[9] The subsequent move to viable planets and life then takes roughly a similar amount of time, so that within say half an order of magnitude the current age of the

---

[5]See e.g. Smith and Szathmáry (1995) and Ward and Brownlee (2000) for theories of the origin of life.

[6]The conservatism—perhaps even lack of imagination—of such scenarios is striking. But science-fiction movies such as *Star Trek, Star Wars, E.T., My Stepmother is an Alien* (not to speak of *Emmanuelle, Queen of the Galaxy*) hardly do better. Conway's *Game of Life* suggests that initial complexity is not at all needed to generate complex structures, which may well include intelligent life in as yet unknown guise.

[7]Reprimanding the late Carl Sagan, who expected intelligent life to exist in millions of places even within our own Galaxy, Ward and Brownlee (2000) claim that whereas this might indeed apply to the most basic forms of life, it is the move to complex (let alone intelligent) life that is extremely rare (because of the multitude of special conditions required), perhaps having been accomplished only on Earth.

[8]See e.g. Rees (1999), Ellis (2007), and Weinberg (2008), at increasing level of technicality.

[9]The reason (which may be baffling on first reading) is that in addition to the light elements formed in Big Bang nucleosynthesis (i.e., about 75 % hydrogen and 25 % helium, with traces of other elements up to Lithium, see Galli and Palla 2013), the heavier elements in the Periodic Table (many of which are necessary for biochemistry and/or the composition of the Earth and its atmosphere) were formed in stars, to be subsequently blown into the cosmos by e.g. supernova explosions. In that way, some of these elements eventually ended up in our solar system, where they are indispensable in constituting both the Earth and ourselves. See Arnett (1996) for a technical account and Ward and Brownlee (2000) for a popular one.

Universe seems necessary to support life. In view of the expansion of the Universe, a similar comment could be made about its size, exaggerated as it might seem for the purpose of explaining life on earth.

2 Evidence for Fine-Tuning

Thanks to impressive progress in both cosmology and (sub) nuclear physics, over the second half of the 20th Century it began to be realized that the above scenario is predicated on seemingly exquisite fine-tuning of some of the constants of Nature and initial conditions of the Universe. We just give some of the best known and best understood cases here.[10]

One of the first examples was the 'Beryllium bottleneck' studied by Hoyle in 1951, which is concerned with the mechanism through which stars produce carbon and oxygen.[11] This was not only a major correct scientific prediction based on 'anthropic reasoning' in the sense that some previously unknown physical effect (viz. the energy level in question) *had* to exist in order to explain some crucial condition for life; it involves dramatic fine-tuning, too, in that the nucleon-nucleon force must lie near its actual strength within about one part in a thousand in order to obtain the observed abundances of carbon and oxygen, which happen to be the right amounts needed for life (Ekström et al. 2010).

Another well-understood example from nuclear physics is the mass difference between protons and neutrons, or, more precisely, between the down quark and the up quark (Hogan 2000).[12] This mass difference is positive (making the neutron heavier than the proton); if it weren't, the proton would fall apart and there would be no chemistry as we know it. On the other hand, the difference can't be too large,

---

[10]See Barrow and Tipler (1986), Leslie (1989), Davies (2006), Ellis (2007), and Barnes (2012) for further examples and more detailed references. Stenger (2011) and Bradford (2011, 2013) attempt to play down the accuracies claimed of fine-tuning, whilst Aguirre (2001) casts doubt on its limited scope.

[11]In order to make carbon, two ^{4}He nuclei must collide to form ^{8}Be, upon which a third ^{4}He nucleus must join so as to give ^{12}C (from which, in turn, ^{16}O is made by adding another ^{4}He nucleus). This second step must happen extremely quickly, since the ^{8}Be isotope formed in the first step is highly unstable. Without the exquisitely fine-tuned energy level in ^{12}C (lying at the ^{8}Be $+^4$He reaction energy) predicted by Hoyle, this formation process would be far too infrequent to explain the known cosmic abundances. Opponents of anthropic reasoning would be right in pointing out that these abundances as such (rather than their implications for the possibility of human life) formed the proper basis for Hoyle's prediction.

[12]Quarks are subnuclear particles that come in six varieties, of which only the so-called 'up' and 'down' quarks are relevant to ordinary matter. A neutron consists of one up quark and two down quarks, whereas a proton consists of two up quarks and one down quark. The electric charges (in units where an electron has charge -1) are 2/3 for the up quark and $-1/3$ for the down quark, making a neutron electrically neutral (as its name suggests) whilst giving a proton charge $+1$. Atoms consist of nuclei (which in turn consist of protons and neutrons) surrounded by electrons, whose total charge exactly cancels that of the nucleus.

for otherwise stars (or hydrogen bombs, for that matter) could not be fueled by nuclear fusion and stars like our Sun would not exist.[13] Both require a fine-tuning of the mass difference by about 10 %.

Moving from fundamental forces to initial conditions, the solar system seems fine-tuned for life in various ways, most notably in the distance between the Sun and the Earth: if this had been greater (or smaller) by at most a few precent it would have been too cold (or too hot) for at least complex life to develop. Furthermore, to that effect the solar system must remain stable for billions of years, and after the first billion years or so the Earth should not be hit by comets or asteroids too often. Both conditions are sensitive to the precise number and configuration of the planets (Ward and Brownlee 2000).

Turning from the solar system to initial conditions of our Universe, but still staying safely within the realm of well-understood physics and cosmology, Rees (1999) and others have drawn attention to the fine-tuning of another cosmological number called Q, which gives the size of inhomogeneities, or 'ripples', in the early Universe and is of the order $Q \sim 0.00001$, or one part in a hundred thousand.[14] This parameter is fine-tuned by a factor of about ten on both sides (Rees 1999; Tegmark et al. 2006): if it had been less than a tenth of its current value, then no galaxies would have been formed (and hence no stars and planets). If, on the other hand, it had been more than ten times its actual value, then matter would have been too lumpy, so that there wouldn't be any stars (and planets) either, but only black holes. Either way, a key condition for life would be violated.[15]

The expansion of the Universe is controlled by a number called Ω, defined as the ratio between the actual matter density in the Universe and the so-called critical density. If $\Omega \leq 1$, then the Universe would expand forever, whereas $\Omega > 1$ would portend a recollapse. Thus $\Omega = 1$ is a critical point.[16] It is remarkable enough that

---

[13]Technically, the fundamental 'pp–reaction' (i.e., proton + proton $\rightarrow$ Deuteron + positron + neutrino), which lies at the beginning of nuclear fusion, would go in the wrong direction.

[14]The Universe is approximately 13.7 billion years old (which is about three times as old as the Earth). Almost 400.000 years after the Big Bang, the Universe (which had been something like a hot soup of elementary particles until then) became transparent to electromagnetic radiation (which in everyday life includes light as well as radio waves, but whose spectrum is much larger) and subsequently became almost completely dark, as it is now. The so-called cosmic microwave background (CMB, discovered in 1964 by Penzias and Wilson), which still pervades the Universe at a current temperature of about 3 K (=−270 °C), is a relic from that era. It is almost completely homogeneous and isotropic, except for the ripples in question, whose (relative) size is given by the parameter Q. This provides direct information about the inhomogeneities of the Universe at the time the CMB was formed, i.e., when it was 400.000 years old.

[15]As analyzed by the Planck Collaboration (2014), variations in the constants of Nature would also affect the value of Q, which is ultimately determined by the physics of the early Universe. Hence its known value of 10^{-5} constrains such variations; in particular, the fine-tuning of Q necessary for life in turn fine-tunes the fine-structure constant α (which controls electromagnetism and light) to within 1 % of its value (1/137).

[16]Roughly, the physics behind this is that at small matter density the (literally 'energetic') expansion drive inherited from the Big Bang beats the gravitational force, which tries to pull matter together.

currently $\Omega \approx 1$ (within a few percent); what is astonishing is that this is the case at such a high age of the Universe. Namely, for Ω to retain its (almost) critical value for billions of years, it must have had this value right from the very beginning to a precision of at least 55 decimal places.[17]

This leads us straight to Einstein's cosmological constant Λ, which he introduced into his theory of gravity in 1917 in order to (at least theoretically) stabilize the Universe against contracting or expanding, to subsequently delete it in 1929 after Hubble's landmark observation of the expansion of the Universe (famously calling its introduction his "biggest blunder"). Ironically, Λ made a come-back in 1998 as the leading theoretical explanation of the (empirical) discovery that the expansion of the Universe is currently accelerating.[18] For us, the point is that even the currently accepted value of Λ remains very close to zero, whereas according to (quantum field) theory it should be about 55 (some even say 120) orders of magnitude larger (Martin 2012). This is often seen as a fine-tuning problem, because some compensating mechanism must be at work to cancel its very large natural value with a precision of (once again) 55 decimal places.[19]

The fine-tuning of all numbers considered so far seems to be dwarfed by a knock-down FTA given by Roger Penrose (1979, 2004), who claims that in order to produce a Universe that even very roughly looks like ours, its initial conditions (among some generic set) must have been fine-tuned with a precision of one to $10^{10^{123}}$, arguably the largest number ever conceived: all atoms in the Universe would not suffice to write it out in full.[20] Penrose's argument is an extreme version of an idea originally due to Boltzmann, who near the end of the 19th Century argued that the direction of time is a consequence of the increase of entropy in the

---

[17]If not, the expansion would either have been too fast for structures like galaxies to emerge, or too slow to prevent rapid recollapse due to gravity, leading to a Big Crunch (Rees 1999). This fine-tuning problem is often called the *flatness problem*, since the Universe is exactly flat (in the sense of Einstein's Theory of General Relativity) when $\Omega = 1$ (otherwise it either has a spherical or a hyperbolic geometry). The fine-tuning problem for Ω is generally considered to be solved by the (still speculative) theory of cosmic inflation (Liddle and Lyth 2000; Weinberg 2008), but even if this theory is correct, it merely shifts the fine-tuning from one place to another, since the parameters in any theory of inflation have to be fine-tuned at least as much as Ω; Carroll and Tam (2010) claim this would even be necessary to ten million decimals. In addition, the flatness problem may not be a problem at all, like the horizon problem (McCoy 2015).

[18]The Physics Nobel Prize in 2011 was awarded to Perlmutter, Schmidt, and Riess for this discovery. The cosmological constant Λ can theoretically account for this acceleration as some sort of an invisible driving energy. Thus reinterpreted as 'dark energy', Λ contributes as much as 70 % to the energy density of the Universe and hence it is currently also the leading contributor to Ω (Planck Collaboration 2015).

[19]The broader context of this is what is called the *naturalness problem* in quantum field theory, first raised by the Dutch Nobel Laureate Gerard 't Hooft in 1980. His claim was that a theory is unnatural if some parameter that is expected to be large is actually (almost) zero, unless there is a symmetry enforcing the latter. This generates its own fine-tuning problems, which we do not discuss here; see Grinbaum (2012).

[20]The number called "googol" that the internet company Google has (erroneously) been named after is 'merely' 10^{100}; Penrose's number is even much larger than a one with googol many zeroes.

future but not in the past,[21] which requires an extremely unlikely initial state (Price 1997; Uffink 2007; Lebowitz 2008). However, this kind of reasoning is as brilliant as it is controversial (Callendar 2004, 2010; Earman 2006; Eckhardt 2006; Wallace 2010; Schiffrin and Wald 2012). More generally, the more extreme the asserted fine-tuning is, the more adventurous the underlying arguments are (or so we think).

To be on the safe side, the fine-tuning of Ω, Λ, and Penrose's initial condition should perhaps be ignored, leaving us with the other examples, and a few similar ones not discussed here. But these should certainly suffice to make a case for fine-tuning that is serious enough to urge the reader to at least make a bet on one the five options listed above.

3   General Arguments

Before turning to a specific discussion of the Design and the Multiverse proposals, we make a few critical (yet impartial) remarks that put the FTA in perspective (see also Sober 2004; Manson 2009). Adherents of the FTA typically use analogies like the following:

- Someone lays out a deck of 52 cards after it has been shuffled. If the cards emerge in some canonical order (e.g., the Ace of Spades down to 2, then the Ace of Hearts down to 2, etc.), then, on the tacit assumption that each outcome is equally (un)likely, this very particular outcome supposedly cannot have been due to 'luck' or chance.
- Alternatively, if a die is tossed a large number of times and the number 6 comes up every time, one would expect the die to be loaded, or the person who cast it to be a very skillful con man. Once again, each outcome was *assumed* equally likely.

First, there is an underlying assumption in the FTA to the effect that the 'constants' of Nature as well as the initial conditions of the Universe (to both of which the emergence of life is allegedly exquisitely sensitive) are similarly variable. This may or may not be the case; the present state of science is not advanced enough to decide between chance and necessity concerning the laws of nature and the beginning of the Universe.[22]

---

[21]This is a technical way of saying that heat flows from hot bodies to cold ones, that milk combines with tea to form a homogeneous mixture, that the cup containing it will fall apart if it falls on the ground, etc.; in all cases the opposite processes are physically possible, but are so unlikely that they never occur.

[22]Our own hunch tends towards necessity, for reasons lying in constructive quantum field theory (Glimm and Jaffe 1987): it turns out to be extremely difficult to give a mathematically rigorous construction of elementary particle physics, and the value of the constants may be fixed by the requirement of mathematical existence and consistency of the theory. For example, the so-called scalar φ^4 theory is believed to be trivial in four (space-time) dimensions, which implies that the relevant 'constant of Nature' must be zero (Fernandez et al. 1992). This is as fine a fine-tuning as anything! Similarly, in cosmology the Big Bang (and hence the initial conditions for the

Second, granted that the 'constants' etc. are variable in principle (in the sense that values other than the current ones preserve the existence and consistency of the theories in which they occur), it is quite unclear to what extent they can vary and which variations may be regarded as 'small'; yet the FTA relies on the assumption that even 'small' variations would block the emergence of life (Manson 2000). In the absence of such information, it would be natural to assume that any (real, positive as appropriate) value may be assumed, but in that case mathematical probabilistic reasoning (which is necessary for the FTA in order to say that the current values are 'unlikely') turns out to be impossible (McGrew et al. 2001; Colyvan et al. 2005; Koperski 2005).[23] But also if a large but finite number of values (per constant or initial condition) needs be taken into account, it is hard to assign any kind of probability to any of the alternative values; even the assumption that each values is equally likely seems totally arbitrary (Everitt 2004; Norton 2010).

Nonetheless, these problems may perhaps be overcome and in any case, for the sake of argument we will continue to use the metaphors opening this section.

4 Critiquing the Inference of Design from Fine-Tuning

The idea that cosmic fine-tuning originates in design by something like an intelligent Creator fits into a long-standing Judeo-Christian tradition, where both the Cosmos and biology were explained in that way.[24] Now that biology has yielded to the theory of Evolution proposed by Darwin and Wallace in the mid 19th Century,[25] the battleground has apparently moved back to the cosmos. Also there,

---

(Footnote 22 continued)

Universe it gives rise to) actually seems to be an illusion caused by the epistemic fact that we look at the quantum world through classical glasses. In this case, the requirement that cosmology as we know it must actually emerge in the classical limit of some quantum theory (or of some future theory replacing quantum mechanics) may well fix the initial conditions.

[23]Suppose some constant takes values in the real axis. In the absence of good reasons to the contrary, any alternative value to the current one should have the same probability. But there is no flat probability measure on the real numbers (or on any non-compact subset thereof). Even if there were such a measure, any finite interval, however large or small, would have measure zero, so that one could not even (mathematically) express the difference between some constant permitting life if it just lies within some extremely small bandwidth (as the FTA has it), or in some enormously large one (which would refute the FTA).

[24]See footnote 1 and refs. therein. Note that a fine-tuning intelligent Creator is still a long shot from the Christian God whom Swinburne (2004), Küng (2005), and Collins (2009) are really after!

[25]The Dutch primatologist De Waal (2013) recently noted how reasonable Creationism originally was: animals known to the population of the Middle East (where Judeo-Christian thought originated) included camels etc. but no primates, and hence all living creatures appeared very different from mankind.

Design remains a vulnerable idea.[26] For the sake of argument we do not question the coherence of the idea of an intelligent Creator as such, although such a spirit seems chimerical (Everitt 2004; Philipse 2012).

First, in slightly different ways Smith (1993) and Barnes (2012) both made the point that the FTA does not claim, or support the conclusion, that the present Universe is *optimal* for intelligent life. Indeed, it hardly seems to be: even granted all the fine-tuning in the world as well as the existence of our earth with its relatively favorable conditions (Ward and Brownlee 2000), evolution has been walking a tightrope so as to produce as much as jellyfish, not to speak of primates (Dawkins 1996). This fact alone casts doubt on the FTA as an Argument of Design, for surely a benign Creator would prefer a Universe optimal for life, rather than one that narrowly permits it? From a theistic perspective it would seem far more efficient to have a cosmic architecture that is robust for its designated goal.

Second, the inference to Design from the FTA seems to rest on a decisive tacit assumption whose exposure sustantially weakens this inference (Bradley 2001). The cards analogy *presupposes* that there was such a thing as a canonical order; if there weren't, then any particular outcome would be thought of in the same way and would of course be attributed to chance. Similarly, the dice metaphor *presupposes* that it is special for 6 to come up every single time; probabilistically speaking, every other outcome would have been just as (un)likely as the given sequence of sixes.[27] An then again, in the case of independently tunable constants of Nature and/or initial conditions, one (perhaps approximate) value of each of these must first be marked with a special label like 'life-permitting' in order for the analogy with cards or dice (and hence the appeal of the FTA) to work. The FTA is predicated on such marking, which already *presupposes* that life is special.

---

[26]In this context, it is worth mentioning that the familiar endorsement of the Big Bang by modern Christian apologists (see footnote 1) as a scientific confirmation of the creation story in *Genesis* 1 seems wishful thinking based on a common mistranslation of its opening line as "In the beginning God created the heavens and the earth" (and similarly in other languages), whereas the original Hebrew text does not intend the "beginning" as an absolute beginning of time but rather as the starting point of the action expressed by the following verb, whilst "created" should have read "separated" (Van Wolde 2009). More generally, the world picture at the time of writing of *Genesis* was that of a disk surrounded by water, the ensuing creation story not being one of *creatio ex nihilo*, but one in which God grounds the Earth by setting it on pillars. This led to a tripartite picture of the Cosmos as consisting of water, earth, and heaven. See also Van Wolde's contribution to this volume, as well as Noordmans (1934). The remarkable *creatio ex nihilo* story introduced by the Church Fathers therefore lacks textual support from the Bible.

[27]Entropy arguments do not improve the case for Design. It is true that although the probability of a sequence of all sixes is the same as the probability of any other outcome, the former *becomes* special if we coarse-grain the outcome space by counting the number of sixes in a given long sequence of throws and record that information only. The outcome with sixes only then becomes extremely unlikely, since it could only have occurred in one possible way, whereas outcomes with fewer sixes have multiple realizations (the maximum probability occurring when the number of sixes is about one-sixth of the total number of throws). The point is that the very act of coarse-graining again *presupposes* that six is a special value.

It is irrelevant to this objection whether or not life is indeed special; the point is that the assumption that life be special has to be made *in addition* to the FTA in order to launch the latter on track to Design. But the inference from the (assumed) speciality of life to Design hardly needs the FTA: even if all values of the constants and initial conditions would lead to a life-permitting Universe, those who think that life is special would presumably point to a Creator. In fact, both by the arguments recalled at the beginning of this section and those below, their case would actually be considerably stronger than the FTA.

In sum, fine-tuning is not by itself sufficient as a source for an Argument of Design; it is the *combination* with an assumption to the effect that life is somehow singled out, preferred, or special. But that assumption is the one that carries the inference to Design; the moment one makes it, fine-tuning seems counter-productive rather than helpful.

Attempts to give the Design Argument a quantitative turn (Swinburne 2004; Collins 2009) make things even worse (Bradley 2002; Halvorson 2014). Such attempt are typically based on *Bayesian Confirmation Theory*. This is a mathematical technique for the analysis and computation of the probability $P(H|E)$ that a given hypothesis H is true in the light of certain evidence E (which may speak for or against H, or may be neutral). Almost every argument in Bayesian Confirmation Theory is ultimately based on *Bayes' Theorem*

$$P(H|E) = P(E|H) \cdot P(H)/P(E),$$

where $P(H|E)$ is the probability that E is true given the truth of the hypothesis H, whilst $P(H)$ and $P(E)$ are the probabilities that H and E are true without knowing E and H, respectively (but typically assuming certain background knowledge common to both H and E, which is very important but has been suppressed from the notation).[28]

In the case at hand, theists want to argue that the Universe being fine-tuned for *Life* makes *Design* more likely, i.e., that $P(D|L) > P(D)$, or, equivalently, that $P(L|D) > P(L)$ (that is, Design favors life). The problem is that theists do not merely ask for the latter inequality; what they really believe is that $P(L|D) \approx 1$, for the existence of God should make the emergence of life almost certain.[29] For simplicity,

---

[28]This implies, in particular, that $P(H|E) > P(H)$, i.e., E confirms H, if and only if $P(E|H) > P(E)$, which is often computable. The probabilities in question are usually (though not necessarily) taken to be *epistemic* or (inter)*subjective*, so that the whole discussion is concerned with probabilities construed as numerical measures of *degrees of belief*. For technical as well as philosophical background on Bayesianism see e.g. Howson and Urbach (2006), Sober (2008), and Handfield (2012).

[29]Swinburne (2004), though, occasionally assumes that $P(L|D) = 1/2$ as a subjective probability (based on our ignorance of God's intentions), which still makes his reasoning vulnerable to the argument below. In Swinburne (2004), arguments implying $P(L|D) > P(L)$ are called *C-inductive*, whereas the stronger ones implying $P(L|D) > 1/2$ are said to be *P-inductive*. Swinburne's strategy is to combine a large number of *C*-inductive arguments into a single overarching *P*-inductive one, but according to Philipse (2012) every single one of Swinburne's *C*-inductive arguments is actually invalid (and we agree with Philipse).

first assume that $P(L|D) = 1$. Bayes' Theorem then gives $P(D|L) = P(D)/P(L)$, whence $P(D) \leq P(L)$. More generally, assume $P(L|D) \geq 1/2$, or, equivalently, $P(L|D) \geq P(\neg L|D)$, where $\neg L$ is the proposition that life does not exist. If (D,L) is the conjunction of D and L, we then have

$$P(D) = P(D,L) + P(D, \neg L) \leq 2P(D,L) \leq 2P(L),$$

since $P(D, \neg L) \leq P(D,L)$ by assumption. Thus a negligible prior probability of life (on which assumption the FTA is based!) implies a hardly less negligible prior probability of Design. This inequality make the Argument from Design self-defeating as an explanation of fine-tuning, but in any case, both the interpretation and the numerical value of $P(D)$ are so obscure and ill-defined that the whole discussion seems, well, scholastic.

5   Critiquing the Inference of a Multiverse from Fine-Tuning

The idea of Design may be said to be human-oriented in a spiritual way, whereas the idea of a Multiverse more technically hinges on the existence of observers, as expressed by the so-called (weak) *Anthropic Principle* (Barrow and Tipler 1986; Bostrom 2002). The claim is that there are innumerable Universes (jointly forming a 'Multiverse'), each having its own 'constants' of Nature and initial conditions, so that, unlikely as the life-inducing values of these constants and conditions in our Universe may be, they simply *must* occur within this unfathomable plurality. The point, then, is that we have to observe precisely those values because in other Universes there simply are no observers. This principle has been labeled both 'tautological' and 'unscientific'. Some love it and some hate it, but we do not need to take sides in this debate: all we wish to do is find out whether or not the FTA speaks in favour of a Multiverse, looking at both an explanatory and a probabilistic level. Thus the question is whether the (alleged) fact of fine-tuning is (at least to some extent) explained by a Multiverse, or if, in the context of Bayesian confirmation theory, the evidence of fine-tuning increases the probability of the hypothesis that a Multiverse exists.[30] To get the technical discussion going, the following metaphors have been used:

---

[30]Although fine-tuning has been claimed (notably by Rees 1999) to provide independent motivation for believing in a Multiverse, the existence of a Multiverse may be a technical consequence of some combination of string theory (Susskind 2005; Schellekens 2013) and cosmological inflation (Liddle and Lyth 2000; Weinberg 2008). Both theories are highly speculative, though (the latter less so than the former), and concerning the 'landscape' idea it is hard to avoid the impression that string theorists turn vice into virtue by selling the inability of string theory to predict anything as an ability to predict everything (a similar worry may also apply to inflation, cf. Smeenk 2014). Let us also note that even if it were to make any sense, the 'emergent multiverse'

- Rees's 'off the shelf' clothes shop has already been mentioned in the Introduction: if someone enters a shop that sells suits in one size only (i.e., a single Universe), it would be amazing if it fitted (i.e., enabled life). However, if all sizes are sold (in a Multiverse, that is), the client would not at all be surprised to find a suit that fits.
- Leslie's (1989) firing squad analogy states that someone should be executed by a firing squad, consisting of many marksmen, but they all miss. This amounts to fine-tuning for life in a single Universe. The thrust of the metaphor arises when the lucky executee is the sole survivor among a large number of other convicts, most or all of whom are killed (analogously to the other branches of the Multiverse, most or all of which are inhospitable to life). The idea is that although each convict had a small a priori probability of not being hit, if there are many of them these small individual probabilities of survival add up to a large probability that *someone* survives.
- Bradley (2009, 2012) considers an urn that is filled according to a random procedure:

  - If a coin flip gives Heads (corresponds to a single Universe), either a small ball (life) or a large one (no life) is entered (depending on a further coin flip).
  - In case of Tails (modeling a 'Binaverse' for simplicity), two balls enter the urn, whose sizes depend on two further coin flips (leaving four possibilities).

Using a biased drawing procedure that could only yield either a small ball or nothing, a small ball is obtained (playing the role of a life-enabling Universe). A simple Bayesian computation shows that this outcome confirms Tails for the initial flip.

Each of these stories is insightful and worth contemplating. For example, the first one nicely contrasts the Multiverse with Design, which would correspond to bespoke tailoring and hence, at least from a secular point of view, commits the fallacy of putting the customer (i.e., life) first, instead of the tailor (i.e., the Universe as it is). The Dostoyevskian character of the second highlights the Anthropic Principle, whose associated selection effects (Bostrom 2002) are also quantitatively taken into account by the third.

Nonetheless, on closer inspection each is sufficiently vulnerable to fail to clinch the issue in favour of the Multiverse. One point is that although each author is well aware of (and the second and the third even respond to) the *Inverse Gambler's Fallacy* (Hacking 1987),[31] this fallacy is not really avoided (White 2000). In its simplest version, this is the mistake made by a gambler who enters a casino or a pub, notices that a double six is thrown at some table, and asks if this is the first roll

---

(Footnote 30 continued)

claimed to exist in the so-called Many-Worlds (or Everett) Interpretation of quantum mechanics (Wallace 2012) is a red herring in the present context, for, as far as we understand, all its branches have exactly the same laws of nature, including the values of the constants.

[31]The *Gambler's Fallacy* is the mistake that after observing say 35 throws of two fair dice without a double six, this preferred outcome has become more likely (than 1/36) in the next throw.

of the evening (his underlying false assumption being that this particular outcome is more likely if many rolls preceded it). Despite claims to the contrary (Leslie 1988; Manson and Thrush 2003; Bradley 2009, 2012), Hacking's analysis that this is precisely the error made by those who favor a Multiverse based on the FTA in our opinion still stands. For example, in Rees' analogy of the clothes shop, what needs to be explained is not that *some* suit in the shop turns out to fit the customer, but that the one *he happens to be standing in front of* does. Similarly, the probability that a *given* executee survives is independent of whoever else is going to be shot in the same round. And finally, the relevant urn metaphor is not the one described above, but the one in which Tails leads to the filling of *two* different urns with one ball each. Proponents of a Multiverse correctly state that its existence would increase the probability of life existing in *some* Universe,[32] but this is only relevant to the probability of life in *this* Universe if one identifies *any* Universe with the same properties as ours with *our* Universe.[33] Such an identification may be suggested by the (weak) Anthropic Principle, but its is by no means implied by it, and one should realize that the inference of a Multiverse from the FTA implicitly hinges on this additional assumption.[34]

Moving from a probabilistic to an explanatory context, we follow Mellor (2002) in claiming that if anything, a Multiverse would make fine-tuning even more puzzling. Taking the firing squad analogy, there is no doubt that the survival of a single executee is unexpected, but the question is whether it may be explained (or, at least, whether it becomes less unexpected) by the assumption that simultaneously, many other 'successful' executions were taking place. From the probabilistic point of view discussed above, their presence should have no bearing on the case of the lone survivor, whose luck remains as amazing as it was. From another, explanatory point of view, *it makes his survival even more puzzling*, since we now know from this additional information about the other executions that apparently the marksmen usually do kill their victims.

---

[32]Bradley (2012, p. 164) states *verbatim* that he is computing the probability that "At least one universe has the right constants for life", other authors doing likewise either explicitly or tacitly.

[33]Bradley (2009) counters objections like Hacking's by the claim that "if there are many Universes, there is a greater chance that Alpha [i.e., our Universe] will exist". This implies the same identification.

[34]We side with Hartle and Srednicki (2007) in believing that the identification in question is solidly wrong: "This notion presupposes that we exist separately from our physical description. But we are not separate from our physical description in our data; we *are* the physical system described (...) It is *our data* that is used in a Bayesian analysis to discriminate between theories. What other hypothetical observers with data different from ours might see, how many of them there are, and what properties they might or might not share with us (...) are irrelevant for this process.".

6   Conclusion

Already the uncontroversial examples that feed the FTA suffice to produce the fascinating insight that the formal structure of our current theories of (sub)nuclear physics and cosmology (i.e., the Standard Model of particle physics and Einstein's theory of General Relativity) is insufficient to predict the rich phenomenology these theories give rise to: the precise values of most (if not all) constants and initial conditions play an equally decisive role. This is a recent insight: even a physicist having the stature of Nobel Laureate Glashow (1999, p. 80) got this wrong, having initially paraphrased the situation well:

> "Imagine a television set with lots of knobs: for focus, brightness, tint, contrast, bass, treble, and so on. The show seems much the same whatever the adjustments, within a large range. *The standard model is a lot like that.*
> Who would care if the tau lepton mass were doubled or the Cabibbo angle halved? The standard model has about 19 knobs. They are not really adjustable: they have been adjusted at the factory. Why they have their values are 19 of the most baffling metaquestions associated with particle physics."

In our view, the insight that *the standard model is not like that at all* is the real upshot of the FTA.[35] Attempts to draw further conclusions from it in the direction of either *Design* or a *Multiverse* are, in our opinion, unwarranted. For one thing, as we argued, at best they fail to have any explanatory or probabilistic thrust (unless they rely on precarious additional assumptions), and at worst fine-tuning actually seems to turn against them.

Most who agree with this verdict would probably feel left with a choice between the options of *Blind Chance* and *Blind Necessity*; the present state of science does not allow us to make such a choice now (at least not rationally), and the question even arises if science will ever be able to make it (in a broader context), except perhaps philosophically (e.g., à la Kant). However, we would like to make a brief case for the fifth position, stating that the fine-tuning problem is *misguided* and that all we need to do is to clear away confusion.

There are analogies and differences between cosmic fine-tuning for life through the laws of Nature and the initial conditions of the Universe, as discussed so far, and Evolution in the sense of Darwin and Wallace. The latter is based on random (genetic) variation, survival of the fittest, and heritability of fitness. All these are meant to apply locally, i.e., to life on Earth. We personally feel that arguments to extend these principles to the Universe in the sense that the Cosmos may undergo some kind of 'biological' evolution, having descendants born in singularities, perhaps governed by different laws and initial conditions (some of which, then, might be 'fine-tuned for life', as in the Multiverse argument), as argued by e.g., Wheeler (in Ch. 44 of Misner et al. 1973) and Smolin (1997), imaginative as they may be, are too speculative to merit serious discussion. Instead, the true analogy seems to be as

---

[35]Callender (2004) understandably misquotes Glashow, writing: "The standard model is *not* like that.".

follows: as far as the emergence and subsequent evolution of life are concerned, the Universe and our planet Earth should simply be taken as given. Thus the fundamental reason we feel 'fine-tuning for life' requires no explanation is this[36]:

Our Universe has not been fine-tuned for life: life has been fine-tuned to our Universe.

Acknowledgement   The author is indebted to Jeremy Butterfield, Craig Callender, Jelle Goeman, Olivier Hekster, Casey McCoy, Herman Philipse, Jos Uffink, and Ellen van Wolde for comments, discussions, and encouragement, all of which considerably improved this paper.

Open Access   This chapter is distributed under the terms of the Creative Commons Attribution-Noncommercial 2.5 License (http://creativecommons.org/licenses/by-nc/2.5/) which permits any noncommercial use, distribution, and reproduction in any medium, provided the original author(s) and source are credited. The images or other third party material in this chapter are included in the work's Creative Commons license, unless indicated otherwise in the credit line; if such material is not included in the work's Creative Commons license and the respective action is not permitted by statutory regulation, users will need to obtain permission from the license holder to duplicate, adapt or reproduce the material.

References

Aguirre, A. (2001). The Cold Big-Bang cosmology as a counter-example to several anthropic arguments. *Physical Review D, 64*, 083508.

Aquinas, T. (1485). *Summa theologica* (written 1268–1274). Basel: M. Wenssler. Written.

Arnett, D. (1996). *Supernovae and nucleosynthesis: An investigation of the history of matter, from the Big Bang to the present.* Princeton: Princeton University Press.

Barnes, L. A. (2012). The fine-tuning of the universe for intelligent life. *Publications of the Astronomical Society of Australia, 29*, 529–556.

Barrow, J. D., & Tipler, F. (1986). *The anthropic cosmological principle.* Oxford: Clarendon Press.

Bentley, R. (1692). *A confutation of atheism from the origin and frame of the world.* London: Mortlock.

Bostrom, N. (2002). *Anthropic bias: Observation selection effects in science and philosophy.* New York: Routledge.

Bradford, R. (2011). The inevitability of fine tuning in a complex universe. *International Journal of Theoretical Physics, 50*, 1577–1601.

Bradford, R. (2013). Rick's critique of cosmic coincidences. Retrieved on April 6, 2015 www.rickbradford.co.uk/Coincidences.html.

Bradley, D. (2009). Multiple universes and the observation selection effects. *American Philosophical Quarterly, 46*, 61–72.

Bradley, D. (2012). Four problems about self-locating belief. *Philosophical Review, 121*, 149–177.

---

[36]From this point of view, what we see as the essential mistake made by those who feel fine-tuning does require an explanation is similar to what Butterfield (1931) famously christened 'Whig History', i.e., "the tendency in many historians to write on the side of Protestants and Whigs, to praise revolutions provided they have been successful, to emphasize certain principles of progress in the past and to produce a story which is the ratification if not the glorification of the present." (quoted from the Preface).

Bradley, M. C. (2001). The fine-tuning argument. *Religious Studies, 37*, 451–466.
Bradley, M. C. (2002). The fine-tuning argument: The Bayesian version. *Religious Studies, 38*, 375–404.
Butterfield, H. (1931). *The Whig interpretation of history*. London: Bell.
Callender, C. (2004). Measures, explanations and the past: Should 'special' initial conditions be explained? *The British Journal for the Philosophy of Science, 55*, 195–217.
Callender, C. (2010). The Past Hypothesis meets gravity. In G. Ernst & A. Hütteman (Eds.), *Time, chance, and reduction: Philosophical aspects of statistical mechanics* (pp. 34–58). Cambridge: Cambridge University Press.
Carroll, S. M., & Tam, H. (2010). Unitary evolution and cosmological fine-tuning. arXiv:1007. 1417.
Collins, R. (2009). The teleological argument: An exploration of the fine-tuning of the universe. In W. L. Craig & J. P. Moreland (Eds.), *The Blackwell companion to natural theology* (pp. 202–281). Chichester: Wiley-Blackwell.
Colyvan, M., Garfield, J. L., & Priest, G. (2005). Problems with the argument from fine tuning. *Synthese, 145*, 325–338.
Davies, P. (2006). *The Goldilocks dilemma: Why is the universe just right for life?* New York: Mariner.
Dawkins, R. (1986). *The blind watchmaker*. New York: Norton.
Dawkins, R. (1996). *Climbing mount improbable*. New York: Norton.
de Spinoza, B. (1677). *Ethica ordine geometrico demonstrata*. Amsterdam: J. Rieuwertsz.
De Waal, F. (2013). *The Bonobo and the atheist*. New York: Norton.
Earman, J. (1987). The SAP also rises: A critical examination of the anthropic principle. *American Philosophical Quarterly, 24*, 307–317.
Earman, J. (2006). The "Past Hypothesis": Not even false. *Studies in History and Philosophy of Modern Physics, 37*, 399–430.
Eckhardt, W. (2006). Causal time asymmetry. *Studies in History and Philosophy of Modern Physics, 37*, 439–466.
Ekström, S., et al. (2010). Effects of the variation of fundamental constants on Population iii stellar evolution. *Astronomy & Astrophysics, 514*, A62.
Ellis, G. G. R. (2007). Issues in the philosophy of cosmology. In J. Butterfield & J. Earman (Eds.), *Philosophy of physics Part B* (pp. 1183–1286). Amsterdam: Elsevier.
Everitt, N. (2004). *The non-existence of God*. London: Routlegde.
Fernandez, R., Fröhlich, J., & Sokal, A. D. (1992). *Random walks, critical phenomena, and triviality in quantum field theory*. Heidelberg: Springer.
Glimm, J., & Jaffe, A. (1987). *Quantum physics: A functional integral point of view*. New York: Springer.
Galli, D., & Palla, F. (2013). The dawn of chemistry. *Annual Review of Astronomy and Astrophysics, 51*, 163–206.
Glashow, S. L. (1999). Does quantum field theory need a foundation? In T. Y. Cao (Ed.), *Conceptual foundations of quantum field theory* (pp. 74–88). Cambridge: Cambridge University Press.
Grinbaum, A. (2012). Which fine-tuning problems are fine? *Foundations of Physics, 42*, 615–631.
Hacking, I. (1987). The inverse gambler's fallacy: The argument from design. The anthropic principle applied to Wheeler universes. *Mind, 96*, 331–340.
Halvorson, H. (2014). A probability problem in the fine-tuning argument. http://philsci-archive. pitt.edu/11004/.
Hartle, J. B., & Srednicki, M. (2007). Are we typical? *Physical Review D, 75*, 123523.
Handfield, T. (2012). *A philosophical guide to chance*. Cambridge: Cambridge University Press.
Hogan, C. J. (2000). Why the universe is just so. *Reviews of Modern Physics, 72*, 1149–1161.
Holt, J. (2012). *Why does the world exist? An existential detective story*. New York: W.W. Norton.
Howson, C., & Urbach, P. (2006). *Scientific reasoning: The Bayesian approach* (3rd ed.). Chicago: Open Court.
Jorink, E. (2010). *Reading the book of nature in the Dutch Golden Age* (1575–1715). Leiden: Brill.

Kenny, A. (1969). *The five ways: St Thomas Aquinas' proofs of God's existence.* London: Routledge & Kegan Paul.

Koperski, J. (2005). Should we care about fine-tuning? *The British Journal for the Philosophy of Science, 56,* 303–319.

Küng, H. (2005). *Der Anfang aller Dinge: Naturwissenschaft und Religion.* München: Piper.

Lebowitz, J. L. (2008). From time-symmetric microscopic dynamics to time-asymmetric macroscopic behavior: An overview. In G. Gallavotti, W. L. Reiter & J. Yngavson (Eds.), *Boltzmann's Legacy* (pp. 38–62). Zürich: European Mathematical Society.

Leslie, J. (1988). No inverse Gambler's fallacy in cosmology. *Mind, 97,* 269–272.

Leslie, J. (1989). *Universes.* London: Routledge.

Leslie, J., & Kunh, R. L. (2013). *The mystery of existence: Why is there anything at all?* Chichester: Wiley-Blackwell.

Liddle, A. R., & Lyth, D. A. (2000). *Cosmological inflation and large-scale structure.* Cambridge: Cambridge University Press.

Manson, N. A. (2000). There is no adequate definition of 'fine-tuned for life'. *Inquiry, 43,* 341–352.

Manson, N. A. (Ed.). (2003). *God and design.* London: Routledge.

Manson, N. A. (2009). The fine-tuning argument. *Philosophy Compass, 4,* 271–286.

Manson, N. A., & Thrush, M. J. (2003). Fine-tuning, multiple universes, and the "this universe" objection. *Pacific Philosophical Quarterly, 84,* 67–83.

Martin, J. (2012). Everything you always wanted to know about the cosmological constant problem (but were afraid to ask). *Comptes Rendus Physique, 13,* 566–665.

McCoy, C. (2015). What is the horizon problem? Preprint.

McGrew, T., McGrew, L., & Vestrup, E. (2001). Probabilities and the fine-tuning argument: A sceptical view. *Mind, 110,* 1027–1038.

Mellor, D. H. (2002). Too many universes. In N. A. Manson (Ed.), *God and design: The teleological argument and modern science* (pp. 221–228). London: Routledge.

Misner, C., Thorne, K., & Wheeler, J. A. (1973). *Gravitation.* San Francisco: Freeman.

Monod, J. (1971). *Chance and necessity: An essay on the natural philosophy of modern biology.* New York: A. Knopf.

Noordmans, O. (1934). *Herschepping.* Zeist: De Nederlandsche Christen-Studenten-Vereeniging.

Norton, J. D. (2010). Cosmic confusions: Not supporting versus supporting not. *Philosophy of Science, 77,* 501–523.

Paley, W. (1802). *Natural theology: Or evidences of the existence and attributes of the deity collected from the appearances of nature.* London: J. Foulder.

Penrose, R. (1979). Singularities and time-asymmetry. In S. W. Hawking & W. Israel (Eds.), *General relativity: An Einstein centenary survey* (pp. 581–638). Cambridge: Cambridge University Press.

Penrose, R. (2004). *The road to reality: A complete guide to the laws of the universe.* London: Jonathan Cape.

Philipse, H. (2012). *God in the age of science? A Critique of religious reason.* Oxford: Oxford University Press.

Planck Collaboration (2014). *Planck* intermediate results. XXIV. Constraints on variation of fundamental constants. *Astronomy & Astrophysics.* arXiv:1406.7482.

Planck Collaboration (2015). Planck 2015 results. XIII. Cosmological parameters. arXiv:1502.01589.

Price, H. (1997). *Time's arrow & Archimedes' point.* Cambridge: Cambridge University Press.

Rees, M. (1999). *Just six numbers.* London: Weidenfeld & Nicolson.

Rundle, B. (2004). *Why there is something rather than nothing.* Oxford: Oxford University Press.

Schellekens, A. N. (2013). Life at the interface of particle physics and string theory. *Reviews of Modern Physics, 85,* 1491–1540.

Schiffrin, J. S., & Wald, R. M. (2012). Measure and probability in cosmology. *Physical Review D, 86,* 023521.

Schilpp, P. A. (1949). *Albert Einstein: Philosopher-scientist* (Vol. 1). La Salle: Open Court.

Smeenk, C. (2014). Predictability crisis in early universe cosmology. *Studies in History and Philosophy of Modern Physics, 46*, 122–133.

Smith, J. M., & Szathmáry, E. (1995). *The major transitions in evolution.* Oxford: Freeman.

Smith, Q., & Craig, W. L. (1993). *Theism, atheism and Big Bang cosmology* (pp. 203–204). Oxford: Clarendon Press.

Smolin, L. (1997). *The life of the cosmos.* Oxford: Oxford University Press.

Sober, E. (2004). The design argument. In W. E. Mann (Ed.), *Blackwell guide to the philosophy of religion* (pp. 117–147). Oxford: Blackwell.

Sober, G. (2008). *Evidence and evolution.* Cambridge: Cambridge University Press.

Stenger, V. (2011). *The fallacy of fine-tuning: Why the universe is not designed for us.* Amherst (NY): Prometheus.

Susskind, L. (2005). *The Cosmic landscape: String theory and the illusion of intelligent design.* New York: Little, Brown.

Swinburne, R. (2004). *The existence of God* (2nd ed.). Oxford: Oxford University Press.

Tegmark, M., Aguirre, A., Rees, M. J., & Wilczek, F. (2006). Dimensionless constants, cosmology, and other dark matters. *Physical Review D, 73*, 023505.

Uffink, J. (2007). Compendium of the foundations of classical statistical physics. In J. Butterfield & J. Earman (Eds.), *Philosophy of physics part B* (pp. 923–1074). Amsterdam: Elsevier.

Van Wolde, E. (2009). Why the verb bāra' does not mean 'to create' in Genesis 1.1–2.4a. *Journal for the Study of the Old Testament, 34*, 3–23.

Wallace, D. (2010). Gravity, entropy, and cosmology: In search of clarity. *The British Journal for the Philosophy of Science, 61*, 513–540.

Wallace, D. (2012). *The emergent multiverse: Quantum theory according to the Everett interpretation.* Oxford: Oxford University Press.

Ward, P., & Brownlee, D. (2000). *Rare earth.* New York: Copernicus Books.

Weinberg, S. (1977). *The first three minutes.* New York: Basic Books.

Weinberg, S. (2007). Living in the multiverse. In B. Carr (Ed.), *Universe or multiverse* (pp. 29–42). Cambridge: Cambridge University Press.

Weinberg, S. (2008). *Cosmology.* Oxford: Oxford University Press.

White, R. (2000). Fine-tuning and multiple universes. *Noûs, 34*, 260–276.

Chance in the Hebrew Bible: Views in Job and Genesis 1

Ellen van Wolde

Abstract There are a variety of views on 'chance' to be found in the Hebrew Bible, or Old Testament. In this chapter we will discuss the Book of Job and the opening chapter in the Book of Genesis, i.e. Genesis 1, both as narratives and as poetic texts and explore the philosophical and theological consequences for a better understanding of the concept of chance. In the prologue of the Book of Job, chance is referred to as the result of a wager between God and the satan, who is described as one of the sons of God. In the dialogue between Job and his friends, bad luck is viewed as a consequence of bad behaviour while good luck is the result of good behaviour. In this sense, chance clearly functions within a moral framework of retribution. At the end of the Book of Job, in God's speech out of the whirlwind, chance is linked to a multifocal view of the universe and understood in terms of position, perspective, and scale. Also the opening chapter of the Book of Genesis offers a non-deterministic view on chance. Chance is not the exception in a causal or necessary chain of events, but it stands out in a framework of non-linear thinking in which totality and instantaneity alternate. With regard to both biblical texts, God's speech in the Book of Job and Genesis 1, chance can be conceived as a disqualifier of this chain of events, and even as an ultimate denial of the existence of necessity.

1 The Prologue of the Book of Job: Chance as a Wager

Job's life is going well, very well indeed. Job is rich, wealthier than anyone in the East. He has a large herd of cattle, a huge number of employees, and a very large household. Above all, he has the family that every rich man desired at that time, namely seven sons and three daughters. Who could wish for more? Because of these blessings, or maybe by choice, Job lives his life as righteously as he can,

E. van Wolde (✉)
Faculty of Philosophy, Theology and Religious Studies, Radboud University, Nijmegen,
The Netherlands
e-mail: e.vanwolde@ftr.ru.nl

© The Author(s) 2016
K. Landsman and E. van Wolde (eds.), *The Challenge of Chance*,
The Frontiers Collection, DOI 10.1007/978-3-319-26300-7_7

treating others as he would wish to be treated. Because he is an honest, upright, and god-fearing man, people respect him. Suddenly, seemingly by chance, bad luck strikes. One day as his sons and daughters socialise with their friends, a terrible storm arose and lifted the roof off smashing it back down onto the group. They are all killed in an instant—no one survives. The servants have to break this dreadful news to their master. Then, barely has one disaster struck when another employee rushes in from the fields to tell Job that cattle-thieves have stolen all his livestock: thousands of oxen, she-asses, sheep, goats and camels are gone. Within the wink of an eye this god-fearing man who had everything has lost everything.

This story of devastating misfortune is told in the book of Job,[1] one of the books in the Hebrew Bible.[2] Surprisingly, chapters 1 and 2 already offer an explanation why this happened. It seems that Job's misfortune, or the shift from fortune to misfortune, was the consequence of a deal made in heaven. Through the description of a meeting by the divine council we find out what lay behind Job's misfortune from heaven's perspective. In this meeting, Yahweh[3] opens the discussion by asking a fellow divine being,[4] one of the sons of God called the satan,[5] the

---

[1] Most scholars today would date the composition of the Book of Job to some point between the seventh and fourth centuries BCE. There are a number of indications in the book that it was not written all at one time, but went through a phases of composition. In the most recent monography on Job (Seow 2013, pp. 40–44), the Book of Job is dated to the late sixth to mid-fifth century BCE. Seow's arguments are based on literary parallels to Deutero-Isaiah and Zechariah 3, as well as to the historical reference to the Chaldeans in Job 1:17.

[2] Some of the most comprehensive and recent monographs on the book of Job are: Habel (1985); Clines (1989–2009); Newsom (2003); Seow (2013).

[3] The notion of 'God' or 'deity' is expressed in Biblical Hebrew by the word 'elōhîm, a plural noun of the singular form 'el or 'eloah, 'God', and this plural noun is commonly used with a singular verb form. The personal name of the God of Israel is yhwh, Yahweh. In the Hebrew Bible sometimes reference is made to the God of Israel by its common name 'elōhîm, other times by the personal name yhwh. In the Book of Job both terms are used to designate the deity (see, e.g., here in Job 1:6: "the sons of 'elōhîm presented themselves before yhwh"; or Job 1:21 where Job says: "yhwh has given, yhwh has taken away"; see also Job 2:10, in which Job says to his wife: "Should we accept good from the hands of the deity (ha-'elōhîm), should we not accept evil?"). Throughout this chapter I will refer to 'elōhîm or yhwh by the term 'God', with the exception of literally quoted verses.

[4] 'Sons of God' or 'divine beings' (in Hebrew benê-'elōhîm or benê-'elîm) figure in several passages in the Hebrew Bible and designate the divine beings that live in heaven and are seen as closely related to Yahweh or to Elyon, 'God, the most high'. The notion of a 'divine council' denotes a formal gathering of these 'sons of God' and this council is viewed as the godly government, which most likely resembled the earthly royal court. The earthly and heavenly councils formally operated in two ways: the first way would be an advisory board for the king/deity regarding matters of state or government; the second way was as a formal judicial court. For extensive discussion, see White (2014). Reference to these divine beings and/or a divine council is made in the Hebrew Bible in: Gen. 1:26; 3:22; 11:7; Exod 15:11; Deut 4:19; 17:3; 32:8; 33:2–3; Judg. 5:20; 1 Kg 22:19–23; Isa 6; 14:13; Jer 8:2; 23:18.22a; Am 8:14; Sach 3; 14:5; Pss 25:14; 29:1–2; 49:20; 58:1–2; 73:15; 82; 89:6–9; 96:4–5; 97:7–9; 148:2–3; Job 1–2; 15:8; 38:7; Dan 7:9–14; Neh 9:6; 1 Chron 16:25 (book order follows the Hebrew canon).

[5] 'The satan' is the translation of ha-saṭān (in Hebrew this is the nominalised form of the participle of the verb saṭan 'accuse', preceded by the definite article ha-) and the definite article indicates that

following question. "Have you noticed my servant Job? There is no one like him on earth, a blameless and upright man who fears God and shuns evil!" The satan replies, "Is it 'for naught' (Hebrew *chinām*) that Job has put his faith in you? You have protected him, all his life." In this sense, the satan argues that the principle of retribution,[6] or 'tit for tat', drives human behaviour, including Job's model behaviour. In other words, the satan claims that Job puts his faith in God only because God protects him and to make sure things go well for him. God takes the opposite position. Simply put, God assumes that Job is pious at the same time as being rich, whereas the satan claims that Job is pious *because* he is rich and wants to stay rich. Challenged by the satan, God places his bets on Job. It is an important question for God: do people fear God unconditionally or do they put their faith in him in order to ensure they stay well off? [7] God cannot test everyone so he puts Job, the epitome of a pious man, to the test. The aim is to answer the following questions: is people's loyalty to God pure, that is to say not driven by self-interest? Are disasters the consequence of bad behaviour or caused by a lack of trust in God? Do human beings who live a good life deserve happiness? Did Job deserve happiness? Is there any rationality behind the alternation of fortune and misfortune on earth? To demonstrate the significance of these questions, the narrator sets the exchange between God and the satan in heaven. Here the discussion between God and the satan can be more open and intense. However, only the readers know about the wager. The character Job knows nothing of this heavenly experiment.

The next scene is set on earth and shows how Job reacts when blow after blow strike. Although deeply miserable and unable to understand what is happening to him, he does not blame God. Instead he says: "Naked I came from my mother's womb, and naked I shall return. Yahweh has given and Yahweh has taken away; blessed be the name of Yahweh" (Job 1:21). The interesting point of this response is that Job does not consider misfortune as mere bad luck or as something inexplicable that happened by accident, but he attributes everything, either good or bad, to God. Job's position, therefore, is one of complete faith or trust.

But then, new disasters strike Job. This time his body is affected and his skin peels away until his body is raw. Eventually he ends up in a rubbish dump covered

---

(Footnote 5 continued)

'the satan' does not express a name, but refers to someone who performs the task of 'accusing'. Figuring in a judicial court, one might translate 'the satan' with 'the (public) prosecutor'. In the divine council operating as a judicial court (see note 4), the various divine beings each play their own role: the satan acts as the public prosecutor, while (the highest) God acts as the judge, and the *malach* (traditionally translated with 'angel') functions as the messenger who brings the divine judgements as messages to the human beings.

[6]The term retribution derives from the Latin *retribuare*, 'to pay, grant, repay'.

[7]The modern terminology 'to believe in God', 'to love God', or 'to have pure faith' does not adequately reflect the idea of 'to fear God'. In the Hebrew Bible, 'to fear God' includes notions like 'trust', 'respect', 'awe' and 'loyalty', which figure in a hierarchical framework of thinking. The adequate human expression of this fear is 'to serve God'. The question in Job's prologue is, therefore, do human beings fear God because they trust and respect God, or, in contrast, because they expect reward and try to avoid punishment?

with loathsome ulcers from the soles of his feet to the crown of his head. He scratches himself with a pot shard but still he utters no reproach. Suddenly Job's wife turns up. Where did she come from? She was not mentioned before.[8] The narrator told us about Job's sons and daughters but never mentioned a wife, and when he lost his offspring there was no reference to her either. In his deepest misery Job says that he is all alone in the world ("naked I came, naked I will go") without mention of a wife—apparently she does not count. Yet, now Mrs. Job enters the picture and challenges her husband: "Do you still keep your integrity? Say good-bye to God ('elōhîm) and die" (Job 2:9). In a way, Job's wife draws the same conclusion that many secular readers would draw under similar circumstances. Embedded in her words are questions such as: "How can you keep on being loyal to God when all this misfortune befalls you? Why are you being targeted? You, my dear husband, do not deserve this. You live an upright life, I can testify to it." Job's wife is motivated by the principle of causality as the steering principle of faith: you place your trust in God since he is the one who made you, supports you, perhaps, even punishes you when you deserve it. There appears to be balance in this God-created universe. But disaster and misery prove that such a balance does not exist, so you might as well give up your loyalty to God. Modern secular people would add: it is not just the fact that this cosmic order does not exist, but the so-called originator and defender of this cosmic order does not exist either. This would be unthinkable in the context of ancient Near Eastern culture. Here in this text the phrase is: 'Say adieu to God'; the existence of God is not at issue. Nevertheless, this farewell to God is what Job's wife proposes and we, as modern readers, are likely to agree with her as we often understand misfortune in individual lives or in nature as evidence of the non-existence of God.

Yet, Job contests this view fiercely. He dismisses his wife's words as foolish: "Should we accept good from the hands of the deity, but should we not accept evil?" (Job 2:10). Still, her words have an effect. By confronting Job with his own death and pointing out to him the choice between blessing God or saying good-bye to God, she forces him to respond. The difference between the wording of Job's first reaction in Job 1:21 ("Yahweh has given and Yahweh has taken away; blessed be the name of Yahweh") and his response to his wife is striking. The first time Job speaks he refers to God by the name Yahweh. Thus Job acknowledges Yahweh as Lord. The second time, immediately following his wife's remarks, Job speaks about God as 'the deity', ha-'elōhîm. Although Job still considers God as the agent or distributor of good and bad luck, this sounds more detached. In addition, whereas

---

[8]Her namelessness, her absence in chapter 1, her short and unclearly presented speech in chapter 2, and her departure after the second chapter of the book of Job never to return in the rest of the book, have aroused interpreters' interest in Job's wife throughout history. From the Greek translation of the Hebrew Bible in the Septuagint (third century BCE) in which a section on her is added to the translation, through the interpretation history of the Bible, to contemporary novels and theatre plays, Mrs. Job has received much more attention than in the biblical book of Job. For a recent survey, see Gravett (2012, pp. 97–125). For a textual analysis of Job 1–2 and of the narrative function of Job's wife, see: van Wolde (1995, pp. 201–221).

the first reaction was a statement, the second is formulated as a question. And Job who first choses to bless God (in Job 1:21) now stops blessing God. His wife introduces the notion of death and this instils doubt in Job. He is no longer sure of anything and begins to ask himself questions. He even starts to reason from a human point of view instead of automatically adopting the perspective of God. His wife's taunts trigger Job to change from an assured believer into someone who asks questions. The responses of an ardent believer would not have provided material for such a dramatic story. The book of Job is made human and lifelike through the doubt and spirit of a man who has to confront his trust in God in the light of the suffering, misery and undeserved and devastating bad luck that has befallen him.

Thus the opening chapters of the book of Job explore the theme of chance through narrative. What seems to be an inexplicable change of fortune on earth is described as the consequence of a wager in heaven. The bet turns out to be a kind of empirical research. God's hypothesis is that people serve him 'for naught'. His is a framework of non-causality. The counterhypothesis, formulated by the satan, is that people serve God in order to secure a better life for themselves. His framework is one of causality. The test is performed on God's model servant on earth, Job. The concept of chance thus figures in the domain of causality. By alternating between scenes on earth and scenes in heaven, the reader is able to view the topic from two perspectives through the characters in the two domains, i.e. God and the satan in heaven, and Job and his wife on earth. By positioning the four characters in a kind of matrix, the narrator reveals his preferences. The narrative strategy of Job 1–2 is to convince readers to share both God's and Job's point of view and agree with them that it is enough to accept that everything (good luck and bad luck) is given or taken away by God. God and Job conclude that the satan and women (not just Job's wife) hold a point of view is seductive but incorrect. However, by introducing these opposing characters, readers are challenged to consider questions such as: Are the concepts of causality and retribution helpful in understanding the incidence of fortune and misfortune in someone's life? Are patterns of regularity, logic and ethical balance sufficient to explain the unexpected disruptions in someone's life or not?

2 Dialogue in the Book of Job: Chance as Proof of Moral Balance

Job's friends know the answers to these questions. In endless discussions (Job chapters 4–37), usually called 'dialogues', Job and his friends defend the view of a moral balance in the world in various degrees.[9]

---

[9]For an extensive description of the dialogue sections, see: van Wolde (2003, pp. 42–106).

Eliphaz is the first of Job's friends to speak and his speech is characterized by dignity, sobriety and reticence. The nub of what he wants to say is: "Is not your fear of God your confidence and the integrity of your life your hope?" (Job 4:6). Job can be reassured precisely because he believes in justice and knows that God guarantees that human beings will be recompensed in accordance with their behaviour, good or bad, says Eliphaz. He thus sketches a hopeful future for Job. The second friend, Bildad, is less optimistic. He calls God a just judge and in his view there is no doubt that God administers law in the right way. Bildad even goes as far as seeing God's justice illustrated by the fate of Job's sons: they partied too much and have sinned so they were punished. In actual fact, his argument is back to front: because Job's sons were punished, they must have sinned. The third friend, Zophar, even goes one step further. He identifies where Job went wrong and explicitly condemns the process that Job is going through, since he understands that Job risks throwing the whole of the traditional doctrine of retribution overboard. Zophar's reaction is caustic and what he says can be summarized as follows: "don't think that you can understand everything by your talk and chatter. It cannot be grasped at all, so submit to the traditional views and know that God's justice is a great mystery". Thus Zophar puts Job's behaviour to shame. In contrast to Eliphaz, who regarded Job as innocent, and Bildad, who regarded the sons as guilty, Zophar now accuses Job outright of sin and calls on him to repent.

Job's friends' views on misfortune and chance clearly function within the moral framework of retribution. David Clines, one of the most prominent Job scholars of our times, offers a fair reflection on their position:

> Now it is very easy to mock the friends' concept of God as the executor of retribution, and to point to the myriad of examples we all know in which reward has been denied the godly and the wicked have escaped punishment. Yet the alternatives to this theology may be worse still: imagine a world in which there is simply no predictable correspondence between act and consequence. How will any parent inculcate right behavior in children, how will any state warn the criminally inclined, if there is no underlying principle of retribution? The attractiveness of the theology is that it is not purely experiential and anecdotal, an accumulation of instances, but a systematic, principled thinking through of the way the world ought to work, should be governed, must be conceived. It posits a fundamental justice at the heart of God's design for the universe. From this perspective, any number of examples, or apparent examples, where it fails to be implemented cannot subvert the principle, for— although it is often stated as an account of what actually happens in the real world—it is not so much a description of reality as a blueprint for it. (Clines 2004, p. 42)

Job, as a man who fears God and shuns evil (Job 1:1), had long accepted the same theology. But since he has experienced a refutation of that theology at first hand, his whole view of God's justice is called into question. He draws the bitter conclusion that there is no retribution and that there is no justice. His personal tragedy has led to disillusionment with God and the whole of the moral universe.

3   God's Answer "Out of the Whirlwind"

Surprisingly, in the book of Job, God's speech out of the whirlwind is presented in the form of an answer to Job: "Then, Yahweh replied to Job out of the whirlwind and said" (Job 38:1). God's answer, which stretches out over four chapters (Job 38–41), is set in a poetic style, with short sentences, fixed rhythms, and multiple series of rhetorical questions, which very often open with the interrogatives 'who', 'when', 'where', 'what', or 'do you know'? Yet, a real answer to the earlier questions posed, it is not. It seems more of a monologue in which God does not really react to Job's questions and cry for justice. God's first words to Job are full of significance: "Who is this who darkens counsel, speaking without knowledge?" (Job 38:2). God reproves Job for setting his own agenda. In his quest for justice, Job obscures the fact that God does nothing to ensure that justice reigns in the world. God speaks about a completely different order, when he continues, "Where were you when I laid the earth's foundations? Speak if you have understanding." (Job 38:4). God not only points out their varying levels of knowledge, but also Job's physical location. God refers to Job's position as well as his implied spatial limits and, accordingly, limited perspective (Joode 2015, pp. 198–199). In fact, God's spatial scale is of a different order. Not only does he know everything about the created universe, he is its architect. "Do you know who fixed its dimensions? Or who measured it with a line? Onto what were the earth's bases sunk? Who set its corner stone? When the morning stars sang together and all the divine beings[10] shouted for joy?" (Job 38:4–7)[11] And God continues: "Have you penetrated to the sources of the sea? Or walked in the recesses of the deep?[12] Have the gates of death been disclosed to you? Have you seen the gates of deep darkness? Have you surveyed the expanses of the earth? If you know of these, tell me" (Job 38:16–18). Again and again, Job is forced to acknowledge his limited position, limited perspective, and, therefore, his limited knowledge.

Later, God carries on asking about all kinds of animals,[13] implying that in their own way they have all the freedom to reproduce and treat their young as they see

---

[10]For divine beings, see footnote 4. In Mesopotamia the stars are conceived as the heavenly manifestations of deities, and hence as divine beings. The same divine beings are at the same time physically present on earth, e.g. in statues (inaugurated after mouth-washing rituals) or in temples. This fluidity of the divine selfhood in Mesopotamia, Canaan and possibly also in the Hebrew Bible is discussed in Sommer (2009).

[11]The translation of the verses in Job 38–39 is the Jewish Publication Society's-translation.

[12]For the tripartite worldview behind these questions (heaven, earth, and teħom or 'the deep'), see below the first paragraph in the section on Genesis 1.

[13]E.g. Job 38:39 and 39:1–4: "Can you hunt prey for the lion? And satisfy the appetite of the king of beasts? They crouch in their dens, lie in ambush in their lairs. Do you know the season when the mountain goats give birth? Can you mark the time when the hinds calve? Can you count the months they must complete? Do you know the season they give birth? When they couch to bring forth their offspring, to deliver their young? Their young are healthy; they grow up in the open. They leave and return no more."

fit. In a long series of rhetorical questions, God reflects on the universe and its inhabitants, showing that he infused all entities and creatures with wisdom so that they would be capable of acting on the own accord. Creatures reproduce, nurture and sustain themselves and their offspring in their own ways and God does not need to know everything they do. He does not watch over the mother ostrich when she decides to hide her eggs, forgetting that other animals could tread on them.[14] He does not get involved in the moral convictions of human beings who want him to share their ideas of justice. There is a universal order, which God upholds, but its principles are not balance and equity, or retribution and equivalence, as Job and his friends seem to think. God's principles are more strategic and focus on intimate knowledge, sustenance and variety (Clines 2004, p. 48). In his discourse, God knows his universe intimately, but he does not tell the stars or the earth's inhabitants what to do and how to behave. The purposes of the universe are infinitely multiple, each of its elements has its own perspective and rules. As for humans, they are merely one part. The world has not been designed just for them. If they want to up hold justice they must to do it themselves, according to their own rules.

Finally, the theology of the divine speech contains an implicit answer to the satan's question: Does Job serve God *chinām*? The satan had suggested that Job was pious because he found it benefitted him to be pious. Job's behaviour in the opening chapters proves he is pious 'without cause' but now, in the divine speech, this question is raised again in a different sense. Since the divine speech denies that there is a causal relationship between deed and consequence, it follows that every deed is done *for free,* without a reason and without reward (Clines 2004, p. 49). There is no principle of retribution at work in the universe. Any system of moral causation, of moral order, will not be from the universe or God, but will be made and maintained by human beings.

4 Chance in the Book of Job

At the start of the Book of Job, readers are confronted with Job's fate and we cannot but feel compassion for him. Yet, as readers we know that what appears to be bad luck for Job on earth is actually a consequence of the wager in heaven between God and the satan. It is this dynamic interaction between the heavenly wager and its impact on earth that makes the risk of good or bad luck acceptable to readers of the Bible. This bi-focal perspective disappears in the dialogues between Job and his friends, since here the friends present their respective mono-focal views, in which chance clearly functions within a moral framework of retribution and is reduced again to a simple balance. However, by the end of the Book of Job when we read about God's speech out of the whirlwind, these simplistic views are replaced by a multifocal view of the universe in which chance is understood in terms of

---

[14]See Job 39:13–16.

perspective, place and scale. In a long series of rhetorical questions God reflects on the universe and its inhabitants, showing every phenomenon's spatial limitedness, bound to each limited perspective. What is considered unacceptable or unjust on one scale may be explicable on another scale, and vice versa. Thus the Book of Job advocates a perspectival and scale dependent view on causal chains and events that cannot be reduced to human explanations and simple schemes of retribution.

This non-deterministic framework we see in the Book of Job has not played a major role in Jewish and Christian theologies. Like the satan, Job's wife and Job's friends, people continue to ground their faith in God on causality and explain life in a deterministic framework. That is to say, people develop causal explanations for the sometimes inexplicable alternation of events, with their brains and rationality, and then they make God responsible for what they consider to be a 'reasonable' or 'necessary' chain of events. They blame God for bad luck, injustice, natural disasters, and in this find a reason to conclude that God does not exist. God's speech in the Book of Job invites its readers to examine their views on the topic of chance as this exposes the human quest for causal explanations as a result of a human need for moral order, logical structure, and a system they can understand. The text teaches us to consider chance as the residue of our quest for necessity, for moral and logical patterns and our desire to call patterns God's design. The Book of Job does not present its teachings through an abstract discourse, a learned essay, or a treatise with generalizations. It offers narrative and poetic material[15] that reflects ambiguity, and uses a matrix of characters' perspectives to challenge us to make up our own minds on the topic of justice, moral and logical order, and chance.

5   From Narrative to Philosophy

If we turn from the literary aspects of the Book of Job with its discussions on the moral balance in the world to the deterministic views on chance that have dominated Orthodox Judaism and Orthodox Christianity, we discover elements that still influence present discussions on rationality and faith. Orthodox Jewish tradition has adhered to a theology that celebrates Yahweh, the God of salvation, who elected his people Israel out of the nations, acted mightily for Israel at the exodus and at the conquest of the land, and gracefully in its offer of a covenant and of the Torah. In return, his people must acknowledge him as the one and only God, serve him and respect him, and live following his laws of covenant. Yahweh then will act as the executor of a system of retributive justice.

Orthodox Christian theology has followed Jewish tradition in this theology of retribution and has at the same time been influenced by Aristotelean ideas of

---

[15]The narrative style of chapters 1 and 2 in the Book of Job, characterized by sequential verbal forms, long sentences, embedded speeches of distinct characters, and an observable narrator's voice, differs greatly from the poetic style of God's speech in chapters 38–41, in which these characteristics are absent.

regularity, causality, and coherence, in which God is the initiator of all changes in events. These religious deterministic worldviews are based on the convictions of a divine cosmic order as well as a divine moral order: God is the initiator and dominant agent behind all entities and the causal chains of events, and God upholds the moral order according to the principle of retribution. According to this retributionary view of God, those who act properly are rewarded with blessings, while wrongdoers are punished.

Today, most people in Western Europe no longer uphold these orthodox traditions. Nevertheless, in modern notions of chance, ideas of regularity and causality, which have their roots in ancient Christian adaptations of the Aristotelean conception of causality (cf. Hulswit 2002), often resurface. Aristotle, in particular, defined an 'efficient cause' as the primary source of change that is brought about for the sake of an end. As part of the Newtonian revolution in science during the seventeenth century, this concept of causality underwent a radical change, in that goals or ends were replaced by initial conditions, and causal relations became instances of deterministic laws. What remains unchanged, however, is the view that *causal relationships* were conceived as if they are *ontologically* there.

From Hume and Kant onwards, this view also started to be questioned. There was an awareness that causality presupposes selection or a predisposition that is created from the perspective of the rule or scientific law that the human mind accepts as such, but which may not be ontological. This development from ontology to epistemology in modern science and philosophy obviously has consequences for in understanding the notion of chance. See, for example, the position taken by Hume described in the first chapter of the present book *The Challenge of Chance*: "The chance or indifference lies only in our judgment on account of our imperfect knowledge, not in the things themselves, which are in every case equally necessary, though to appearance not equally constant or certain" (Hume, *Treatise* part 3, Sect. 1). Notice also Hume's conclusion that, "one should not suppose that the attributes of God have any analogy or likeness to the perfections of a human creature. (…) We ascribe to God Wisdom, Thought, Design, Knowledge (…) because these words are honourable among men, forgetting that He is infinitely superior to our limited view and comprehension" (Hume 1779, 46).

Surprisingly, this view expressed by Hume, and similarly by Kant, is not so very different from the position ascribed to God in the Book of Job. As shown above, God in his speech presented in Job 38–41 advocates a perspective and scale dependent view on causal chains and events that cannot be reduced to human explanations and simple schemes of retribution. God's position is non-deterministic and embedded in a framework where every living creature is responsible for his, her, or its own decisions that are necessarily limited in scale, time, place and position. The Book of Job does, therefore, not make God responsible for the chain of events. Even Leibniz, Clarke, or Hume would not dare to speak of God's decisions in terms of a betting game. Yet, the openings chapters of the Book of Job do talk about God's actions in this way. So Einstein's words that "God does not play dice" is in a way countered by Job. We could even conclude that the European philosophical tradition does not consist of "a series of footnotes to Plato" (as A.N. Whitehead famously held), but to footnotes to Job as well.

6 From Philosophy Back to Narrative: Genesis 1

Does everything that exists, have a beginning? Does everything that begins to exist have a cause? For aeons Christian theology offered an answer to these two questions by means of a notion commonly known as *creatio ex nihilo*: God 'created out of nothing', which contrasts with *creatio ex materia* 'creation out of some pre-existent, eternal matter'. Christian theology posited that all things, which have a beginning, must also have a source or cause, and that, because the universe has an apparent beginning, it must also have a transcendent cause. The idea of a beginning demands a creator who existed without a beginning and prior to and outside the universe. Currently, the general public (as can be seen on many websites) link the common phrase *creatio ex nihilo* to Genesis 1:1. This verse is considered to be a description of the first act by God through which everything came into being. The implication is that before this instant of creative action there was nothing: God did not make the universe from pre-existing material, but he started from scratch.

In the twentieth century, it became accepted in biblical scholarship that this idea of 'creation out of nothing' was not based on texts from the Hebrew Bible but on Greek texts (especially on 2 Maccabees 7 in the Septuagint) and on later interpretations influenced by Hellenistic philosophy. Over the last years, more nuanced studies have been written on 2 Maccabees 7, the Septuagint, and Hellenistic Jewish and Christian texts, showing that the idea of 'creation out of nothing' was not present in 2 Maccabees 7 and not elsewhere in the Septuagint, but was developed in the second century CE.[16] In addition, new studies on creation texts in the Hebrew Bible have been published that demonstrate how these texts were conceived in a completely different intellectual framework than the later Jewish and Christian traditions.[17] Yet, it is not the original ancient texts that influence the notion of 'creation out of nothing' in our times (the 21th century CE), but the reception and transformation of these texts by Christian traditions from the Early Middle Ages up to today. It turns out that texts in the Hebrew Bible never presupposed the concepts that lie at the heart of the *creatio ex nihilo*-theory, namely the concepts of nothingness and of material origins.

I will now focus on Genesis 1 to explain the cognitive framework of the ancient Near East in which Genesis 1 originated, a framework that differs from the Greek and Hellenistic framework and from medieval Jewish and Christian traditions.

---

[16]See Schmuttermayr (1973), O'Neill (2002), Niehoff (2013).

[17]See three of the most recent comprehensive studies of Genesis 1: Smith (2010); Walton (2011); Batto (2013).

Subsequently, I show how this framework is different from the common understanding of Genesis 1 in modern non-academic and ecclesiastical circles that have
been greatly influenced by these later Jewish and Christian traditions. Upon a
sketch of the worldview this text presupposes, a short textual analysis of Genesis
1:1–3 will follow in order to elucidate the view this text offers of the beginnings of
the universe. Finally, some of its consequences for our understanding of the notion
of 'chance' will be drawn.

7 Worldview in the Hebrew Bible

"There is more between heaven and earth", in this and other everyday conversations
it seems natural to make a distinction between heaven and earth. However, this is
not as self-evident as it appears. The endless universe is, in fact, continuous and not
split up into a heaven and an earth. Although the word string "heaven and earth" is
used in the Hebrew Bible as a merism to express the totality of all and everything,
biblical texts share the ancient Near Eastern view that the cosmos consists of at least
three layers: heaven, earth and the netherworld.[18] This tripartite cosmic view serves
as a backdrop for all the texts in the Hebrew Bible.

The tripartite cosmic view is immediately apparent in the three opening verses of
Genesis 1. God performs an action with respect to two direct objects, 'heaven' and
'earth'. The two nouns hāšāmayîm, heaven(s), and hā'āreṣ, earth, reflect the
worldview that the universe consists of at least two components or levels. The
following two verses presuppose another level in the universe below the earth,
namely tĕhōm: the netherworld or abyss that is filled with water. What did the
ancient Israelites think of when they spoke of tĕhōm? The Biblical material allows
us to construe an inventory of the possible concepts underlying tĕhōm: it is conceived as (1) a spatial realm under the earth, (2) a vertical depth, (3) a large expanse
of water expanded vertically and horizontally, (4) a container of water that is the
source of springs, wells, fountains, and rivers on earth, and (5) a layer on which the
earth rests. Based on the first two concepts, the semantic content of tĕhōm is
considered in terms of depth and translated into English as 'the deep' or 'the abyss'.
Based on concepts 3 and 4, the semantic content of tĕhōm is considered in terms of
huge volumes of water and translated into English as 'waters' or '(primeval) ocean'.
In short, in the Hebrew Bible, the tĕhōm is clearly conceived as the lowest tier in the
tripartite cosmos—a deep container filled with water.

Heaven is the highest tier, and biblical texts including Genesis (1:6–8) present
the idea that heaven is made of solid vertically arranged material that holds volumes

---

[18]Cf. Cornelius (1994); Horowitz (1998); Pongratz-Leisten (2001); Keel and Schroer (2002);
Walton (2011).

of water in place.[19] The function of this heavenly vault is to prevent the waters above the vault from falling down on the earth. The *tĕhōm* or the spatial realm beneath the earth is also filled with water. Earth occupies a central position in the tripartite view of the cosmos sandwiched between heaven above and *tĕhōm* below. Ancient maps, such as the Babylonian map in the British Museum and Greek maps drawn by Anaximander and Herodotus, share the belief that the inhabited world was a disk of earth surrounded by water. Biblical texts also conceive the earth as a single, disk-shaped continent surrounded by an ocean of water. The Hebrew word *tebel* refers specifically to this 'earth-disk', i.e. the earth as a single entity. The word *'eres* is the more general term referring either to the (dry) land or ground, or to the whole earth. Genesis 1:9–10 states that the *'eres* was formed as the result of the waters moving horizontally outwards to leave dry land behind at the centre. This produced two spatial domains on earth, namely land in the middle and waters surrounding the land. Many texts in the Hebrew Bible from Prophets to Psalms and Job describe how the earth comes into existence when God establishes the earth by setting it on pillars to prevent the earth-disk from sinking beneath the waters of *tĕhōm*—the underworld ocean.

8   Genesis 1:1–3

Genesis 1:1–3 tells us how this world came into being and this is commonly translated "In the beginning God created the heaven and the earth. And the earth was void and bare, darkness was over the abyss, and God's spirit moved over the waters. God said: Let there be light and light was." A more detailed analysis shows the flaws of this translation. The very first word, *bĕrē'šît* 'in the beginning of', marks not an absolute starting point in time, but expresses the starting point of the action expressed by the following verb. The meaning of this verb *bārā'* has been widely discussed. In van Wolde (2009) the hypothesis is presented that this verb designates 'to separate' as a purely spatial term, a view that was further explained and substantiated in van Wolde and Rezetko (2011).[20] Based on comprehensive linguistic studies, the conclusion is that the verb *bārā'* functions in the cognitive domain of space and designates [SEPARATION], [DIVISION], or [SETTING APART]. Dependent on the context it can be translated as, 'to divide, separate, set apart, spread out, disconnect.' Hence, Genesis 1:1 should be translated: "In the beginning

---

[19]Genesis 1:6–8 uses the term *rāqîa'*, 'vault', which refers to the result of either a gold/silversmith who beats out metal/solid plates or of someone who spreads out a plate or other solid material. Based on a metaphorical structuring of this concept, God's making of the heaven is conceptualized in terms of the beating out or spreading of solid plates of the heavenly vault in the endless water expanse.

[20]van Wolde (2009, pp. 3–23), van Wolde and Rezetko (2011, pp. 2–39).

when God separated the heaven and the earth, ...".[21] The implication is that the sentence is not concluded in verse 1, but continues in verse 2. It marks the starting point of the divine action of separation over and against the situation described in verse 2.

In verse 2a, the following two pictures are painted (1) the earth as *tōhû wa-bōhû*, which creates an image of the earth not yet set on pillars, hence, not yet visible and still covered with the pre-cosmic waters of the *tĕhōm*, and (2) a vast darkness over this primeval ocean or *tĕhōm*. Verse 2b describes how God's *ruach* or wind/breath hovers over and faces these waters. In this sense, verse 2a depicts an endless expanse of water stretching out in all directions, covered in complete darkness, whereas verse 2b describes God's spatial movement and actions with regard to the waters.

This is a powerful image of what happened when God began to act in a universe that, till then, had only consisted of water. Verse 1 describes the beginning of this action and qualifies it as separation: when God began to separate the heaven and the earth. Verse 2a continues with the situation that the earth is covered with water and darkness covers the abyss of waters, that is, it zooms in on the condition of the heaven and earth referred to as direct object in verse 1. However, verse 2b zooms in on God's act of separation described in verse 1. Consequently, verse 2b shows that it is God's breath (or the wind) that separated the primordial waters to make a spatial realm between heaven and earth. In this deep, dark and watery context, verse 3 uses only two Hebrew words to evoke God's first act of creation: "And God said: *wayyehi 'ōr*, "Let light be", followed by the immediate result: "And light was". The proposed translation is therefore:

1. In the beginning when God separated the heaven and the earth
2a. The earth was ungrounded and without foundation
    and darkness covered the abyss filled with waters
2b. God's breath/wind was moving/blowing over the waters,
3a. And God said:
3b. Let light be.
3c. And light was.

Genesis 1:1–3 uses imagery to convey how God's breath or wind transforms a world filled with water. This divine act of dividing by breathing shows us that God does not fill a void (the classical idea of *creatio ex nihilo*), but rather that he splits the oneness of the primordial waters open to create a new reality.

---

[21]The syntactic structure of this verse is: "In the beginning of (the act by which) God separated the heaven and earth", which in English becomes "in the beginning when God separated the heaven and the earth". Cf. Holmstedt (2008, pp. 353–359).

9 The Framework of Non-Linearity in Genesis 1

Genesis 1 is usually read from a chronological and causal perspective. Therefore, the text is understood as a temporal arrangement in which the first thing told happened first and the next thing told happened later. In a chronological reading, the opening verse represents the beginning and the subsequent days show what happens next. In a causal reading, the same text is read as a causal chain: the first element told not only happens first, but it is also the cause of the second element, which is therefore an effect of the first event. In this reading causality will be the story's main theme: everything originates from the creative actions of God. He is the initiator and the created phenomena are the effects of his actions. Causal relationships also occur between the created elements themselves. Water and light are created first and only then can the earth can bring forth plants. These plants in their turn must be created before the animals as they are necessary for the animals to live on. A causal understanding of the text has important consequences because the last element told is considered the most valuable or important (at least this is the way it is interpreted in history). In the Jewish tradition this has led to the conclusion that the seventh and last day is the climax of the story. In the Christian tradition many people infer that humans are the pinnacle of creation, and the 6th day of creation is considered to be the story's climax. On the 6th day God created the human being and with this final creature, creation reaches its culmination, possibly even its goal. According to this causal conception, Genesis 1 is understood as an explanation of the special position of the human being within the created universe: heaven is made for the benefit of the earth, the earth for the benefit of humans, while plants and animals are made to provide the necessary conditions for the human beings to live on earth. However, if this causal approach is applied to Genesis 2 and the story of paradise, a woman ('Eve') is the last creature to be made so we would have to infer that creation reached its climax and ultimate goal when the human female was created. Illogically, the opposite conclusion is usually drawn.

This linear interpretation of Genesis 1 rivals the scientific view, because it understands causality in the same way as science does in the sense that they both provide a linear explanation of the actual causal relations between objects and events (see section above on causality as ontology). However, does this linear arrangement actually apply to the text of Genesis 1? Some shortcomings can easily be detected. If linearity were the fundamental device, how can it be explained that God made the light in verse 4, and the sun and moon much later on, in verses 14–16? How God could have possibly made the heaven and the earth in verse 1? And do the earth and the heaven exist now, according to the story at this stage? If they do, it is inexplicable that God in verses 6–8 creates a vault in the middle of the waters and calls it 'heaven', and that in verses 9–10 God separates the waters from the dry land and he calls the waters 'seas' and the dry land 'earth'. Does God create them twice? The most striking problem with a linear reading is that Genesis 2 is positioned immediately after Genesis 1. It seems as though Genesis 2:4b–7 starts from the beginning again. "On the day Yahweh God made earth and heaven, the earth was without plants and human beings … and he made human beings from dust of the earth." Yet, this had already happened before, as was described in Genesis 1:26–28. People who read linearly are completely baffled. But what if this linear conception is too limited a view?

In a non-linear reading the text can be explained as follows. The first action narrated is marked as an action by God through which he alters an existing situation or totality by separating the expanses of water into heaven and an earth, and from this point onwards God (and the narrator in the text) focus on the various elements. Over and over again we see the non-linear pattern return. The starting point is totality, and then the text zooms in on the making of one or more of its elements. For example, verse 3 tells us about God creating light, but only later on, in verses 14–17, does the text mention God making the sun, moon and stars. Or, in verses 9–10 God makes the earth as a whole, while zooming in in verses 11–13 on the plants and trees on earth. Or, verse 1 describes the separation of the waters into heaven and earth, while later on, in verses 6–8, we read that the heavenly vault was created to keep the waters apart. This non-linear form of narration and conceptualisation can also be seen in the two entire stories presented in the first three chapters of the Book of Genesis, namely Genesis 1 telling the story of creation, and Genesis 2–3 telling the story of paradise. In the first story, we are told how God made the human beings on the 6th day (1:26–28). This is a kind of overarching view of what occurs in the second story, when Yahweh God first makes a male human being and then from him makes a female human being while describing the details of their new existence. In other words, the story of paradise refers back to what has previously been told through images of an overarching summary of creation. This non-linear arrangement shows some similarities with fractal structures. The starting point is like a fractal image at the highest level, from where the text zooms in on one element, which in itself exhibits a fractal structure, too. Over and over again, new elements are specified that form new smaller fractals.

10   The Non-linear Arrangement in Genesis 1 and the Concept of Chance

In the opening chapter of the Book of Genesis we discover various non-linear arrangements in which the text first introduces the elements in one big picture and, subsequently focuses on one of these elements in detail. Depending on the perspective chosen (i.e., the level of detail zoomed in on) a new kind of 'realm' is revealed. In each realm, the species have their own organisation and responsibilities.

For example, verses 11–12 relate to the plants and trees on earth. God instigates the earth to produce plants and trees, each with its own seeds and fruit in order to reproduce distinct species. The words, 'the seed' is repeated six times in these verses, three times with regard to plants and three times with regard to trees. The causative sense of the verb in verse 11 indicates that the plants are conceived as producing the seed, and the seeds themselves are responsible for the process of germination and production of new life in the ground. In verse 12, the fruits of the trees are described as seed containers. The notion that each plant and tree should bring forth new life according to its own species is repeated three times. In this way the text emphasizes both the activity of the plants themselves and their system for maintaining the necessary distinctions between their offspring.

Another example is the animal kingdom. In verses 20–23, God addresses first (in verse 20) two groups of animals: the animals that swarm the seas, and the birds that are characterized in relation to earth and heaven. And the swarming sea animals are blessed and encouraged to be fruitful and multiply and fill the waters of the seas whereas the birds are also blessed but are only told to multiply. However, in verse 21, also a third group of animals are mentioned: the *tannînîm*, the inhabitants of the *tĕhôm* or the abyss. They are considered to have existed prior to God's creative activities and to differ from the other animals in their origin and procreative abilities. They are not asked to reproduce themselves. In contrast, the sea animals (the second group of animals mentioned in verse 21) are presented as having been brought forth by the waters and they are asked to reproduce themselves in order to swarm the sea. The birds are described as flying over the earth across the sky; they are still related to the earth and to the aerial realm below the heavenly firmament. God assigns each party to its own life sphere, which they have to fill with their own species of animate life, with the exclusion of the *tannînîm* who are not recorded as reproducing new life.

In a cultural framework dominated by a non-linear way of thinking, the concepts of necessity and chance also function differently. In a non-linear perspective, the concept of chance is not understood in terms of a break in a causal or deterministic chain of events, but it stands out in a framework of thinking in which totality and instantaneity alternate. Because Genesis 1 alternates between scales, it does not represent a temporal sequence or a causal arrangement. Thus the reader is made aware of a new sense of coherence at each and every level or scale, and, more importantly, challenged with the lack of necessity for sequence between the various levels. Because of the absence of a causal chain of events, the text of Genesis 1 opens our eyes and shows us the fractal structure of the universe. Chance—so often conceived as the opposite of necessity—turns out to be present in every event depending on the scale and the perspective chosen. In this sense, Genesis does not differ from the view presented by God in his represented speech in the book of Job.

11 Conclusion: Views on Chance in Job and Genesis 1

God's speech out of the whirlwind in the Book of Job and the opening chapter of the Book of Genesis both offer a non-deterministic view on chance. Chance is not the exception in a causal or necessary chain of events, but is scale dependent. The view is unmasked that causal relationships have to be conceived as if they are *ontologically* present. In his speech in the Book of Job, God invites its readers to examine their views on the topic of chance as this exposes the human quest for causal explanations that results from the human need for moral order, logical structure, and an understandable system. The text teaches us that chance accompanies our quest for necessity, for moral and logical patterns and our desire to call patterns God's design. In addition, chance is linked to a multifocal view of the universe and understood in terms of position, perspective, and scale. Moreover, the opening chapter of the Book of Genesis offers a non-deterministic view on chance. In Genesis 1, chance is not an exceptional event that disrupts some causal or deterministic chain of events, but rather it is highlighted within a framework of non-linear thinking where totality and instantaneity alternate. In a world, where God zooms in or out on various lower-level components, any claims for completeness or order can no longer be made. In sum, in both Job and Genesis 1, chance is presented as a disqualifier of causal chains and even as an ultimate denial of necessity.

Acknowledgments I would like to thank Lut Callaert, Ruti Vardi, Klaas Landsman, Chris Mollema, and Christoph Lüthy for their comments on earlier drafts of this chapter.

Open Access This chapter is distributed under the terms of the Creative Commons Attribution-Noncommercial 2.5 License (http://creativecommons.org/licenses/by-nc/2.5/) which permits any noncommercial use, distribution, and reproduction in any medium, provided the original author(s) and source are credited. The images or other third party material in this chapter are included in the work's Creative Commons license, unless indicated otherwise in the credit line; if such material is not included in the work's Creative Commons license and the respective action is not permitted by statutory regulation, users will need to obtain permission from the license holder to duplicate, adapt or reproduce the material.

References

Batto, B. F. (2013). *In the beginning: Essays on creation motifs in the ancient Near East and the Bible.* Winona Lake, IN: Eisenbrauns.

Clines, D. J. A. (1989–2009). *Job. Word biblical commentary* (3 vols.: Job 1–20 publ. in 1989; Job 21–37 in 2006; Job 38–42 in 2009). Vols. 1 and 2: Dallas: Word Press; Vol. 3: Nashville: Thomas Nelson Publishers.

Clines, D. J. A. (2004). Job's God. In E. van Wolde (Ed.), *Job's God* (Concilium 2004/4) (pp. 39–51). London: SCM Press.

Cornelius, I. (1994). The visual representation of the world in the ancient Near East and the Hebrew Bible. *Journal of Northwest Semitic Languages, 20,* 193–218.

De Joode, J. (2015). *Landscapes of hurt and healing. An Exploratory cognitive-linguistic analysis of spatial metaphors in the book of job.* Dissertation, Catholic University Leuven.

Gravett, E. O. (2012). Biblical responses: Past and present retellings of the enigmatic Mrs. Job. *Biblical Interpretation, 20,* 97–125.

Habel, N. C. (1985). *The book of Job. A commentary (Old Testament Library).* London: SCM Press.

Holmstedt, R. D. (2008). The restrictive syntax of Genesis 1:1. *Vetus Testamentum, 22,* 353–359.

Horowitz, W. (1998). *Mesopotamian cosmic geography (Mesopotamian civilizations)*. Winona Lake, IN: Eisenbrauns.

Hume, D. (1779). *Dialogues concerning natural religion* (2nd ed.). London.

Hulswit, M. (2002). *From cause to causation. A Peircean perspective*. In Philosophical Studies Series (Vol. 90). Dordrecht/Boston/London: Kluwer.

Keel, O., & Schroer, S. (2002). *Schöpfung: Biblische Theologien im Kontext altorientalischer Religionen*. Göttingen/Freiburg: Vandenhoeck & Ruprecht.

Newsom, C. A. (2003). *The book of Job. A contest of moral imaginations*. Oxford: Oxford University Press.

Niehoff, M. R. (2013). The emergence of monotheistic creation theology in hellenistic judaism. In L. Jenott & S. K. Gribetz (Eds.), *Jewish and Christian cosmogony in late antiquity* (pp. 85–106). Mohr Siebeck: Tübingen.

O'Neill, J. C. (2002). How early is the doctrine of creation Ex Nihilo? *Journal of Theological Studies, 53*(2), 449–465.

Pongratz-Leisten, B. (2001). Mental map und Weltbild in Mesopotamien. In B. Janowski & B. Ego (Eds.), *Das biblische Weltbild und seine altorientalischen Kontexte* (pp. 261–279). Mohr Siebeck: Tübingen.

Schmuttermayr, G. (1973). Schöpfung aus dem Nichts' in 2 Makk 7,28? Zum Verhältnis von Position und Bedeutung. *Biblische Zeitschrift, 17*, 203–228.

Seow, C. L. (2013). *Job 1–21: Interpretation and commentary*. Grand Rapids: Eerdmans.

Smith, M. S. (2010). *The Priestly vision of Genesis 1*. Minneapolis: Fortress.

Sommer, B. (2009). *The bodies of God and the world of ancient Israel*. Cambridge: Cambridge University Press.

van Wolde, E. (1995). The development of Job: Mrs. Job as catalyst. In A. Brenner (Ed.), *A Feminist companion to wisdom literature* (pp. 201–221). Sheffield, UK: Sheffield Academic Press.

van Wolde, E. (2003). *Mr. and Mrs. Job* (pp. 42–106). London: SCM Press.

van Wolde, E. (2009). Why the verb ברא does not mean 'to create' in Genesis 1.1–2.4a. *Journal for the Study of the Old Testament* 34.1, 3–23. http://radboud.academia.edu.ellenvanwolde.

van Wolde, E., & Rezetko, R. (2011). Semantics and the semantics of ברא. A Rejoinder to the Arguments Advanced by B. Becking and M. Korpel. *Journal of Hebrew Studies 11*(9), 2–39. http://radboud.academia.edu.ellenvanwolde.

Walton, J. H. (2011). *Genesis 1 as ancient cosmology*. Winona Lake, IN: Eisenbrauns.

White, E. (2014) *Yahweh's council* (Forschungen zum Alten Testament 2. Reihe, Vol. 65). Tübingen: Mohr Siebeck.

Happiness and Invulnerability from Chance: Western and Eastern Perspectives

Johannes M.M.H. Thijssen and David R. Loy

Abstract Since the beginning of Western philosophy, thinkers have discussed how one might lead a good, i.e. a happy, life and what role luck plays in flourishing. According to one dominant Ancient Greek tradition, life's circumstances are not relevant for our happiness, and, moreover, they fall outside of our control. What is up to us is how we respond to life's circumstances and adversities. Christianity, however, rejected ancient tradition and moved happiness to a new home: heaven. Because Adam and Eve were disobedient in Paradise, God punished the human species with a 'genetic' defect which made life miserable for each and every individual. Chance or (bad) luck is an inevitable ingredient of human suffering. Buddhism also perceives chance or luck as intrinsic to life, but locates it into the sphere of human control. It is not the gods, but we, who, through our own actions, are responsible for what happens to us. This is called the law of *karma*: we reap what we have sown. There are striking parallels between the Greek methods to train our mental responses to (bad) luck and the Buddhist analysis of unwholesome actions and corresponding advice to improve our karma. Both traditions are still helpful today in our attempts to secure happiness in the face of chance adversity.

1 Introduction

On 17 July 2014, Malaysian Airlines flight MH 17 departed from Amsterdam. It never reached its destination in Kuala Lumpur; the plane crashed in the Ukraine, not far from the Russian border, claiming 298 lives. A few days after the crash, a Dutch newspaper ran a story about a family that had 'miraculously' missed flight

J.M.M.H. Thijssen (✉) · D.R. Loy
Faculty of Philosophy, Theology and Religious Studies, Radboud University, Nijmegen
The Netherlands
e-mail: h.thijssen@ftr.ru.nl

D.R. Loy
e-mail: davidrobertloy@gmail.com

© The Author(s) 2016
K. Landsman and E. van Wolde (eds.), *The Challenge of Chance*,
The Frontiers Collection, DOI 10.1007/978-3-319-26300-7_8

151

MH 17. The family had arrived slightly late at Schiphol Airport and, as the flight was overbooked, had been transferred to another flight.

In the summer of 2005, Hurricane Katrina swept the Gulf of Mexico coast from central Florida to Texas. At least 1833 people were killed, most of them in New Orleans. The storm caused $108 billion in damage, depriving many people of their homes and possessions.

These examples are just two of many others that people tend to associate with 'luck', 'chance', or 'karma', by those familiar with the Buddhist term. Initially, the family that arrived late would feel they were unlucky in missing flight MH 17. But after the crash the reverse was true; apparently they had been extremely lucky or they had good karma Likewise, the people whose homes lay in the path of hurricane Katrina and lost their homes and lives were clearly unfortunate while those who lived a mile out of the path were just lucky.

What do we mean when we attribute a plane crash or a hurricane to chance or (bad) luck? It means that we feel these events are random, unlikely and the result of conjunctions of causes that are unknown to us. These events are unpredictable: there is no apparent purpose or plan which includes all causal chains. Moreover, and more importantly for this chapter, these events appear to be beyond our control. Chance events expose the vulnerability of our happiness and even our lives. One moment we are prosperous and happy, and the next moment our lives are disrupted and our happiness is shattered.

In some languages, the words happiness and luck have an etymological connection. For example, in German and Dutch, the same word is used for 'chance' and 'happiness,' thus implying that it takes some measure of luck (*Glück*; *geluk*) to achieve happiness (*Glück*; *geluk*). In English, there is also an etymological connection. In Middle-English 'hap' means 'luck' or 'chance,' and also occurs in 'happiness.' Yet is happiness only a matter of luck? How much good luck can be expected and how much bad luck must be endured in attempts to lead a good life? How insecure is our happiness?

Socrates and Siddharta Gautama, who later came to be known as the Buddha, were near contemporaries. The philosophical traditions they initiated were anchored in existential questions about how to live well, i.e. how to end suffering and be happy. Interestingly, early Indian Buddhist philosophers as well as early Western philosophers reflected on the relationship between chance and leading a happy life. We look at ancient Greek and Buddhist philosophies and Christianity to explore how they developed ways of thinking to get to grips with the terrifying notion of life based on chance.

According to contemporary Western thinking, which is heavily influenced by Greek notions, Hurricane Katrina and flight MH 17 appear to be chance events or occurrences of (bad) luck. Even though there is a connection between certain actions and certain consequences, for instance, between having bought a ticket and being on flight MH 17, the actions and interactions are far too many and far too complicated to be able to distinguish any direct causal chains. This is what, for instance, Aristotle meant by *tuchê*, by luck: when a man goes to the marketplace and runs into a debtor that he wished, but did not expect to find, their meeting is a result, not of a determinate cause, but of luck (*Physics* II.5). Similarly, the travellers did not board flight MH 17 expecting to die. That it happened was the outcome of (bad) luck.

In her now classic *The Fragility of Goodness*, Martha Nussbaum has drawn attention to many Greek philosophers' preoccupation with luck.[1] Their concern was to find out how a person might lead a good, i.e. happy, life and become immune to bad luck. As will be explained below, the Western approach is to learn how to cope mentally with the undesirable results of luck in the pursuit of happiness.

Buddhist philosophy would take a different view of a plane crash or a hurricane and attribute the devasting effects to the karma of the victims: as if the misfortune happened through the victims own agency, instead of, just happening. A western response to this view might be: "What terrible deeds have the victims done in the past to deserve Katrina or flight MH 17?" It is true that the term 'karma' is central to the Buddhist analysis of human action and seems to suggest a responsibility for whatever is happening to us. Karma, however, is *not* a calculus of rewards and punishments for one's actions. In the words of Jay Garfield, it is not "a cosmic bank account".[2] Nevertheless, many of the events in one's life are seen to be due to one's karma, i.e. they are the result of previous actions, whether in this or a former life. According to traditional Buddhist teachings, the quality of an action is determined both by the intentions behind it and by its consequences. This includes many, perhaps most, of the good and bad things that happen to us, Thus, mere chance is abolished and, although it may be delayed, we retain a way to control what happens to us by behaving in a wholesome way. In this chapter, we try to compare and contrast Western Greek and Asian Buddhist attitudes towards chance and how chance events impact on happiness.

2   Ancient Greek Philosophy as a Way of Life: The Pursuit of Happiness

The distinguished British philosopher Bernard Williams once quipped: "The legacy of Greece to Western philosophy is Western philosophy"![3] And indeed, there is a continuity between ancient and contemporary philosophy. Not only does philosophy continually refer to themes that originated in antiquity, but philosophers today remain fascinated by the views and theories of ancient Greeks and Romans. Yet, at the same time, Williams's aphorism overlooks one crucial aspect of ancient Western philosophy that has disappeared from contemporary academic philosophy. Philosophy today is mainly a theoretical and conceptual discourse, whereas ancient philosophy crucially involved a way of life.

---

[1]Nussbaum (1986), although her angle is different from the one taken here.
[2]Garfield (2015), 284. See also Loy (2008).
[3]Williams (2006), 3.

Ancient philosophy covers a lengthy period of time, which conventionally runs from the appearance of Thales in the sixth century B.C. until 529 C.E. when Emperor Justinian, under pressure from a local Christian group, closed the philosophical school in Athens. Durig this long period of twelve centuries, philosophy was initially situated within a Greek culture and then continued within a Roman context. Greek was the main language of the philosophers. When we study the earliest Western philosophy, we tend to see it through the lens of contemporary philosophical perspectives, and thus discover an ancient science, logic, ethics, metaphysics and even a philosophy of mind. Due to this fragmentation, we tend to miss the overall character of philosophy at that time, which runs through the diversity and heterogenity of views and schools. In his *Philosophies for sale*, the satirist Lucian (c. AD 125–180) wittily captures the heart of ancient philosophy.[4] In a marketplace, Hermes is setting up stalls for selling philosophies, with prices that vary considerably. Each philosopher represents a specific school along with the life-style, the *bios*, that comes with it, and is loudly advertised by Hermes. These philosophies are attractve to buyers because they are guides for living a good life.

Yet what does this mean about philosophy as the search for wisdom and truth, an inquiry into all kinds of topics and problems as well as the art of analysis and argumentation? The first Western philosophers were engaged in those activities as well, but in addition, and in contrast to philosophers nowadays, they also *lived* their philosophy. They operated on the (tacit) assumption that philosophy can save your life. Philosophy is an authoritative guide on how to live, since the knowledge of the world and your place in it will motivate you to live your life accordingly. In sum, the philosophical life is a life based on reasoning, but it is about more than reasoning.

The French scholar Pierre Hadot, more than anyone else, has drawn attention to the ancient conception of philosophy as a way of life, and has emphasized its existential and spiritual dimensions.[5] Philosophical discourse was an integral part of a specific way of life. It was meant to justify and disseminate the way to live, both among followers and opponents. At the same time, philosophical discourse also expressed a way of life. And finally, it functioned as a type of mental exercise or spiritual practice.

With Plato, this conception of philosophy came to be firmly established in all the different schools. The most prominent among them were Plato's own Academy, Aristotle's Lyceum, the Stoa, Epicurus's Garden and the Skeptics.[6] The last school criticized the other schools for clinging to theories and statements (*dogmata*), whose truth remained open to doubt, and hence to further investigation. According

---

[4]References to ancient texts are according to the standard system. Unless otherwise stated, the editions of the *Loeb Classical Library* (Cambridge, MA-London) have been used for Greek and Latin texts and their English translations. Lucianus' edition of his *Biôn Praksis* and its translation are in vol. II, 450–511 of his works.

[5]Hadot (1995), which has been translated into several languages, among which English in Hadot (2002).

[6]A convenient recent introduction to the philosophical schools in antiquity, which, moreover, has been inspired by Hadot's studies is Cooper (2012).

to the Skeptics, the most sensible view is to assume that truth is beyond our powers, a claim that in itself should not be taken as a dogmatic statement. The schools were not just groups of pupils or followers who identified with a particular teacher, but were also physically located in certain places: for instance, at a gymnasium outside the walls of Athens, at a painted collonade in the marketplace, or in a garden.

In his studies, Hadot has emphasized the role of 'spiritual exercises' (*askêsis*, *meletê*) in each school. The earliest Western philosophers were mental athletes who through their practical exercises, which were part of their way of life, tried to transform themselves spiritually. Hadot took his inspiration from the title of Ignace of Loyola's famous *Spiritual exercises* (*Exercitia spiritualia*), which in his view were nothing but a Christian continuation of Greek and Roman practices.[7] Mental training takes place according to a method that is independent from any theory, and hence is applicable to any theory. The purpose of the method is to 'digest' the specific doctrines, and thus prepare the practitioner for a life-change. Among the exercises that Hadot has explored are diet, meditation on the breath, dialogue and discussion between master and pupil, the study of maxims, self-examination and self-mastery. One such exercise, familiar to anyone who ever studied for an exam, consists of writing summaries or lists of key concepts and memorising them. Epicurus wrote special summaries (*epitomai*) for the sake of his pupils. The Stoic Epictetus compared the process of digesting such material with the mastication of food. In sheep, the digestion of food will produce milk and wool, whereas the digestion of philosophical propositions will lead to a change of behaviour (*Encheiridion*, 46). Marcus Aurelius claimed that from the repetition of Stoic views the soul, like a garment, will receive a new color (*Meditationes*, 5.16). Epicurus's encouragement to become accustomed to the idea that death is nothing to us, since we will not be aware of our own death, is also a type of spiritual exercise (*Letter to Menoikeus*, 124). It can diminish our desire to be immortal and hence help us to enjoy our mortal life.

In view of Hadot's heightened awareness of ancient philosophy as a way of life, as a program for-self-improvement-through-exercise, his virtually complete omission of the *purpose* of philosophy is remarkable. The philosophical schools had the ambition to contribute to the happiness (*eudaimonia*) of their adherents.

Ancient philosophers had a keen eye for the human propensity to seek happiness.[8] As Plato points out: "We all strive to be happy" (*Euthydemus* 282a). The desire to be happy is so evident that it does not make sense to ask "*Why* do you wish to be happy?" (*Symposium* 205a). Happiness is an end-in-itself and hence does not need further justification. Moreover, being happy means that you are doing well (*eu prattein*), and finally, happiness implies the presence of good things and the absence of bad things in your life. All these characteristics of *eudaimonia* are taken up in the traditions after Plato and further elaborated–in particular the question

---

[7]Hadot (2002).

[8]Throughout this article, *eudaimonia* has been translated as 'happiness,' and *eudaimôn* as 'happy.' See also Long (1996), 181–84 and (2001), 33–34. The Greek texts also use *makarios* (happy) and *makariotês* (happiness), which recur in the New Testament, and are often translated as blessed. See for instance *The Gospel of Matthew*, 5, 2–10. See also note 17.

which ingredients in a human life contribute to happiness and which ones are obstacles. Ancient philosophy is motivated by the concern to help us lead a life that is worth living, a life that flourishes (*eu prattein*). In the words of Aristotle:

> .....let us say what it is that we say political expertise seeks, and what the topmost of all achievable goods is. Pretty well most people are agreed about what to call it: both ordinary people and people of quality say 'happiness', and suppose that living well and doing well are the same thing as being happy (*Ethica Nicomachea* 1095a17–20).[9]

Once you realize that the goal in your life is to become happy, you have to bring order to your life and make important choices. You do not want to end up with a life that is 'unlived', which consists of merely killing time. So what is the best possible life? In other words, how can one really become happy? Each school offered its own vision of the nature of the world and the human condition, and built its own way of life upon those insights.

Human beings are vulnerable: not only the playthings of desires and emotions, but also exposed to social and physical circumstances beyond their control, such as untrustworthy rulers, wars, poverty, disease and obscurity that can all be described as 'bad luck'. All these factors can be obstacles to happiness and thus a source of suffering. How much do such circumstances affect one's quest for happiness? Although different philosophical schools provided different anwers, their approaches were all based on the revolutionary idea that human beings are (or can be) masters of their own happiness. Happiness is achievable for anyone. Happiness is what you think, and you can learn to think by doing philosophy!

In one of his tragedies, Euripides (480–406 B.C.) asks the following important question:

> "O Zeus, what should I say? That you watch over men? Or that you have won the false reputation of doing so, while chance (*tuchê*) in fact governs all mortal affairs?" (*Hecuba* II, 488–91)

The play is about the former queen of Troy. After its fall, Hecuba had become a Greek slave and as the story unfolds she will learn about the death of her two children. The idea that we are governed by gods may seem disconcerting to some, but the idea that we are living in a world that was not made for us, ruled by random chance and constant change, may be even more frightening.

3   Immunising Against Luck: Ancient Greek Approaches

Since ancient times, Greek poets and philosophers have struggled with the role of *tuchê*, luck or chance, in human life, including how to avert bad *tuchê*: how to avoid one's life turning into a tragedy. Seeking support from the gods seemed one sensible strategy. Yet the gods behave in erratic ways and are difficult to control.

---

[9]*Aristotle. Nicomachean Ethics* (2002).

Furthermore, our misfortunes may have been caused not by us, but by what our ancestors did. In the story of Pandora's box, the poet Hesiod explains how the miseries and misfortunes of humankind originated. Before Pandora lifted the lid of the storage jar "the tribes of men used to live upon the earth entirely apart from evils, and without grievous toil and distressful diseases, which gave death to men" (Hesiod, *Works and Days*, 90–93).

The philosophical response to Euripides' question was offered by Plato. He gave an entirely new twist to an already extant term: *eudaimonia* and, moreover, was the first to bring up whether chance (*tuchê*) is also one of the good things that we need in order to become happy (*Euthydemus*, 279c). The original meaning of *eudaimonia* is that one is favored by the gods. A person who is *eudaimôn* has a 'good (*eu*) *daimon*,' usually an identified Olympic God, and hence is in possession of the good things that such a *daimon* is supposed to provide. Yet how can one guarantee to be favored by the gods, who in the myths appear to be as capricious as human beings? One can try to please them with sacrifices and prayers, but in the end we still have to surrender to the disconcerting idea that our happiness, our *eudaimonia*, depends on (good) luck (*tuchê*): we have no control over the gods.

Plato's brilliant move was to internalize the *daimon*.[10] Within us, we have a godlike capacity: our reason. By putting oneself under the rule of reason, one still submits to a god, though now an internal one. Be master of your own life by following your reason. Only by using our rationality can we become happy, i.e. temporarily becoming like a god (*homoiôsis theôi*).[11] If you live according to reason, life does not have to turn into a tragedy, run by blind luck and change. That is the novel powerful reply that Plato gives to the literary tradition.

By making happiness dependent upon our rational capacities, Plato opens the door for reconsidering the influence of external circumstances that seem to depend on luck, and that is exactly what the various schools did. Philosophers since Plato have not taken recourse to pacifying the gods, but have instead developed other ways of thinking to make themselves immune to contingencies and the inherent vulnerability of human existence. The ancient schools developed strategies to eliminate the power of ungoverned *tuchê*, of the impact of external circumstances beyond our control. In this chapter, the emphasis will be on those schools that are based on the insight that, although we cannot change the world, we can change our mental attitude if we are willing to commit to a certain way of life.[12] In what follows, we shall briefly focus on three such strategies that were meant to make our

---

[10]See Long (2001) and also Mikalson (2002).

[11]Plato, *Timaeus* 90a–c and also *Theaetetus* 176a–b. That the gods are happy, is mentioned by Plato, *Symposium* 202c7. See Sedley (1999) for a fundamental discussion of becoming like a god.

[12]In his *Nicomachean Ethics* Aristotle discusses the two extreme views about the influence of luck on human happiness. According to one position, being happy is a matter of good luck; it is a gift from the gods. The other position claims that the factors relevant to happiness are within the agent's control. The strategy of the advocates of this view is to make happiness invulnerable to luck. Aristotle himself takes a middle course. See Nussbaum (1986): 318–342.

happiness safe from luck. Unfortunately, there is not the space here to elaborate all the theoretical details and intricacies, so we will confine ourselves to an outline.[13]

One common obstacle to happiness is not getting what we desire, for success in acquiring what we want is never guaranteed. So we need to be careful about what we desire. According to Epicurus, human beings are led by pleasure (*hedonê*) and pain (*lupê*). These 'instincts' determine what we choose and what we avoid. Unfortunately, human beings are often confused in their judgments about what they want. Epicurus provides an intelligent classification of human desires, and an analysis of the beliefs upon which they depend. Very few desires turn out to be natural and necessary, such as those for food, drink and sex. Most desires are unnecessary, because we are too much affected by habitual preferences. It is not really necessary to eat filet mignon every day. And finally, some desires are empty, because they are based on wrong ideas, such as the wish to become famous or wealthy. These desires are not important for our pleasure and happiness. Those who are capable of satisfying their natural desires are free from pain (*aponia*) and mental distress (*ataraxia*), and, as a consequence, they are happy (*eudaimôn*).

A second obstacle to happiness is emotional distress caused by our reactions to (random) circumstances. The Stoics developed strategies to manage our emotions (*pathê*). Their basic idea is that the happy life is a life of virtuous activity, i.e. a life in which one's actions and behaviour are an expression of the virtues (such as justice, magnanimity, temperance, courage). According to the Stoic conception, our usual emotions are often a result of social conditioning, and are, in fact, ways of feeling born out of ignorance. The Stoics are particularly concerned about unskillful emotions. Not fear in response to *real* danger, but anxiety, desire, anger, grief, obsessive love and jealousy are the targets of their therapy, because those emotions disturb our lives, and, consequently, threaten our happiness. They are erroneous value judgments. By revising or 'unlearning' these value judgments, we can learn to see things differently.

According to the Stoics, we are caught in a dualism between the pursuit of what we believe is good and the avoidance of what we think to be bad. However, only the virtues are really good, and only the vices are really bad. The persons or situations that give rise to emotions are actually not, on Stoic theory, important for our happiness. We tend to judge them as 'good' or 'bad', whereas they are 'indifferent', or neutral. The benefits equal the harms. The Stoics encourage us with various exercises that keep our emotions from getting a hold on us. It is "up to us" how we interpret and respond to whatever happens to us. In this way, Stoic philosophy can shield us from misfortune: we learn not to be affected by whatever happens. We are free from emotions (*apatheia*). How the Stoic immunisation against bad luck works can be seen, for instance, in the following advice from Epictetus:

---

[13]The following studies are extremely useful for understanding these aspects of ancient philosophy: Annas (1993), Cooper (2012), Long (1996) and (2006), Nussbaum (1994), Sorabji (2000), Tsouma (2009) and Warren (2009).

Some things are up to us and others are not. Up to us are opinion, impulse, desire, aversion, and, in a word, whatever is our own action. Not up to us are body, property, reputation, office, and, in a word, whatever is not our own action. The things that are up to us are by nature free, unhindered and unimpeded; but those that are not up to us are weak, servile, subject to hindrance, and not our own. Remember, then, that if you suppose what is naturally enslaved to be free, and what is not your own to be your own, you will be hampered, you will lament, you will be disturbed, and you will find fault with both gods and men. But if you suppose only what is your own to be your own, and what is not your own not to be your own (as is indeed the case), no one will ever coerce you, no one will hinder you, you will find fault with no one, you will accuse no one, you will not do a single thing against your will, you will have no enemy, and no one will harm you because no harm can affect you (*Encheiridion*, 1)[14]

Epictetus's advise is neatly summarized in the following well-known, but hard-gained advice:

Do not ask things to happen as you wish, but wish them to happen as they do happen, and your life will go smoothly (*Encheiridion*, 8)

A third obstacle is addressed by the Skeptics. They also wish to free us from the dualism of good and bad. We think that we are struck by bad circumstances, and we pursue the things that we believe are good, but which we lack. Once we have acquired these so-called good assets, we are afraid to lose them, and, as a consequence, experience troubles "For those who hold the opinion that things are good or bad by nature are perpetually troubled" (Sextus Empiricus, *PH* 1.27).[15] Their strategy was to carefully investigate (*skepsis*) the various arguments and theories of the different schools. Since this inquiry remained inconclusive, it led to a suspension of judgement about the 'real' nature of things. It is not possible to affirm or deny anything about a matter under investigation. We can talk only about appearances, without arriving at the truth. Nevertheless, such skeptical inquiry has beneficial effects: "But those who make no determination about what is good and bad by nature neither avoid nor pursue anything with intensity; and hence they are tranquil" Sextus Empiricus, *PH* 1.28).

By not entertaining fixed views about the nature of things or a situation, the level of one's anxiety is not unnecessarily raised. The Skeptic experiences hunger and thirst, yet does not add the value judgement that it is really unfortunate that this is happening to her, of all people.[16] To use a Zen metaphor: she does not place another head upon her own head. The goal of the Skeptic is to attain peace of mind or tranquility (*ataraxia*) towards situations that are a matter of opinion or appearance, and maintain composure (*metriopatheia*) towards situations that are inevitable (*PH* 1.25). Once we have suspended judgement, freedom from confusion will follow "as a shadow follows a body" (*PH* 1.29).

---

[14]The Epictetus translations are taken from Epictetus (1995).

[15]The Sextus Empiricus translations are taken from Annas and Barnes (1994).

[16]See also Sextus Empiricus, *PH* 3.235–238 and *M* 11.110–167.

In such ways these earliest Western philosophers responded to the human motivation to become happy. Moreover, they were all concerned to make happiness immune from chance events beyond our control. Their important message is that it is a matter of ignorance to think that to live a happy life is due to circumstances beyond one's control. Rather, it is a matter of how we deal with those circumstances. This insight, and living according to this insight, requires training. We cannot eliminate suffering, but it is up to us whether it will make us unhappy. It all depends on the perspective that we have on the world and on ourselves. Philosophy has an important role to play in providing us with this perspective, as Epictetus points out:

> Philosophy does not promise to secure anything external for man, otherwise it would be admitting something that lies beyond its proper subject-matter. For as the material of the carpenter is wood, and that of the statuary bronze, so the subject-matter of the art of living is each person's own life (*Dissertationes* I.15.2)

The terms *a-taraxia*, *a-ponia* and *a-patheia* are significant. The schools promise that after a thorough training, they can free us from several kinds of mental suffering: from confusion, from pain, and from unskillful emotions. In this way, our happiness will become invulnerable to the world. The Skeptics are concerned to free us from the suffering that arises when we get entangled in opposing views and theories; the Epicurians teach us to learn what we really want, to analyse our desires and not to desire more than you need; the Stoics help us to see our emotional responses for what they really are: upheavals of thought that alternate between the poles of attraction and aversion.

4   A Christian Perspective: The Myth of the Fall

Christianity brought a total change of scene. In particular Augustine does not believe in the human capacity to achieve long-term happiness here and now and in the role which ancient philosophers claimed to help achieve it.[17]

> But such is the stupid pride of these men who suppose that the supreme good is to be found in this life and that they can be the agents of thir own happiness, that their wise men,–I mean the man whom they describe as such with astounding inanity,– whom, even if he be blinded and grow deaf and dumb, lose the use of his limbs, be tortured with pain, and visited by every other evil of the sort that tongue can utter or fancy conceive, whereby he is driven to inflict death on himself, they do not scruple to call happy (*De civitate dei*, XIX, 4).

Augustine presents Christianity as an alternative philosophy in the ancient sense of a way of life. Becoming Christian now comes to be the sure route to happiness, though not in this life. In one of his most famous works, *The City of God*

---

[17]Augustine uses *beatus* (happy) and *beatitudo* (happiness), which are translations of *makarios* and *makariotês*, respectively.

(*De civitate dei*), written between 412 and 426/27, Augustine presents his complex vision of earthly life and contrasts it with eternal life in the heavenly Jerusalem.[18] Book 19 is devoted to the philosophers' pathetic attempts to attain happiness within the misery of human life (*De civitate dei* 19.1).

Those who think that there is any happiness in this world, reveal their astonishing lack of understanding. According to Augustine, even the rhetorically gifted are not able to describe life's miseries to any extent. This does not prevent Augustine from offering a page-long complaint about human suffering due to not getting what we want, losing what we have, ailments, decay, mental illness, and the incessant strife between virtue and evil. The best we can do in this life, is to foster hope for happiness in the future, i.e. after our death (*De civitate dei* 19.4). We should not overlook that Augustine's keen eye for human suffering was sharpened by a civil war and the invasion of Germanic tribes. In 410, Alaric and his Goths sacked Rome, the eternal city. Its impact was much greater than that of 9/11 in the West. From the Augustinian perspective, bad luck is just part of human life. However, from the perspective of God, there is no luck or randomness. God is all-powerfull, just, has complete knowledge, and hence, is in total control. So, the question of how to deal with luck did not arise for Augustine. His concern rather is to explain the miseries of human life in view of a God who is neither weak, nor unjust.

So how do we explain and deal with humankind's misfortunes? Augustine offers an ingenious interpretation of the *Book of Genesis*, which becomes a fundamental Christian doctrine in both its Catholic and Protestant versions.[19] The only explanation that Augustine can think of is that our suffering in this life is a punishment from God. A punishment not for something we did, but a punishment for the disobedience of Adam and Eve in the Garden of Eden. Augustine's story is based on his reading of Genesis 2:18–3:24. God had explicitly forbidden the first humans to eat from the tree of knowledge of good and evil. However, a fallen angel, using a snake as its instrument, started with "the weakest link of the human couple" and seduced Eve to eat from its fruit; and Eve offered the fruit to Adam. Obviously, God discovered their disobedience and punished them with expulsion from Paradise, and hence from eternal life and happiness. Suddenly, mankind found itself in a hostile world, in which it had to toil for a living and was inflicted with bodily decay and death. The blissfull order between soul and body was destroyed. The disobedience of Adam and Eve to God has been punished with another corresponding disobedience: the human body is no longer under control of the will, as is clear both from inconvenient sexual temptation and from unwanted failure to perform (*De civitate dei* 14.17). God has punished us with *concupiscentia carnis*, with carnal desire. It is

---

[18]See, for instance, Van Oort (1991).

[19]Nisula (2012) is the most fundamental recent study on the topic, which focuses on sexual desire (*concupiscentia*) as the key concept in Augustine's theory. Augustine's theory is also discussed in Nelson (2011). Among the many studies published about Augustine, see further Brown (1969) Chadwick (2009) and Rist (1994) for details about his life and the broader intellectual background of his views.

this disobedience of Adam and Eve to God, which tainted them with a weakness that has been passed on to future generations. One contemporary opponent, Julian of Aeclanum, consistently described Augustine's position as *peccatum naturale*, a natural defect or sin. The disorder of sexual desire (*concupiscentia*) disseminates itself, so to speak, in the off-spring and thus becomes 'genetic'.[20]

As in Hesiod's story, Augustine too believes that our misfortunes are caused by what our ancestors did. There is no way to escape our miserable life on earth. Only after it ends may we become eternally happy, if we follow the Christian way of life and if God grants us his grace. In the hands of Augustine, Christianity's solution to the indifference of chance came to be its abolishment: God has total control and complete knowledge. At the divine dimension, there is no contingency, whereas at the human level, chance or (bad) luck are part of human suffering and have to be accepted as God's severe, but just punishment for Adam and Eve's disobedience. They are part of God's plan.

5   The Asian Buddhist Perspective: Karma Rather than (Bad) Luck

Buddhism, lacking ruling gods or a creator God, removes the intermediary between our moral actions and their results. Karma (Pali, *kamma*) is understood as an impersonal law of the cosmos: our intentional acts are causes that have direct effects, sooner or later, in that what we do rebounds back onto us.[21] Again, however, as in Christianity, the horrifying specter of mere chance is abolished. Although the consequences of our actions may be delayed, we have a handle on what will happen to us in the future. Insofar as we continue to be reborn, our present circumstances are a result of what we have done earlier, and our future circumstances will be a result of what we are doing now. The doctrine of karma offers an explanation for the repeated suffering of human beings. It stretches out the cause and effect process over several lifetimes and thus makes acceptable that the vicious are not punished immediately and the virtuous may suffer like Job in this life. However, not original sin, but a spiritual ignorance lies at the origin of suffering. Nothing happens to us by chance or luck, but as the result of our karma. According to the Buddhist view, we are 'heir' to our actions, as Peter Harvey puts it. We reap what we have sown, although not everything that happens to us is caused by karmic actions in the past.[22]

For Augustine, happiness here on earth is not possible, yet if we obey God's will we can hope for an eternity in heaven after we die. But what can we do according to the Buddhist view to diminish our suffering and to contain what seems to happen to

---

[20]See Nisula (2012), chapter three and especially 127-134 with the relevant texts in the footnotes.
[21]See also Loy (2008) and the excellent introductions in Carpenter (2014) and Harvey (2013).
[22]Harvey (2013), 39–40.

us by (bad) luck? By following the Buddha's teachings, we can end our ignorance and improve our karma. The foundation of these teachings is the doctrine of the "Four Noble Truths" and the related Buddhist analysis of the roots of unskilful or unwholesome actions. In what follows, we will present a brief overview of these crucial elements of Buddhist thought

For early Buddhism the ultimate goal is *nirvana* (*nibbana* in Pali), but the nature of that goal is less clear. This world of *samsara* is a realm of suffering (Sanskrit *duhkha*, Pali *dukkha*), craving, and delusion; nirvana signifies the end of them, because it is the end of rebirth and karmic retribution. According to the earliest texts we have, in the Pali Canon, Sakyamuni the historical Buddha stated that he taught only *dukkha* and how to end it, but apparently he offered few positive descriptions of the goal.[23] Then is someone who has attained nirvana *happy*? Despite occasional references to *sukha* (the Pali term that corresponds most closely to the English term *happiness*, but which also can be translated as *comfort* or *ease*), the emphasis in the Buddhist tradition has been more on serenity and peace of mind.

Perhaps it is not surprising, then, that lay Buddhists have often been less interested in attaining nirvana–which requires thousands of lifetimes of hard practice, according to the common understanding—than in "merit-making" that will lead to a more favorable (i.e., more enjoyable) rebirth. In popular practice, the Buddha's nuanced teachings about karma have been simplified and commodified into a one-dimensional emphasis on generosity: by making offerings (usually food and money), especially to monastics and temples, you accumulate merit (Sanskrit *punya*, Pāli *puñña*) that will improve your circumstances, if not in this life then in your next one. There is a curious parallel here with the commodification of sin that led to the sale of indulgences by the medieval Church: *merit* is positive, something to be sought, while *sin* is negative, something that needs to be absolved, yet in both cases the belief benefits the religious institution, which therefore has little incentive to correct it.

This shared preoccupation with *what happens after we die* should not, however, distract us from more important similarities between the pre-Christian Western philosophical traditions and the main teachings of Buddhism, regarding *how to live now*. In fact, the parallels are so striking that we are led to reflect on the possibility of historical influence, a topic that has recently received much scholarly attention.[24] Because we normally describe Epicureanism, Stoicism, and Skepticism as philosophies, but view Buddhism as a religion, we do not usually think to compare them. Yet if we suspend any judgement about the transcendent nature of nirvana, the similarities become truly remarkable.

Buddhist teachings focus on two basic causes of *dukkha* (suffering): craving (Pali *tanha*, Sanskrit *trisna*) and ignorance (Pali *avijja*, Sanskrit *avidya*, literally "not seeing"). *Tanha* is the origin of dukkha, according to the second of the four noble truths believed to have been taught by the Buddha in his very first teaching

---

[23]In both the *Alagadduupama Sutta* and the *Anuradha Sutta.*

[24]See, in particular, McEvilley (2001).

(as preserved as the *Dhammacakkappavattana Sutta*) after his awakening. The third noble truth asserts that there is an end to our dukkha (when our craving ceases), and the fourth noble truth gives the eightfold path that leads to its cessation: right view, right intention, right speech, right action, right livelihood, right effort, right mindfulness, and right concentration (or meditation).

Noticeably absent from this list is any reference to ascetic practices, which the Buddha reputedly tried before rejecting them in favor of mindfulness and meditation. The Buddhist path is a "middle way" between hedonism and asceticism, emphasizing not only ethical behavior but most of all realizing the way things really are (including oneself): hence the term enlightenment or, more literally, "awakening" ("the Buddha" means "the awakened one"). Although all eight parts of the path are important, there is nonetheless special emphasis on the last two, which involve the mind-control and personal transformation that is also the main focus of pre-Christian philosophies.

Other similarities with classical Epicureanism, Stoicism, and the Skepticism are hard to miss. The Buddhist path emphasizes nonattachment, so Buddhist monastics live according to rules that clearly regulate what they are allowed to own, and what desires they are able to satisfy. In the Theravada tradition, the basic possessions of monks are three robes, a belt, sewing needle, razor, and water filter; they may also have some incidentals such as toiletries (but not perfumes), a mosquito net, medicines, dharma books, etc. They are mendicant and beg for their food, normally eating only once a day, before noon. They must abstain from all sexual activity and intoxicants such as alcohol. Of course, this lifestyle assumes that, as Epicurus also realized, attempting to satisfy incessant desires is not the way to become truly happy.

Even as the Skeptics were concerned about the dogmatism of fixed views, so the Buddha emphasized that his teachings were heuristic: rather than offering a metaphysical position to identify with, they are helpful for discovering something for ourselves. Two well-known stories illustrate this. One tells of a dialogue between the Buddha and the monk Malunkyaputta, who is troubled by the Buddha's silence regarding fourteen questions, including the finitude or infinitude of the universe, and what happens to a Buddha after he dies. In response to his declaration that he will leave the monastic order if the Buddha does not answer his questions, the Buddha offers a parable:

> Suppose, Mālunkyāputta, a man were wounded by an arrow thickly smeared with poison, and his friends and companions, his kinsmen and relatives, brought a surgeon to treat him. The man would say: 'I will not let the surgeon pull out this arrow until I know whether the man who wounded me was a noble or a brahmin or a merchant or a worker.' And he would say: 'I will not let the surgeon pull out this arrow until I know the name and clan of the man who wounded me; ... until I know whether the man who wounded me was tall or short or of middle height; ... until I know whether the bow that wounded me was a long bow or a crossbow...
>
> The questions go on and on ...
> All this would still not be known to that man and meanwhile he would die. So too, Mālunkyāputta, if anyone should say thus: 'I will not lead the holy life under the Blessed One until the Blessed One declares to me: "the world is eternal" ... or "after death a

Tathāgata neither exists nor does not exist,"' that would still remain undeclared by the Tathāgata and meanwhile that person would die (*Culamalunkya Sutta, Majjhima Nikaya* 63)[25]

As Thich Nhat Hanh glosses, "The Buddha always told his disciples not to waste their time and energy in metaphysical speculation.... Life is short."[26]

Even more famous is the simile comparing the Buddha's teaching to a raft that a man might use to cross a "great expanse of water, whose near shore was dangerous and fearful and whose further shore was safe and free from fear".

> ... Then, when he had got across and had arrived at the far shore, he might think thus: 'This raft has been very helpful to me, since supported by it and making an effort with my hands and feet, I got safely across to the far shore. Suppose I were to hoist it on my head or load it on my shoulder, and then go wherever I want.' Now, bhikkhus, what do you think? By doing so, would that man be doing what should be done with that raft?"
> "No, venerable sir."
> ... 'Suppose I were to haul it onto the dry land or set it adrift in the water, and then go wherever I want.' Now, bhikkhus, it is by so doing that that man would be doing what should be done with that raft. So I have shown you how the Dhamma is similar to a raft, being for the purpose of crossing over, not for the purpose of grasping" (*Alagadupama Sutta, Majjhima Nikaya* 22).[27]

A common Zen metaphor admonishes us not to take the finger for the moon. The finger is pointing at something, which cannot be grasped conceptually. As the Skeptics might say, the goal is not to discover the correct view—a precise set of concepts—that we should fixate on, but to understand our inquiry as a path that seeks other beneficial effects.

Like Stoicism, Buddhism is particularly concerned about "afflictive emotions" (Sanskrit *klesa*, Pali *kilesa*) such as anger, pride, jealousy, and grief, which can lead us to act in ways that we regret later. The Buddha used the metaphor of two darts to emphasize the difference between pain and our emotional reaction to it:

> When an untaught worldling is touched by a painful (bodily) feeling, he worries and grieves, he laments, beats his breast, weeps and is distraught. He thus experiences two kinds of feelings, a bodily and a mental feeling. It is as if a man were pierced by a dart and, following the first piercing, he is hit by a second dart.... Having been touched by that painful feeling, he resists (and resents) it. ... He is fettered by suffering, this I declare.
> But in the case of a well-taught noble disciple, O monks, when he is touched by a painful feeling, he will not worry nor grieve and lament, he will not beat his breast and weep, nor will he be distraught. It is one kind of feeling he experiences, a bodily one, but not a mental feeling. It is as if a man were pierced by a dart, but was not hit by a second dart following the first one.... Having been touched by that painful feeling, he does not resist (and resent) it. Hence, in him no underlying tendency of resistance against that painful feeling comes to underlie (his mind) *Sallatha Sutta (Samyutta Nikaya* 36.6).[28]

---

[25]The translation is from Bhikkhu Nanamoli and Bhikkhu Bodhi (1995), 534–35.

[26]Thich Nhat Hanh (1974), 42.

[27]*See* Bhikkhu Nanamoli and Bhikkhu Bodhi (1995), 228–29.

[28]The translation is from Bhikkhu Bodhi (2000), 1264–65.

The issue of emotional reactions brings us back to the Buddha's understanding of karma, which emphasizes *why* we do what we do.

Although karma and rebirth were already widely accepted in pre-Buddhist India, Brahminical teachings understood karma mechanistically: performing a Vedic sacrifice in the proper fashion would sooner or later lead to the desired consequences. The Buddha transformed this ritualistic approach into a moral principle by focusing on *cetana*, which literally means "volitions" or "motivations." The beginning of the *Dhammapada* makes this point:

> Experiences are preceded by mind, led by mind, and produced by mind. If one speaks or acts with an impure mind, suffering follows even as the cart-wheel follows the hoof of the ox…. If one speaks or acts with a pure mind, happiness follows like a shadow that never departs.[29]

The term *karma* literally means "action." Focusing on the eventual consequences of our actions puts the cart (effect) before the horse (action), and loses the revolutionary implications of the Buddha's innovation. Emphasizing the initial act yields a different insight: that my life-situation can be transformed by transforming the motivations of my actions right now. Just as my body is composed of the food eaten and digested, so "I" am (re)constructed by my habitual mental attitudes. By choosing to change what motivates me, I can change the kind of person that I am.

Buddhist teachings say little about evil per se, but a lot about what are sometimes called the three "roots of evil" (also known as the three fires, or the three poisons) that often motivate our actions: greed, ill will, and delusion. We are encouraged to transform them into their positive counterparts: generosity, loving-kindness, and the wisdom that realizes our interdependence with others.

From this perspective, karma does not need to be taken as a cosmological law comparable to Newton's second law of motion. It can be understood more psychologically, in a way that accords with Stoic insights into the happiness of a virtuous life: we experience karmic consequences not so much for what we have *done* as for what we have *become*, because what we intentionally and habitually do make us what we are: I become the kind of person who does that sort of thing. In other words, we are "punished" not *for* our "sins" but *by* them. And from the other side, as Spinoza declares at the end of the *Ethics*: happiness is not the reward of virtue, but is virtue itself (*Ethics*, Part V, Proposition XLII).

In other words, to be motivated differently is to become a different kind of person, and to become a different kind of person is to experience the world in a different way. When we respond differently to the challenges and opportunities the world presents to us, the world tends to respond differently to us, because our ways of acting involve feedback systems that incorporate other people. The more I am motivated by greed, ill will, and delusion, the more I must manipulate the world to get what I want, and consequently the more separate I feel from others, and the more alienated others feel when they realize what is happening.

---

[29]*Dhammapada* (2010).

On the other side, the more my actions are motivated by generosity, loving-kindness, and the wisdom that acknowledges our interdependence, the more I can relax and open up to the world. The more I feel genuinely connected with other people, the less I will be inclined to use and abuse them, and consequently the more inclined they will be to trust and open up to me. In such ways, transforming my own motivations not only transforms my own life; it also tends to affect those around me, since, as Buddhism emphasizes, we are not really separate.

This naturalistic understanding of karma does not exclude the possibility of more mysterious possibilities regarding the consequences of our actions, such as their effects on one's rebirth, as traditional Buddhism emphasizes. Whether or not that happens, however, karma as how-to-transform-my-life-situation-by-transforming-my-motivations-right-now is not a fatalistic doctrine but an empowering teaching, with many similarities to pre-Christian philosophies of life. Instead of passively accepting the problematic circumstances of our lives, we are encouraged to improve our situations by addressing them with generosity, loving-kindness and wisdom.

Of course, this approach does not make me invulnerable to external events beyond my control, but focuses instead on training my mental ability to respond to them. Whether or not karma is a cosmic law, whether or not there is rebirth, whether or not nirvana transcends the reality of this world, such teachings have enormous implications for how happily we are able to live here and now, day-to-day.

6 Protection Against Luck: West and East

'Luck' or 'chance' on the one hand, and 'karma' on the other seem, at first glance, opposing concepts. If something happens by luck, it is beyond the agent's control. Hence, the main concern of some ancient philosophical schools has been to make our happiness immune against luck. Karma, however, implies that the agent has a great deal of (indirect) control over what happens to her. Thus, luck or chance has been eliminated. Yet, as this chapter has attempted to demonstrate, ancient grapplings with luck and Buddhist discussions about karma, respectively, address the same salient concerns of human existence. What should we do in order to become happy, or, approaching the same question from the other side of the spectrum, what should we do to end our suffering? Both philosophical traditions indicate ways of how we should respond to oscillations of our experience, caused by internal and external events that seem beyond our control. Both philosophical traditions believe that the invulnerability of our happiness against luck depends upon a mental transformation. The Western tradition has focused more on coping with the emotional effects of bad luck: disappointed desires and expectations, anger, fear, anxiety, grief. The Buddhist tradition, on the other hand, has focused its mental training much more on the agent's motivations. Even though these approaches are quite different, the curative methods offered are aimed to change our experience of the world and are still helpful today in our attempts to secure happiness in the face of chance adversity.

Open Access This chapter is distributed under the terms of the Creative Commons Attribution-Noncommercial 2.5 License (http://creativecommons.org/licenses/by-nc/2.5/) which permits any noncommercial use, distribution, and reproduction in any medium, provided the original author(s) and source are credited. The images or other third party material in this chapter are included in the work's Creative Commons license, unless indicated otherwise in the credit line; if such material is not included in the work's Creative Commons license and the respective action is not permitted by statutory regulation, users will need to obtain permission from the license holder to duplicate, adapt or reproduce the material.

References

Annas, J. (1993). *The morality of happiness*. Oxford: Oxford University Press.
Aristotle (2002) *Nicomachean ethics*. Translation (with historical introduction) by Christopher Rowe; Philosophical introduction and commentary by Sara Broadie. Oxford: Oxford University Press.
Brown, P. (1969). *Augustine of Hippo: A biography*. Berkeley: University of California Press.
Carpenter, A. D. (2014). *Indian Buddhist philosophy*. London: Routledge.
Chadwick, H. (2009). *Augustine of Hippo: A life*. Oxford: Oxford University Press.
The Connected Discourses of the Buddha (2000). A Translation of the *Samyutta Nikaya* by Bhikkhu Bodhi. Somerville, MA.
Cooper John, M. (2012). *Pursuits of wisdom: Six ways of life in Ancient philosophy from Socrates to Plotinus*. Princeton: Princeton University Press.
Dhammapada: The Way of Truth (2010). Translated from the Pali by Sangharakshita. Cambridge, UK: Windhorse.
Epictetus (1995). *The discourses, the handbook, the fragments*. In C. Gill, (Ed.), (Trans. revised R. Hard). London.
Garfield, J. L. (2015). *Engaging Buddhism. Why it matters to philosophy*. Oxford: Oxford University Press.
Hadot, P. (2002). *Exercises spirituels et philosophie antique*. Paris: Albin Michel.
Hadot, P. (1995). *Qu'est-ce que la philosophie antique?* Paris: Gallimard.
Hadot, P. (2002). *What is Ancient philosophy* (M. Chase, Trans.). Cambridge, MA.
Harvey, P. (2013). *An introduction to Buddhism. Teachings, history and practices* (2nd ed.). Cambridge: Cambridge University Press.
Long, A. A. (2006). *From Epicurus to Epictetus: Studies in Hellenistic and Roman philosophy*. Oxford: Clarendon Press.
Long, A. A. (2001). Philosophy's hardest question: "What to make of oneself?". *Representations, 74*, 19–36.
Long, A. (1996). *Stoic studies*. Cambridge: Cambridge University Press.
Loy, D. R. (2008). How to drive your Karma. In *Money, Sex, War, Karma. Notes for a Buddhist revolution* (pp. 53–65). Boston: Wisdom Publications.
McEvilley, T. (2001). *The shape of Ancient thought: Comparative studies in Greek and Indian philosophies*. New York: Allworth Press.
Nelson, D. R. (2011). *Sin: A guide for the Perplexed*. London: T&T Clark.
The Middle Length Discourses of the Buddha (1995). (B. Nanamoli & B. Bodhi, Trans.). Boston: Wisdom Publications.
Mikalson, J. D. (2002). The daimon of eudaimonia. In J. F. Miller, C. Damon & K. S. Myers (Eds.), *Vertis in Usum: Studies in honor of Edward Courtney* (pp. 250–259). München: K.G. Saur.
Nisula, T. (2012). *Augustine and the functions of concupiscence*. Leiden: Brill.
Nussbaum, M. C. (1986). *The Fragility of goodness. Luck and ethics in Greek tragedy and philosophy*. Cambridge: Cambridge University Press.

Nussbaum, M. C. (1994). *The therapy of desire. Theory and practice in Hellenistic ethics*. Princeton: Princeton University Press.

Rist, J. M. (1994). *Augustine: Ancient thought baptized*. Cambridge: Cambridge University Press.

Sedley, D. (1999). The ideal of godlikeness. In G. Fine (Ed.), *Plato 2: Ethics, Politics, Religion and the Soul* (pp. 309–328). Oxford: Oxford University Press.

Sextus Empiricus Outlines of scepticism (1994). (J. Annas & J. Barnes, Trans.). Cambridge.

Sorabji, R. (2000). *Emotion and peace of mind: From Stoic agitation to Christian temptation*. Oxford: Oxford University Press.

Thich Nhat Hanh (1974). *Zen keys*. New York: Anchor Press.

Tsouma, V. (2009). Epicurean therapeutic strategies. In J. Warren (Ed.), *The Cambridge companion to Epicureanism* (pp. 249–265). Cambridge: Cambridge University Press.

Van Oort, J. (1991). *Jerusalem and Babylon. A study into Augustine's City of God and the sources of his doctrine of the two cities*. Leiden: Brill.

Warren, J. (2009). Removing fear. In J. Warren (Ed.), *The Cambridge companion to Epicureanism* (pp. 234–248). Cambridge: Cambridge University Press.

Williams, B. A. O. (2006). The legacy of Greek philosophy. In B. Williams (Ed.), *The sense of the past. Essays in the history of philosophy*; Edited with an introduction by Myles Burnyeat. Princeton and Oxford: Princeton University Press.

The Experience of Coincidence: An Integrated Psychological and Neurocognitive Perspective

Michiel van Elk, Karl Friston and Harold Bekkering

Abstract In this chapter, we focus on psychological and brain perspectives on the experience of coincidence. We first introduce the topic of the experience of coincidence in general. In the second section, we outline several psychological mechanisms that underlie the experience of coincidence in humans, such as cognitive biases, the role of context and the role of individual differences. In the third and final section we formulate the phenomenon of coincidence in the light of the unifying brain account of predictive coding, while arguing that the notion of coincidence provides a wonderful example of a construct that connects the Bayesian brain to folk psychology and philosophy.

1 Prelude

This book concentrates on the topic of coincidence. In this chapter, we focus on psychological and brain perspectives on the phenomenon of coincidence. Humans frequently experiences coincidences in life in the sense of the Oxford dictionary: A remarkable concurrence of events or circumstances without apparent causal connection. To shed light on this issue, we will first introduce the topic of coincidence in general. In the second section, we outline several psychological attri-

M. van Elk
Department of Psychology, University of Amsterdam, Amsterdam, The Netherlands

M. van Elk
Amsterdam Brain & Cognition Center, Amsterdam, The Netherlands

K. Friston
Wellcome Trust Centre for Neuroimaging, Institute of Neurology,
University College London, London, UK

H. Bekkering (✉)
Donders Institute for Brain, Cognition and Behavior, Radboud University,
Nijmegen, The Netherlands
e-mail: h.bekkering@donders.ru.nl

© The Author(s) 2016
K. Landsman and E. van Wolde (eds.), *The Challenge of Chance*,
The Frontiers Collection, DOI 10.1007/978-3-319-26300-7_9

butions that underlie the experience of coincidence in humans like cognitive biases, the role of context and the modulation of the experience of coincidence as a consequence of individual differences. In the third and final section we formulate the phenomenon of coincidence in the light of the unifying brain account of predictive coding, i.e., the assumption that brains are essentially prediction machines supporting perception and action by constantly attempting to match incoming sensory inputs with top-down expectations and predictions. In particular, we will show how the experience of coincidence can be understood as an example of Bayes-optimal model selection.

2 Introduction

In 2011 the newspapers reported the remarkable case of Joan Ginther from Texas.[1] Over several years she won four times a multi-million dollar jackpot, by buying scratch-off lottery tickets. It started in 1993 when she won $5.4 million, followed by $2 million in 2003, $3 million in 2005 and in 2010 she won a $10 million dollar jackpot.[2] Such an extraordinary pattern of wins cries out for an extraordinary explanation. Residents of the town of Bishop were convinced that Joan was born under a lucky star or that God was behind it. Statisticians estimated that the chances of winning such prizes four times in a row were 1 in 18 septillion.[3] Combined with the discovery that Joan had earned a Ph.D. in mathematics at the University of Stanford, this led to the suggestion that Joan had figured out the algorithm behind lotteries. Joan always bought her tickets at the same mini mart in Bishop. By figuring out the algorithm that determines the winner and the schedule by which lottery tickets are distributed across Texas, Joan could have predicted when to buy the winning ticket. Joan further contributed to the mystery, by refusing any interview.

In general, humans are remarkably bad at estimating chances and probabilities (Tversky and Kahneman 1974). As a consequence, coincidental events (i.e. a chance concurrence of events without apparent causal connection) are often imbued with special meaning and result in the search for an ultimate explanation (Brugger et al. 1995). In the case of Joan, the explanation turned out to be less extraordinary than initially thought: the first win was likely based on chance, as the number of the winning ticket matched the date of her birthday. The money that was won may have enabled Joan to buy large quantities of lottery tickets, up to tens of thousands of tickets

---

[1] We would like to thank our colleagues Bastiaan Rutjens & Frenk van Harreveld for bringing this example to our attention in their book on 'Coincidence'.

[2] http://www.dailymail.co.uk/news/article-2023514/Joan-R-Ginther-won-lottery-4-times-Stanford-University-statistics-PhD.html.

[3] http://www.philly.com/philly/news/lottery/How_outrageous_were_the_odds_lottery_legend_Joan_Ginther_beat.html.

a year.[4] Given these large quantities the odds of winning a prize become less unlikely than initially thought. In addition, this strategy also explains the fact that Joan (and a friend with whom she collaborated) won a large number of smaller prizes that passed unnoticed by the media.

In this chapter we focus on the experience of coincidence, which can be defined as the remarkable co-occurrence of two events (e.g. being called by a friend you were just thinking about). In some cases the experience of coincidence results in the inference that a common cause underlies the two events (e.g. some unknown 'force' causing you to think about a friend and causing your friend to call you). In other cases, the co-occurrence of events is attributed to chance. The experience of coincidence thus implies a meta-cognitive perspective, in which the most likely explanation for the events being observed is inferred. The experience of coincidence likely underlies a wide range of human behaviors and beliefs, ranging from belief in conspiracy theories, magic and superstition to belief in faith healing and ultimately belief in supernatural agents, like God. National surveys indicate that the tendency to experience coincidence and to engage in superstitious behavior are widespread, with a prevalence of 26 up to 74 percent in the UK for instance, even among scientists (Wiseman 2003).

3   The Psychology of Coincidence

In this section we will discuss basic psychological mechanisms that underlie the experience of coincidence. First, we will argue that the experience of coincidence is related to the over-generalization of predictive models, which in turn are based on fundamental cognitive biases that may actually confer an adaptive advantage. Next, we will focus on the role of context and individual differences in the experience of coincidence.

3.1   Cognitive Biases and Predictive Models

The experience of coincidence may be considered a specific example of the idea that humans construct a predictive model of the world (Friston and Kiebel 2009). This idea, first articulated by Helmholtz assumes that agents perform inference based on a generative model of the world (Clark 2013; Friston 2010; Friston et al. 2012; Gregory 1980; Rao and Ballard 1999; Schwartenbeck et al. 2013). Such models incorporate associations, which can be used to predict future events (e.g. learning that dark clouds often predict rain) and to predict the consequences of our

---

[4]http://www.philly.com/philly/news/lottery/Lotterys_luckiest_woman_Joan_Ginther_bet_flabber gasting_sums_on_scratch-offs.html.

own and others' actions (e.g. learning how to throw a ball in a basket). Psychological experiments have shown that in many cases, these models are based on fast and frugal heuristic processes, that may be advantageous in specific limited circumstances, but that may be difficult to generalize across different domains (Gigerenzer 2012). Furthermore, it has been suggested that predictive models may come to dominate perception, such that reality is perceived in accordance with the constraints imposed by the model, rather than that the sensory input determines the updating of the model. An extreme example of the dominance of predictive models over perception can be found in research on hypnosis, in which proneness to and acceptance of suggestibility manipulations can result in an altered perception of the environment (Raz et al. 2005). Similarly, it has been suggested that an over-reliance on predictive models and a failure to update these models in accordance with the available sensory evidence may be the basis of illusion in normal perception and delusions and hallucinations in psychopathology (Adams et al. 2013; Corlett and Fletcher 2012).

At a very basic level the experience of coincidence and the construction of a predictive model may be related to basic principles of reinforcement learning and classical conditioning. The behaviorist Skinner already noted that pigeons, when food was presented at a random reinforcement schedule, tended to display superstitious-like behavior (Timberlake and Lucas 1985). The co-occurrence of a specific behavior (e.g. pecking at the wall of the cage) with a specific consequence (e.g. receiving food) resulted in the subsequent reinforcement of that behavior—as if it resulted in the presentation of the food. Similar principles of random reinforcement learning likely play a role in human experiences of coincidence and superstitious behavior as well. For instance, imagine buying a lottery ticket at a specific shop and at a specific time of the day and winning a prize. The next time when you buy a lottery ticket, you may be inclined to buy the ticket at the same shop at the same time—even though you know that the chances of winning at this specific shop are as low as buying a ticket somewhere else.

An over-generalization of the principles of reinforcement learning may often be adaptive, as it enables the learning of novel action-effect contingencies. The so-called 'false positives' generated by learning illusory contingencies based are relatively harmless. Evolutionary psychologists have thus argued that the emergence of superstitious behavior and the belief in coincidence is the consequence of adaptive cognitive biases (Foster and Kokko 2009). In a relatively stable and predictable environment, failing to detect a specific contingency between two events (e.g. knowing that smoke often signals fire) is typically more costly than erroneously inferring a relation between two unrelated events (e.g. believing that drumming causes rain). The evolution of superstition is a specific example of the error management principle (Haselton and Nettle 2006), according to which if there is an asymmetrical distribution between type I errors (i.e. a 'false positive') and type II errors (i.e. a 'false negative'), a bias develops toward committing the least costly error. The experience of coincidence may be related to the overestimation of contingencies in a predictive model. As long as the environment is relatively stable such a model is adaptive, but it may become maladaptive in a different context.

For instance, in young children at home an over-estimation of the amount of control over the environment may be adaptive, as they still need to learn which aspects of their environment can be controlled, but may become maladaptive during adolescence, leading to increased risk taking (Heckhausen and Schulz 1995). Similarly, it has been pointed out that in games of chance, people often rely on the over-generalization of principles of skill and practice: they approach a dice throwing or gambling task for instance with a skill-oriented approach, as if their specific movements or choices influence an outcome that is in fact uncontrollable (Langer 1975). Such a bias is adaptive as long as the losses are small and the potential gains are relatively high, but in specific contexts (e.g. casinos) this behavior may become maladaptive, leading to risky gambling and excessive risk taking.

In psychological research, many other cognitive, reasoning, social, memory and attentional biases have been described that may directly contribute to the experience of coincidence and the construction of mental models that influence subsequent decision making (for an overview, see Kahneman 2011). The *self-attribution bias* reflects the general tendency to over-attribute positive outcomes to oneself and negative outcomes to external factors (Mezulis et al. 2004). The self-attribution bias underlies the experience of coincidence, by incorrectly attributing two unrelated events to a common cause (i.e. oneself). For instance, when throwing a dice or when performing a card guessing game, people tend to take credit for positive outcomes, while they externalize negative outcomes (van Elk, Rutjens and van der Pligt 2015). A well-known example of the self-attribution bias can be observed in John McEnroe, a famous tennis player in the nineteen-eighties who attributed wins on a match to his own capacity and training methods, but losing to bad performance of the umpire. Basically, it has been argued that the self-attribution bias reflects a distorted perceptual process, which is driven by the need to maintain and enhance self-esteem. As such, the selective and biased perception of the world has a strong motivational significance, by avoiding people from becoming passive (e.g. 'learned helplessness'). It has even been argued that an over-optimistic perception of one's own capabilities and the amount of control that can be exerted over the environment, may be adaptive and psychologically healthy (Taylor and Brown 1988).

In formal treatments of the predictive or Bayesian brain, it is fairly straightforward to show that the self-attribution bias is, mathematically, Bayes optimal. This self-attribution bias, also known as optimism bias (Sharot 2012), is a natural consequence of making inferences about the state of the world generating sensory information (Friston et al. 2014). In active (Bayesian) formulations of decision making and choice behavior, we act to realize preferred outcomes by sampling from beliefs about the way that we will behave. Usually, these beliefs are informed by sensory evidence. However, when that evidence is ambiguous the most likely state of the world is the state that is consistent with our ongoing behavior (Friston et al. 2014). Because we believe our behavior will lead to preferred outcomes (that actions can fulfill), this necessarily implies that inferences in an uncertain world are optimistic and are inherently biased by beliefs about our purposeful behavior (FitzGerald et al. 2014). A formal (mathematical) treatment of this issue can be found in FitzGerald et al. (2014) and Friston et al. (2014). In this treatment, the neurobiological correlates of the confidence in beliefs about policies are associated with dopaminergic discharges in the brain—a theme that we will return to later.

Also, it has become quite clear that we do not perceive the world as it is. Above all, the information provided at any moment in time is so abundant that we have to be selective in what we attend to. The question how people are able to attend to the most important information, while ignoring other sources of information has been widely studied in psychology and is typically labeled *selective attention*. Donald Broadbent started his investigations of this phenomenon after working with air-traffic controllers during the second world war (Broadbent 1958). In that situation numerous competing messages from departing and incoming aircraft are arriving continuously, all requiring attention. His basic finding was that air traffic controllers can only deal effectively with one message at a time and so they have to decide which is the most important. Based on his and other findings, cognitive scientist argued that we must have a kind of sensory buffer and the input has to be selected based on the physical characteristics for further cognitive processing. However, this bottom-up approach to information processing was challenged, and for example the attenuation model of Anne Treisman suggested that although we can indeed only limitedly process multiple sensory inputs at once, attention is attenuating specific sensory information rather than applying an early filter on the non-attended sensory information (Treisman 1964). The next step in attention research continued this line of thinking and actually argued that attention is able to select information at a very late stage of processing. MacKay (1973) presented participants information via both ears with a specific instruction, which ear to attend. He found that shadowed ambiguous passages with information on the unattended channel that clarified the ambiguity (ear 1—bank; ear 2—river or money) helped the subsequent memory test regarding the relevant channel; participants were better in recalling sentences for which the un-shadowed word was meaningful, thereby further challenging the bottom-up nature of attention. The research of MacKay nicely illustrates that attention is serving a goal—in his experiment acquiring information from any source available to predict the information relevant for the task. Thus, perception is subjective by nature and the feeling of coincidence based on cognitive biases can be considered in the light that we selectively attend to certain stimuli in the context given while ignoring other information available.

This form of selective attention can also be cast in terms of hypothesis selection. In other words, we are compelled to select among a number of competing hypotheses and search out confirmatory (or dis-confirmatory) sensory evidence for those hypotheses. Clearly, the evidence or stimuli that we attend (or ignore) will be highly sensitive to the current hypothesis entertained by the brain: Humans are biased to selectively attend and recall information that is highly salient or informative (Mcdaniel et al. 1995). In addition, people often rely on representativeness and availability heuristics when judging the likelihood of situational descriptions (Tversky and Kahneman 1974) and may use counterfactual thinking to regulate affect in response to unexpected positive or negative outcomes (Roese 1997).

In general, people are characterized by a misperception of chance events (Tversky and Kahneman 1974), as shown for instance by the tendency to perceive an 'irregular' coin-toss sequence like 'H-T-H-T-T-H' as more likely than a regular sequence like 'H-H-H-T-T-T'. In this example, chance events are considered as a self-corrective process and on each consecutive toss of the coin people take into account the past history of 'heads' and 'tails'—even though the coin obviously has no memory. The latter bias is another good example of the general tendency to construct predictive models of the world—even in cases when such a model is not applicable or appropriate (or in which the model should classify the coin toss as a 'chance event').

In sum, we argue that the experience of coincidence may be considered a specific instance of the tendency to construct and rely on predictive models of the world. These models may often be based on adaptive biases or prior beliefs to detect contingencies (Foster and Kokko 2009) and/or may be supported by other domain-specific biases that confer an adaptive advantage (i.e. heuristics) in specific settings. An over-reliance on internal models and the over-generalization of models to contexts in which they do not apply, may contribute to the experience of coincidence.

3.2 Context and Model Adjustment

In the preceding section we have argued that perception of events in the world is subjective and that cognitive biases at the personal level may result in the experience of coincidence. Specific situations or a given situational context, may also alter your perception of the world dramatically. As has been argued before (FitzGerald et al. 2014), agents have to determine what model to use in the first place and secondly to make inferences about hidden variables to evaluate the likelihood of a model and the precision of the parameters of any plausible model. A given situational context is likely to affect both aspects: which model to use and/or how to weight the parameters within the specific models.

A famous example was demonstrated in a Candid Camera television show in the 1960s (the example is also mentioned in Liebermann 2007). An uninformed individual enters an elevator filled with multiple confederates working with the show. These confederates stand all collectively facing the back of the elevator rather than facing the front. Almost all individuals would look quickly around at the others and then change their orientation in order to stand in line with the confederates. This example is presented in social psychology as one of the fundamental insights of social cognition: "people look to the social environment and external context to guide their behavior, particularly when the appropriate course of action is ambiguous or undefined." This example nicely illustrates how our behavior is context-dependent, but it also nicely illustrates how different models compete for different inferences. Relying on previous knowledge of elevators, you have learned that the door that opened for you when you entered the elevator is also likely the door that will open again when you need to leave the elevator. However, occasionally, you find elevators with two doors, one entrance and one exit door typically at the opposite side of the elevator. The fact that all others are facing the back might strike you as too obvious to be coincidence. Thus, multiple inferences are produced by your brain; the elevator model which activates probabilities about potential door locations that you might perceive to open to allow you to exit the elevator, but also the social model, i.e., the probability that several people all face one direction that is likely going to be the direction at which relevant information will appear. In other words, in a causal model of the world, you expect other agents to anticipate what will happen next, and thus you assume they are directed to the location they expect the door to open—or, you could even infer the candid camera model, i.e., how likely is it that people are making a joke on me. Based on the precision of these different models in terms of what is the best inference on what I can perceive next, most people might make an active inference and turn their side in alignment with the others. Interestingly, this Bayesian approach on a social phenomenon like this emphasizes "the power of the situation" as much as many other well-studied concepts in social psychology, like the confirmation bias (Asch 1956), or the famous obedience to authority phenomenon (Milgram 1965), from a unified framework, predictive coding. Depending on the precision of parameters from different models in your mind you infer what you will perceive next based on the (social) context you are in. Again, we see the emergent theme of selecting among plausible hypotheses that explain the sensory evidence at hand. Above, we have discussed this in terms of perceptual inference, very much along the lines of perception as hypothesis testing (Gregory 1980). Here, the same notion emerges in the context of social inference. We will return to the central role of selecting hypotheses and Bayesian model selection below.

In ambiguous and uncertain contexts, the need for predictive models and the need for making predictions including situational constraints increases. In line with this suggestion, the experience of coincidence and the engagement of superstitious behavior are often strongly related to significant life events that have important consequences, such as well-being, illness or death. It has been found for instance that belief in luck and coincidence increased during times of stress and in potentially threatening situations (Keinan 1994, 2002). Similarly, superstitious behavior is quite prevalent among the performing arts and in sports, and the occurrence of superstitious acts typically increases with the importance of the outcome (e.g. playing the finals; cf. Burger and Lynn 2005). Interestingly, large cultural differences exist in the experience of coincidence and in probabilistic thinking (Wright et al. 1978): Asians compared to westerners typically engage less in probabilistic thinking in terms of 'cause-and-effect' and this may be related to the 'fate-oriented' view in Eastern religion and philosophy. These findings highlight the role of context in the experience of coincidence. Again, these findings make sense in a broader evolutionary framework, according to which the detection of (illusory) contingencies and the need for predictive models is especially important in potentially ambiguous or threatening situations.

Specific contexts may trigger an over-reliance on internal models and a failure to update these models in accordance with the available sensory information, may cause the experience of 'coincidence'. An extreme example of a failure to update one's cognitive model may be found in the phenomenon of *cognitive dissonance* (Festinger et al. 2008). In his seminal work, Festinger describes a religious sect believing that the earth would be flooded and that they would be rescued by extraterrestrials in a flying saucer. When the critical time had passed and the prophecy did not come true, rather than giving up their beliefs, the sect became even more fervent in their faith. Many psychological studies have shown that, rather than changing one's model based on new evidence, humans respond to cognitive dissonance by discarding the evidence or assimilating the evidence to one's current model (Elliot and Devine 1994). For instance, many believers put their trust in a religious leader, who in turn imposes their views on his followers. An increased reliance on religious authority results in a reduced process of error monitoring and a failure to update one's model based on the available evidence. Recently it has been argued that religious rituals are specifically aimed at reducing the process of error monitoring, thereby enhancing people's willingness to uncritically adopt a prevailing worldview (Schjoedt et al. 2013). In line with this suggestion, it has been found for instance that believers are characterized by a reduced activation of the frontal executive monitoring network when listening to a religious authority (Schjoedt et al. 2011). In such contexts, a failure to update one's model may result in the experience of coincidence, as observed for instance during faith healing in which a common cause is inferred (e.g. 'God') for two scientifically unrelated events (e.g. prayer by the religious authority and the (often) temporary recovery of illness).

3.3 Individual Differences and Precision

In addition to contextual effects, individual differences in personality traits and beliefs also play an important role in the experience of coincidence. Some people may prefer more certainty and precision in their predictions than others. In addition some people may more strongly rely on their predictive models than others and may be characterized by systematic biases with respect to taking sensory information into account.

It has been found that the tendency to perceive coincidences is related to the individual trait of need for control (Hladkyj 2001). People scoring high on the need for control (and likely requiring a higher precision in their prediction models) were more likely to experience unusual coincidences as personally significant (c.f., the self-attribution and optimism bias above). In addition, belief in a meaningful world and the imbuement of random events with meaning has been associated with a stronger visual attention capture (Bressan et al. 2008): this finding could reflect that the tendency to perceive coincidences as meaningful is related to a process of error detection of information that is conflicting with one's cognitive schema's.

Several studies have suggested that individual differences in the reliance on internal predictive models of the world are also related to the experience of coincidence. Participants scoring high on schizotypal personality traits are characterized by an increased reliance on internal predictive models and by difficulties to update their model based on new sensory evidence (Corlett and Fletcher 2012). In addition, a relation has been suggested between schizotypy, the perception of coincidence, magical ideation and paranormal beliefs (Williams and Irwin 1991). It has been found, for instance, that people scoring high on schizotypy and magical ideation are more prone toward detecting illusory contingencies (Brugger and Graves 1997). In this task, participants were required to discover the rule whereby navigating a virtual mouse through a maze would result in a reward. In fact, the reward was directly coupled to the amount of time spent navigating: if the participants spent more than three seconds in the maze, they would receive the reward, whereas if they spent less time no reward was provided. Many participants developed beliefs in illusory contingencies (i.e. the belief that moving the mouse repetitiously along a specific path would result in the reward) and the amount of illusory hypotheses that were believed were directly related to magical ideation. In another study using a dice throwing task it was found that the perception of chance events as meaningful is related to a tendency for repetition-avoidance e.g. in guessing outcomes (Brugger et al. 1995). Interestingly, in the same study it was found that the tendency to avoid semantically related guesses was associated to a stronger belief in extrasensory perception. Finally, it has been reported that paranormal believers show fallacies in probabilistic reasoning task and tend to underestimate the likelihood of chance events (Rogers et al. 2009). In addition, paranormal believers are more prone to reporting frequent experiences of coincidence during their life (Bressan 2002). These findings illustrate that individual differences in model selection and the reliance on internal models can have a strong effect on the experience of coincidence.

In summary, when we use internal models to make inferences about the causes of our sensations, we are in the difficult game of carefully balancing the precision of, or confidence in, sensory evidence relative to prior beliefs. In hierarchical models (with multiple levels of abstraction), each level is equipped with a precision that determines how much it predominates over other levels. Crucially, the precision at each and every level of the hierarchy has to be optimized. This optimization itself depends upon biases or priors about expected precision (or expected uncertainty) that can lead to very different inferences and behavior. This may be manifest as normal intersubject variation in cognitive biases or, indeed, provide a formal explanation for false inference in psychopathology (Adams et al. 2013).

4 Predictive Coding and Coincidence

We have defined the experience of coincidence as an inference about the remarkable co-occurrence of two events (Brugger et al. 1995). To conclude, we present a more theoretical view on how Bayesian models, implemented in our brain, can lead to the experience of coincidence. The experience is labeled as a coincidence, when our explanation appeals to the notion of a 'coincidence', as opposed to some underlying common cause. When a causal inference is made, the experience is labeled as coincidence; in contrast, 'non-causal' inference makes the concurrence coincidental. This means that we must have the capacity to infer that an improbable (remarkable) concurrence was or was not causally mediated. This entails the capacity to postulate two concurrent hypotheses (improbable events may or may not have a common cause), and we must also have a (meta-representational) concept of this inferential dilemma.

In this section, we turn to a formal treatment of coincidences from the perspective of the Bayesian brain. To set the scene, it would be useful to rehearse the simplicity of the formal perspectives we have been appealing to. The most general principle guiding action and perception is presumed to be a maximization for the evidence of models used to explain the sensorium. The inverse or complement of model evidence is surprise, prediction error or a quantity called variational free energy. This means that the brain is trying to minimize prediction error (or maximize model evidence). A popular scheme for implementing this minimization is predictive coding, for which there is a substantial amount of circumstantial evidence in terms of neuroanatomy and neurophysiology (Friston and Kiebel 2009). So what does it mean to maximize model evidence? To understand this, we have to appreciate that model evidence has two components:

$$\text{Log evidence} = \text{accuracy} - \text{complexity}$$

Where, mathematically:

$$\text{Log evidence} = \ln \Pr(\text{consequence}|\text{hypothesis})$$
$$\text{Accuracy} = E\left[\ln \Pr(\text{consequence}|\text{cause, hypothesis})\right]$$
$$\text{Complexity} = D[\Pr(\text{cause}|\text{consequence, hypothesis})| \, |\Pr(\text{cause}|\text{hypothesis}))$$

where E[] denotes an expectation or average and D[] the relative entropy or Kullback-Leibler divergence. This mathematical formulation of the goodness of fit of a model is interesting because it says that complexity is the divergence between our prior beliefs (i.e., cognitive biases and preconceptions) and the (posterior) beliefs adopted after seeing sensory information.

Crucially, a high model evidence requires a parsimonious but accurate explanation for sensory consequences (of inferred causes). Generally, these explanations rest upon internal or generative models with a deep hierarchical structure (possibly reflecting the hierarchical organization of cortical areas in the brain). This deep structure is particularly important from the point of view of coincidences, because appealing to a common cause adds an extra level or depth to the hierarchical explanation that can minimize its complexity (and maximize model evidence). To see this clearly, we need to see why complexity is so important.

If we explained all our sensations with a multitude of independent causes, we would have a very accurate (low prediction error) explanation; however, the complexity of this explanation or hypothesis would be very high. This is because complexity increases with the degrees of freedom or number of causes invoked to explain data (the divergence above). The problem with complex but accurate models is that they do not generalize to other situations—a problem known as over-fitting in statistics. This means a good model should also be parsimonious and use the smallest number of causes to explain (sensory) consequences. In turn, this means we are compelled to construct unifying hypotheses about common causes that reduce the cardinality of the causes of our sensory explananda.

It is therefore entirely Bayes-optimal to select hypotheses or models that ascribe a common cause to coincident events; particularly those that are generated by some agency (e.g., oneself, a deity or the CIA). In fact, several studies have shown that the tendency to attribute coincidental events to external agents is universal and may underlie supernatural and conspiracy beliefs (Banerjee and Bloom 2014; Imhoff and Bruder 2014). It is at this point we see the utility of 'coincidence' as an alternative hypothesis for the co-occurrence or succession of coincident events. To make this concrete, consider a situation where you are meeting a friend for coffee and he arrives at exactly the same time as you. This coincidence is surprising and will call for an explanation in your (Bayesian) brain. This is because surprise has to be minimized. There will be a number of competing hypotheses; for example, your friend has been waiting for you, your friend knew exactly when you would arrive because he has been spying on you, you both caught the same tram to the café, the meeting was ordained by God and, finally, it was a coincidence. All of these competing hypotheses or models provide an accurate explanation for the events you have witnessed; however, they differ profoundly in terms of their complexity as scored by the number of (implausible) deviations from your prior beliefs. As we have noted above, selecting the best hypothesis corresponds to accepting the model with the greatest evidence (this is known as Bayesian model selection in statistics). This will be the hypothesis with the minimum complexity; namely the explanation that requires the least divergence from your prior beliefs. In other words, an a priori plausible explanation is most likely inferred (e.g., you arrived on the same tram). However, if there are no tram stops near the café, then the most plausible hypothesis could be a coincidence; provided you believe, a priori, coincidence is plausible. The hypothesis you select will determine whether coincident events (in the real world) are experienced as a coincidence.

The key insight provided by the above treatment is that we are equipped with the hypothesis or heuristic that things can be explained by 'coincidences'. This is a constructive explanation—as opposed to simply ignoring co-occurrences. If this is true, then the way that we deal with (real-world) coincidences depends strongly on our prior disposition to 'coincidence' as a causal explanation. The very fact that we have this hypothesis at hand to explain surprising contingencies is a testament to the sophistication of our hierarchical generative models and may not be seen in lower animals (like pigeons). It also may provide one perspective on the formation of delusional systems in psychosis, where the coincidence hypothesis is simply not available.

There are some other interesting predictions that follow from our line of argument. Above, we have noted that the confidence in our beliefs about chosen outcomes may be signaled by dopamine in the brain. This stands in contrast to alternative explanations based upon dopamine discharges reporting rewards or preferred outcomes. Coincidences may offer an interesting resolution to the competing explanations for dopamine responses. If coincidences resolve surprise, then realizing something is a coincidence should resolve uncertainty and increase precision resulting in elevated dopamine firing. Conversely, if dopamine reports preferred outcomes, even when they are surprising, dopamine should show a response to unexpected rewards that are entirely coincidental.

5 Conclusions

In this chapter we have provided an analysis of the experience of coincidence from a psychological and neurocognitive perspective. As humans we construct predictive models of the world that enable us to generate predictions and to minimize surprise. The experience of coincidence may result from cognitive biases, such as the self-attribution bias and attentional biases, which are Bayes-optimal. Thereby the notion of coincidence provides a wonderful example of a construct that connects the Bayesian brain to folk psychology and philosophy.

Open Access This chapter is distributed under the terms of the Creative Commons Attribution-Noncommercial 2.5 License (http://creativecommons.org/licenses/by-nc/2.5/) which permits any noncommercial use, distribution, and reproduction in any medium, provided the original author(s) and source are credited. The images or other third party material in this chapter are included in the work's Creative Commons license, unless indicated otherwise in the credit line; if such material is not included in the work's Creative Commons license and the respective action is not permitted by statutory regulation, users will need to obtain permission from the license holder to duplicate, adapt or reproduce the material.

References

Adams, R. A., Stephan, K. E., Brown, H. R., Frith, C. D., & Friston, K. J. (2013). The computational anatomy of psychosis. *Frontiers in Psychiatry, 4*, 47.
Asch, S. E. (1956). Studies of independence and conformity: I. a minority of one against a unanimous majority. *Psychological Monographs: General and Applied, 70*(9), 1–70.
Banerjee, K., & Bloom, P. (2014). Why did this happen to me? Religious believers' and non-believers' teleological reasoning about life events. *Cognition, 133*(1), 277–303.

Bressan, P. (2002). The connection between random sequences, everyday coincidences, and belief in the paranormal. *Applied Cognitive Psychology, 16*(1), 17–34.

Bressan, P., Kramer, P., & Germani, M. (2008). Visual attentional capture predicts belief in a meaningful world. *Cortex, 44*(10), 1299–1306.

Broadbent, D. (1958). *Perception and communication.* London: Pergamon Press.

Brugger, P., & Graves, R. E. (1997). Testing vs. believing hypotheses: Magical ideation in the judgement of contingencies. *Cognitive Neuropsychiatry, 2*(4), 251–272.

Brugger, P., Regard, M., Landis, T., & Graves, R. E. (1995). The roots of meaningful coincidence. *Lancet, 345*(8960), 1306–1307.

Burger, J. M., & Lynn, A. L. (2005). Superstitious behavior among American and Japanese professional baseball players. *Basic and Applied Social Psychology, 27*(1), 71–76.

Clark, A. (2013). Whatever next? Predictive brains, situated agents, and the future of cognitive science. *Behavioral and Brain Sciences, 36*(3), 181–204.

Corlett, P. R., & Fletcher, P. C. (2012). The neurobiology of schizotypy: fronto-striatal prediction error signal correlates with delusion-like beliefs in healthy people. *Neuropsychologia, 50*(14), 3612–3620.

Elliot, A. J., & Devine, P. G. (1994). On the motivational nature of cognitive-dissonance—dissonance as psychological discomfort. *Journal of Personality and Social Psychology, 67*(3), 382–394.

Festinger, L., Riecken, H. W., & Schachter, S. (2008). *When prophecy fails.* Chicago: Pinter & Martin Publishers.

FitzGerald, T. H. B., Dolan, R. J., & Friston, K. J. (2014). Model averaging, optimal inference, and habit formation. *Frontiers in Human Neuroscience, 8*, 1–11.

Foster, K. R., & Kokko, H. (2009). The evolution of superstitious and superstition-like behaviour. *Proceedings of the Royal Society B-Biological Sciences, 276*(1654), 31–37.

Friston, K. J. (2010). The free-energy principle: A unified brain theory? *Nature Reviews Neuroscience, 11*(2), 127–138.

Friston, K. J., Adams, R. A., Perrinet, L., & Breakspear, M. (2012). Perceptions as hypotheses: Saccades as experiments. *Frontiers in Psychology, 3*, 151.

Friston, K. J., & Kiebel, S. (2009). Predictive coding under the free-energy principle. *Philosophical Transactions of the Royal Society of London Series B-Biological Sciences, 364*(1521), 1211–1221.

Friston, K. J., Schwartenbeck, P., FitzGerald, T., Moutoussis, M., Behrens, T., & Dolan, R. J. (2014). The anatomy of choice: Dopamine and decision-making. *Philosophical Transactions of the Royal Society of London Series B-Biological Sciences, 369*(1655).

Gigerenzer, G. (2012). Heuristic decision-making. *International Journal of Psychology, 47*, 114.

Gregory, R. L. (1980). Perceptions as hypotheses. *Philosophical Transactions of the Royal Society of London Series B-Biological Sciences, 290*(1038), 181–197.

Haselton, M. G., & Nettle, D. (2006). The paranoid optimist: An integrative evolutionary model of cognitive biases. *Personality and Social Psychology Review, 10*(1), 47–66.

Heckhausen, J., & Schulz, R. (1995). A life-span theory of control. *Psychological Review, 102*(2), 284–304.

Hladkyj, S. (2001). The narrative emplotment of chance events, desire for control and tolerance of ambiguity in the experience of "meaningful coincidence".

Imhoff, R., & Bruder, M. (2014). Speaking (Un-)truth to power: Conspiracy mentality as a generalised political attitude. *European Journal of Personality, 28*(1), 25–43.

Kahneman, D. (2011). *Thinking fast and slow.* New York: Farrar, Straus and Giroux.

Keinan, G. (1994). Effects of stress and tolerance of ambiguity on magical thinking. *Journal of Personality and Social Psychology, 67*(1), 48–55.

Keinan, G. (2002). The effects of stress and desire for control on superstitious behavior. *Personality and Social Psychology Bulletin, 28*(1), 102–108.

Langer, E. J. (1975). The illusion of control. *Journal of Personality and Social Psychology, 32*(2), 311.

Lieberman, M. D. (2007). Social cognitive neuroscience: A review of core processes. *Annual Review of Psychology, 58*, 259–289.

Mackay, D. G. (1973). Aspects of the theory of comprehension, memory and attention. *The Quarterly Journal of Experimental Psychology, 25*(1), 22–40.

Mcdaniel, M. A., Einstein, G. O., Delosh, E. L., May, C. P., & Brady, P. (1995). The bizarreness effect—its not surprising, its complex. *Journal of Experimental Psychology-Learning Memory and Cognition, 21*(2), 422–435.

Mezulis, A. H., Abramson, L. Y., Hyde, J. S., & Hankin, B. L. (2004). Is there a universal positivity bias in attributions? A meta-analytic review of individual, developmental, and cultural differences in the self-serving attributional bias. *Psychological Bulletin, 130*(5), 711–747.

Milgram, S. (1965). Some conditions of obedience and disobedience to authority. *Human Relations, 18*(1), 57–76.

Rao, R. P. N., & Ballard, D. H. (1999). Predictive coding in the visual cortex: A functional interpretation of some extra-classical receptive-field effects. *Nature Neuroscience, 2*(1), 79–87.

Raz, A., Fan, J., & Posner, M. I. (2005). Hypnotic suggestion reduces conflict in the human brain. *Proceedings of the National Academy for Sciences of the United States of America, 102*(28), 9978–9983.

Roese, N. J. (1997). Counterfactual thinking. *Psychological Bulletin, 121*(1), 133–148.

Rogers, P., Davis, T., & Fisk, J. (2009). Paranormal belief and susceptibility to the conjunction fallacy. *Applied Cognitive Psychology, 23*(4), 524–542.

Schjoedt, U., Sørensen, J., Nielbo, K. L., Xygalatas, D., Mitkidis, P., & Bulbulia, J. (2013). Cognitive resource depletion in religious interactions. *Religion, Brain & Behavior, 3*(1), 39–55.

Schjoedt, U., Stodkilde-Jorgensen, H., Geertz, A. W., Lund, T. E., & Roepstorff, A. (2011). The power of charisma-perceived charisma inhibits the frontal executive network of believers in intercessory prayer. *Social Cognitive and Affective Neuroscience, 6*(1), 119–127.

Schwartenbeck, P., FitzGerald, T., Dolan, R. J., & Friston, K. J. (2013). Exploration, novelty, surprise, and free energy minimization. *Frontiers in Psychology, 4*, 710.

Sharot, T., Guitart-Masip, M., Korn, C. W., Chowdhury, R., & Dolan, R. J. (2012). How dopamine enhances an optimism bias in humans. *Current Biology, 22*(16), 1477–1481.

Taylor, S. E., & Brown, J. D. (1988). Illusion and well-being—a social psychological perspective on mental-health. *Psychological Bulletin, 103*(2), 193–210.

Timberlake, W., & Lucas, G. A. (1985). The basis of superstitious behavior—chance contingency, stimulus substitution, or appetitive behavior. *Journal of the Experimental Analysis of Behavior, 44*(3), 279–299.

Treisman, A. M. (1964). Selective attention in man. *British Medical Bulletin, 20*, 12–16.

Tversky, A., & Kahneman, D. (1974). Judgment under uncertainty: Heuristics and biases. *Science, 185*(4157), 1124–1131.

van Elk, M., Rutjens, B., & van der Pligt, J. (2015). The development of the illusion of control and sense of agency in 7-to-12-year old children and adults. *Cognition, 145*, 1–12.

Williams, L. M., & Irwin, H. J. (1991). A study of paranormal belief, magical ideation as an index of schizotypy and cognitive-style. *Personality and Individual Differences, 12*(12), 1339–1348.

Wiseman, R. (2003). UK Superstition Survey. Psychology Department, University of Hertfordshire (Publication no. http://www.richardwiseman.com/resources/superstition_report).

Wright, G. N., Phillips, L. D., Whalley, P. C., Choo, G. T., Ng, K., Tan, I., et al. (1978). Cultural differences in probabilistic thinking. *Journal of Cross-Cultural Psychology, 9*(3), 285–299.

When Chance Strikes: Random Mutational Events as a Cause of Birth Defects and Cancer

Han G. Brunner

Abstract Faithful and stable inheritance of DNA is coupled with occasional random errors of replication that lead to a change in the DNA code known as mutation. Mutations can be considered as "good" because they are the fuel that drives evolution of species. On the level of the individual they are mostly harmful. In fact, the majority of severe intellectual disabilities derives from such random mutational events. In my experience, the tendency to ascribe all events to definite causes is still highly prevalent. Against this background of presumed guilt, parents who are confronted with the birth of a severely handicapped child tend to take solace form the knowledge that the condition was not their "fault". Our recent understanding that severe handicaps may strike anyone, may well lead to the acceptance of a more universal offer of prenatal diagnosis than previous strategies which were based on the identification of high risk groups.

1 Fascination

For as long as we know, people have been devastated and fascinated by the birth of a child with severe malformations or disabilities. Collecting malformed foetuses was a popular pastime for the elite during the 17th Century. Rich and educated men built up sizable private collections of curiosities. One such anatomical collection was sold in its entirety to Czar Peter the Great in 1717 by Frederik Ruijsch from Amsterdam (Baljet and Oostra 1998). An anatomical collection from the 18th century that has been preserved and maintained as a museum is that of Willem and Gerard Vrolik. This is now in the AMC hospital in Amsterdam. People with malformations or other visible developmental defects were put on display in "freak shows" and exhibitions. In the 19th century, PT Barnum in the USA and Tom Norman in the UK traveled widely around their respective countries, with shows of

H.G. Brunner (✉)
Faculty of Medical sciences, Radboud University, Nijmegen, The Netherlands
e-mail: han.brunner@radboudumc.nl

© The Author(s) 2016
K. Landsman and E. van Wolde (eds.), *The Challenge of Chance*,
The Frontiers Collection, DOI 10.1007/978-3-319-26300-7_10

supposed freaks of nature. Quite probably, malformations will continue to scare and excite us forever. Certainly, our fascination with physical abnormality has not ceased in the 20th century. The 1980 movie "the elephant man" directed by David Lynch relates the story of John Merrick whose malformations were exploited by the owner of such a freak show. The 1985 movie "Mask" was based on the life of Roy Lee Dennis who died at age 16 from craniodiaphyseal dysplasia, a progressive deforming bone disease of the skull. Another contemporary movie about malformation is Edward Scissorhands (Tim Burton 1990). The image of a boy born with scissors for hands is clearly inspired by inherited ectrodactyly or "lobster claw malformation" where the middle fingers are missing at birth. A fascination with malformations can further be found in many literary tales, notably Homer's Cyclops in the Odyssey.

2   Divinity and Sorcery

Beyond fascination is the need to find explanations for personal disasters such as the birth of a malformed or handicapped child. In antiquity, and in societies around the world, congenital abnormalities were regarded as omens, or punishment from the gods (Warkany 1959; Beckwith 2012). For example, Tigay (1997) mentions the Babylonian Omen series (Izbu) which lists the predicted significance of women giving birth to children with a wide variety of malformations. "If a woman gives birth (and the child) has two heads: there will be a fierce attack against the land and the king will give up his throne" (Izbu, II, 20 h32) (Pangas 2000). Although divinity was not generally considered a plausible cause after the middle ages, witchcraft and other supernatural phenomena remained serious possibilities until relatively recently. A case cited by Brent and Fawcett (2007) concerns the trial of one George Spencer from Connecticutt, who had a glass eye. When a one-eyed piglet was born on the farm, he was charged with bestiality. He was duly sentenced to death in New England in 1642 for having sired the abnormal pig. George Spencer was hanged. The sow was put to death by the sword.

3   Maternal Impressions

One common belief about malformations which originated very early and appears pervasive in many different cultures is the concept that events and images witnessed by a pregnant woman may somehow imprint themselves on the foetus (e.g. Warkany and Kalter 1962). A positive example of this is the advice given to pregnant women in the Greek city of Sparta, to admire statues of well-formed human beings. The converse idea, that viewing an abnormality can leave an imprint on the developing foetus by some sort of "photographic" effect, remained common until the late 19th century (Fisher 1870). In his book on medical curiosities Jan

Bondeson (1997) extensively discusses these so-called maternal impressions. Bondeson relates the story of the Danish anatomist Bartholin who saw a girl with a cat's head on a visit to Holland in 1738. The explanation given to Bartholin by the locals, was that a cat hiding in her mother's bed, had dashed out unexpectedly and startled the pregnant woman. Bartholin and his colleague Jaccobaeus were influential at the Danish court. On their advice, King Frederik IV ruled that invalid and malformed people should be kept out of sight in a special hospital in Copenhagen. This was not out of pity for the poor and crippled, but to prevent pregnant women from bearing children exactly like them (Bondeson 1997). The last serious description of maternal impression ("Verzien" in Dutch) as a cause for malformation in the Dutch National Journal of Medicine occurred almost exactly 100 years ago (Formijne 1915). Occasional supporters of the concept remain among those who believe in parapsychology.

4 Infections and Teratogens

The discovery by Gregg in the early 1940s (Gregg 1947) that congenital rubella infection causes cataract, deafness, and other abnormalities and the description of severe malformations due to Thalidomide in the early 1960s by McBride in Australia (1961) and Lenz in Germany (1962), in conjunction with experimental work by Warkany in Cincinnati amongst others established the science of teratology, which studies the influence of harmful substances and infections on the foetus (Warkany and Nelson 1940). This concept of the foetus as a vulnerable developing human being inspired dramatic and effective improvements in prenatal care. It is now generally accepted that prenatal factors are responsible for malformations and handicaps in at most of 5 % in newborns in developed countries. In spite of the apparent rarity of teratogenic causes, all mothers of children with severe abnormalities or disabilities feel guilty. Many consider the possibility that something happened during pregnancy that harmed their child, which should have been avoided. In the case of intellectual disability, it is sometimes assumed that a lack of oxygen during delivery was responsible. However, it would seem that this is also rare and that it cannot begin to account for most cases of intellectual disability in the population at this time.

5 Inherited Factors

Inbreeding is an important factor for malformations, and intellectual disability. This reflects recessive inheritance where a child is affected because it received an abnormal gene from both parents. Because most deleterious gene variants are rare, the chance of these coming together in a child is very low, unless the parents are related. Thus, recessive inheritance has an important role in causing malformations and intellectual disability in countries with a high consanguinity rate. A recent study from the UK suggests that the risk of a baby having a malformation is approximately doubled from 3 to 6 % if the parents are first cousins (Sheridan et al. 2013). A recent study from Germany based on prenatal ultrasound scans came to much the same conclusion but the increase was about 3-fold, from 2.8 to 8.5 % for offspring from first-cousin marriages (Becker et al. 2015). No good estimates are available on their frequency, but there is good evidence for recessive inheritance of intellectual disability from populations with high rates of consanguinity such as Iran (Najmabadi et al. 2011).

The frequency of consanguinity varies enormously across the world, from less than 1 % of all marriage unions in the USA and Russia to over 50 % in Sudan and Pakistan (Romeo and Bittles 2014). This variation is tightly linked with customs and existing religious rules. Notably in Europe, the Roman Catholic church generally prohibited first-cousin marriages, while the protestants took a more liberal view. In the UK, following the marriage of Henry VIII first to his sister in law, Catherine of Aragon, and then to Anne Boleyn who was a cousin of his executed second wife, the church of England decided to legalize all first-cousin marriages (Bittles 2009). A dispute about the possible adverse effects of first-cousin marriage in Great Britain in the late 19th century was settled when George Darwin (son of Charles Darwin who married his first cousin Emma Wedgwood) produced evidence that the negative effects of first-cousin marriages were likely small (Darwin 1875; cited in Bittles 2009). Indeed we find that in outbred populations, the contribution of recessive inheritance to intellectual disability appears of modest importance (Gilissen et al. 2014; Deciphering Developmental Disorders Study 2015).

6   De Novo Mutations in Human Genetic Disease

Mutations are sudden changes in the genetic material. Mutations are the fuel of evolution, and therefore beneficial to the adaptation of species to changes in their environment (Crow 2000). Nonetheless, most mutations are either of no effect to the individual (neutral) or detrimental to health and survival. Truly beneficial mutations are clearly exceptionally rare events. Mutations can involve chromosomes, parts of chromosomes, or single genes.

Chromosome abnormalities have been recognized as a cause of severe intellectual disability for many years at least since the discovery of trisomy 21 in Down syndrome 50 years ago. Chromosomal abnormalities are an important cause of severe intellectual disability and explain about 20 % of the total frequency. Techniques for the investigation of chromosomes have become better over time. Still, most individuals with severe intellectual disability have normal chromosomes even when studied by the best available techniques. Patients come from a normal pregnancy, normal birth and from normal families. For these reasons the most common answer to the question why a child has intellectual disability is "I don't know". The possibility to characterize the complete DNA sequence at the single base level by whole genome sequencing has radically changed this situation. It now turns out that most people with a severe intellectual disability do not have abnormalities of whole chromosomes. Some have very small chromosomal changes, but most have an abnormal single gene which has mutated (Gilissen et al. 2014). Similar findings have been reported for autism and schizophrenia but in a lower percentage. Analysis of the affected child and both parents demonstrates that the abnormality has arisen spontaneously in the child by a mutation of a single nucleotide in the DNA. This has important implications since DNA mutations are spread more or less equally across the genome, and occur at a relatively fixed rate of one per 100 million nucleotides per generation. Mutations represent random errors of replication during the formation of our germ cells. Thus, the majority of all instances of severe intellectual disability and a large proportion of other diseases such as autism, schizophrenia and birth defects are due to what seem to be essentially random events (Veltman and Brunner 2012).

7   The Randomness of Mutations

It has now been firmly established that the number of DNA mutations in a newborn child is approximately 100. Of these 100 mutations, on average 1 or 2 hit a gene. Since there are 20,000 genes, the impact of the single gene mutation that every newborn child has will be determined by the nature of the gene that was hit, and by the severity of the mutational event. Both of these factors are random. We may say, that the more we improve our lives, our habits, and our pregnancy care, the more the decision to start a family becomes similar to taking part in a genetic lottery. This comes as no great surprise to most parents. We all know and accept that each pregnancy carries risks. On the other hand, we do want explanations when a severely handicapped child is born. In my experience, the information that a disability is due to a chance event is perceived as good news by parents because it absolves them of feelings of guilt and insufficiency about how they handled their pregnancy.

8   Why Mutations Happen

There are two main causes of new mutations, insufficient DNA repair and random errors during DNA replication. DNA repair is necessary, because the DNA in our cells is under constant attack from external factors that may damage it. External damaging factors include radiation, chemicals, as well as various toxic substances that are generated by the cell itself such as oxygen radicals. To protect our existence, our cells have developed an elaborate system of DNA damage protection and especially DNA repair. This means that the large majority of DNA mutations is immediately corrected and repaired. Our germ cells seem to be especially good at preventing or repairing DNA damage. It was a striking and unexpected result from studies that were performed after the Nagasaki and Hiroshima atom bombs during World War II that there was only limited evidence for an increase of inherited genetic mutations. This is not to say that external factors are not relevant to new mutations. They are obviously very important but at the current level of exposure to noxious influences, they do not seem to be the determining factor whether or not a mutation ensues in a child. In fact, studies of the frequency of new mutations in children suggest a random distribution around the mean of 1–2 gene mutation per newborn individual. The driving force for the generation of new mutations is in the replication of DNA when our germ cells are created. Copying DNA is the essence of creating sperm cells and egg cells. All DNA nucleotides need to be copied with very high fidelity. Viewed like this, it is perhaps surprising that the total number of errors in a newborn is just 100 out of the 3 billion nucleotides of DNA that need to be copied. Mutations are a part of all life.

9   Can We Prevent Mutations?

If we view mutations as copy errors, then we must accept that it will not be easy to prevent them from happening. Consequently, it becomes quite difficult to further reduce the occurrence of severe handicaps and diseases. Once we have minimized the negative influences of DNA damaging substances and radiation, the remaining mutations are due to copy errors that reflect an intrinsic function of our cellular machinery. There may be a practical solution however. We may try to reduce the number of cellular divisions in the germ-line as much as possible. More de novo gene mutations happen during spermatogenesis than during oogenesis. This is because sperm cells continue to copy and then divide over a man's lifetime while the egg cells are already completed by the time a girl is born. In fact, the mutation rate in the child is strongly dependent on the age of the father (Risch et al. 1987; Goriely and Wilkie 2012). While it is probably not practical to try and convince men to have their families young, it is a practical possibility to freeze and store sperm samples at a young age, and then use these later in life. While the impact on an individual may not be immediately apparent, it is clear that if this policy were universally adopted in the face of an increasing age at which men and women start their families, a society could reduce the burden of severe handicaps and autism by a large fraction. Whether this is acceptable or desirable is a different matter and will invite a vigorous societal debate.

10   Accepting Risks

Each pregnancy carries risks and this is a generally accepted fact. Because we cannot prevent mutations from happening, we cannot reduce or eliminate all risk, even if we live healthy lives and provide the best possible pregnancy care.

Ultimately, early detection by prenatal diagnosis may be the only real option if we want to prevent severe handicaps. Whether this is acceptable in the form of universal prenatal diagnosis is again a matter for societal debate. It is clear that such discussions carry tremendous societal, ethical and emotional and even personal connotations and that they cannot be solved from the respective perspectives of biology, medicine or genetics. I believe that such a debate will take place over the coming years. In this respect, it may be instructive to read some of the reactions to a recent paper by cancer expert Vogelstein that suggests that most cases of cancer in Western populations are due to random mutations and that their risk is strongly related to the number of cell divisions per tissue (Tomassetti and Vogelstein 2015a). The authors concluded from their findings, that it is probably more worthwhile for society to try to detect cancers at an early stage than it is for society to invest in cancer prevention. Several commentators objected to this generalization, and partly for good scientific reasons. Nonetheless, the perceived dichotomy between external factors (and inherited predispositions) which we can avoid or ameliorate, and the randomness of mutations which strike from nowhere also seems to have inspired some of these comments. Or as Tomassetti and Vogelstein put it in their response: "Replicative mutations are unavoidable. They are in a sense a side-effect of evolution, which cannot proceed without them. That they play a larger role in cancer than previously believed has important scientific and societal implications." (Tomassetti and Vogelstein 2015b).

All in all, the recent recognition that spontaneous mutations are an important driver of severe illnesses, such as intellectual disability, autism, schizophrenia, and cancer is likely to fuel another nature-nurture debate where random mutation events are contrasted with bad influences from the environment. Nature-nurture debates are never fully solved because the opposing sides are not ready to compromise. Still, such debates are always interesting and instructive, and in the end genome sequencing will provide us with real scientific data to weigh these two respective forces. At the end of the day, we need to come to terms with randomness as an integral part of our biology. This include accepting limits to the extent to which we can and cannot manage our existence.

11   Are Mutations a Necessary Part of Our Existence?

It is often argued that because mutations are the drivers of evolution, we should welcome them as a good thing. In general terms, advantageous mutations may indeed drive improved species adaptation and promote evolution. Nonetheless, since mutations may easily destroy the capacity of the organism to reproduce, there must be an upper limit to the number of random mutations a species can endure. In fact, in humans, the total number of copy errors in a newborn is just 100 out of the 3 billion nucleotides of DNA.

So is there an optimum rate for random mutational events, and how is this determined? First of all, it is clear that the answer to this question varies. In fact, frequency of random mutation can vary 100-fold between species, and each species has its own specific mutation rate. This species-specific mutation rate is not random, as it appears strongly dependent on the size of the genome, with bacteria having the lowest mutation rate and mammals having the highest mutation rate. All this suggests that for each species, there is a relatively constant and likely optimized error-rate of DNA replication.

So if our mutation rate is fixed, why is it what it is? In the absence of a divine plan, we may consider the following possibilities. First, it may be that our current human rate of evolution exactly matches the requirement for adaptation to a changing environment. If this were true then one would expect that there should be some variability of mutation rate within a species over evolutionary time. Simply put: In order to cope with changes in the selection regime, populations should evolve mechanisms that tune the rate of mutation, amongst other things, in order to increase their long-term adaptability (Carja et al. 2014). There is currently not a lot of evidence to support this idea, although it has recently been argued that there are data to support that the rate of human mutation may not be stable over time (Harris 2015).

Another possibility is that the mutation rate is as low as our species can afford. Keeping mutation rates low through high fidelity of DNA replication and reliable repair of mutations, is clearly a strategy that involves considerable cost to the organism. Since resources are limited, there may be a point where it becomes much more rewarding to species overall survival to stop investing in mutation prevention and repair, and rather divert resources and energy to other ways to promote survival and fitness. One weak spot in replication that has not been fixed by evolution, is to do with the defective proofreading capacity of polymerase alpha during replication (Reijns et al. 2012).

12 Conclusion

There may be an inherent tension between the interest of the individual and that of the species it belongs to as to the allowing of randomness. If we go by the "Adapt or die" paradigm, then we need random mutational events to survive as a species. But at the same time such random mutations may kill us before we reproduce. We need a bit of randomness in our existence otherwise our species cannot survive. But we need to dose this randomness very carefully or the resulting chaos will destroy us.

Open Access This chapter is distributed under the terms of the Creative Commons Attribution-Noncommercial 2.5 License (http://creativecommons.org/licenses/by-nc/2.5/) which permits any noncommercial use, distribution, and reproduction in any medium, provided the original author(s) and source are credited. The images or other third party material in this chapter are included in the work's Creative Commons license, unless indicated otherwise in the credit line; if such material is not included in the work's Creative Commons license and the respective action is not permitted by statutory regulation, users will need to obtain permission from the license holder to duplicate, adapt or reproduce the material.

References

Baljet, B., & Oostra, R. J. (1998). Historical aspects of the study of malformations in The Netherlands. *American Journal of Medical Genetics, 77*(2), 91–99.

Becker, R., Keller, T., Wegner, R. D., Neitzel, H., Stumm, M., Knoll, U., et al. (2015). Consanguinity and pregnancy outcomes in a multi-ethnic, metropolitan European population. *Prenatal Diagnosis, 35,* 81–89.

Beckwith, J. B. (2012). Congenital malformations: From superstition to understanding. *Virchows Archiv, 461,* 609–619.

Bittles, A. H. (2009). Commentary: The background and outcomes of the first-cousin marriage controversy in Great Britain. *International Journal of Epidemiology, 38,* 1453–1458.

Bondeson, J. (1997). *A Cabinet of Medical Curiosities.* Ithaca: Cornell University Press.

Brent, R. L., & Fawcett, L. B. (2007). Developmental toxicology, drugs, and fetal teratogenesis. In E. Albert Reece & J. C. Hobbins (Ed.), *Clinical obstetrics: The fetus and the mother.* London: Wiley-Blackwell.

Carja, O., Liberman, U., & Feldman, M. W. (2014). Evolution in changing environments: Modifiers of mutation, recombination, and migration. *Proceeding of National Academy of Science United States of America, 111*(50), 17935–17940.

Crow, J. F. (2000). The origins, patterns and implications of human spontaneous mutation. *Nature Reviews Genetics, 1,* 40–47.

Deciphering Developmental Disorders Study (2015). Large-scale discovery of novel genetic causes of developmental disorders. *Nature, 519,* 223–228.

Darwin, G. H. (1875). Marriages between first cousins in England and Wales and their effects. *Fortnight Review, 24,* 22–41 (2009) (Reprinted in International Journal of Epidemiology *38,* 1429–1439).

Fisher, G. J. (1870). Does maternal mental influence have any constructive or destructive power in the production of malformations or monstrosities at any stage of embryonic development? *American Journal of Insanity, 26*(24), 1–295.

Formijne, A. J. (1915). Het verzien van zwangeren. *Nederlandsch Tijdschrift voor Geneeskunde, 59*(1876–1), 877.

Goriely, A., & Wilkie, A. O. (2012). Paternal age effect mutations and selfish spermatogonial selection: Causes and consequences for human disease. *American Journal of Human Genetics, 2*(90), 175–200.

Gilissen, C., Hehir-Kwa, J. Y., Thung, D. T., van de Vorst, M., van Bon, B. W., Willemsen, M. H., et al. (2014). Genome sequencing identifies major causes of severe intellectual disability. *Nature, 511,* 344–347.

Gregg, N. M. (1947). Congenital defects associated with maternal rubella. *The Australian Hospital, 14*(11), 7–9.

Harris, K. (2015). Evidence for recent, population-specific evolution of the human mutation rate. *Proceedings of the National Academy of Sciences United States of America, 112*(11), 3439–3444.

Lenz, W., & Knapp, K. (1962). Thalidomide embrypopathy. *Deutsche Medizinische Wochenschrift, 15*(87), 1232–1242.

McBride, W. G. (1961). Thalidomide and congenital abnormalities. *Letter to the Editor. The Lancet, 2,* 1358.

Najmabadi, H., Hu, H., Garshasbi, M., Zemojtel, T., Abedini, S. S., Chen, W., et al. (2011). Deep sequencing reveals 50 novel genes for recessive cognitive disorders. *Nature, 478,* 57–63.

Pangas, J. C. (2000). Birth malformations in Babylon and Assyria. *American Journal of Medical Genetics, 91*(4), 318–321.

Reijns, M. A., Rabe, B., Rigby, R. E., Mill, P., Astell, K. R., Lettice, L. A., et al. (2012). Enzymatic removal of ribonucleotides from DNA is essential for mammalian genome integrity and development. *Cell, 149*(5), 1008–1022.

Risch, N., Reich, E. W., Wishnick, M. M., & McCarthy, J. G. (1987). Spontaneous mutation and parental age in humans. *American Journal of Human Genetics, 41*, 218–248.

Romeo, G., & Bittles, A. H. (2014). Consanguinity in the contemporary world. *Human Heredity, 77*, 6–9.

Sheridan, E., Wright, J., Small, N., Corry, P. C., Oddie, S., Whibley, C., et al. (2013). Risk factors for congenital anomaly in a multiethnic birth cohort: An analysis of the born in Bradford study. *The Lancet, 382*, 1350–1359.

Tigay, J. (1997). "'He Begot a Son in His Likeness after His Image' (Gen. 5:3)," in *Tehillah le-Moshe. Biblical and Judaic Studies in Honor of Moshe Greenberg* (pp. 139–147). In M. Cogan, B. L. Eichler & J. H. Tigay (Eds.). Winona Lake, Indiana: Eisenbraun's.

Tomasetti, C., & Vogelstein, B. (2015a). Cancer etiology. Variation in cancer risk among tissues can be explained by the number of stem cell divisions. *Science, 347*, 78–81.

Tomasetti, C., & Vogelstein, B. (2015b). Cancer risk: Role of environment—response. *Science, 347*, 729–731.

Veltman, J. A., & Brunner, H. G. (2012). De novo mutations in human genetic disease. *Nature Reviews Genetics, 13*(8), 565–575.

Warkany, J., & Kalter, H. (1962, December). Maternal impressions and congenital malformations. *Plastic and Reconstructive Surgery and the Transplantation Bulletin, 30*, 628–637.

Warkany, J. (1959). Congenital malformations in the past. *Journal of Chronic Diseases, 10*(2), 84–96.

Warkany, J., & Nelson, R. C. (1940). Appearance of skeletal abnormalities in the offspring of rats reared on a deficient diet. *Science, 92*(2391), 383–384.

Chance, Variation and the Nature of Causality in Ecological Communities

Hans de Kroon and Eelke Jongejans

It's a coincidence, it is not scientific.
Major Walsh in **Close Encounters of the Third Kind**
(Steven Spielberg, director; Columbia Pictures, 1977)

Abstract Chance is pervasive in nature. Erratic events such as storms and fires can cause major damage to an ecosystem. Rare successful long distance dispersal events like a viable seed landing in just the right habitat can form the stepping stone for range expansion of a plant species. Illustrated with two examples we argue that in ecology chance events are scale-dependent. We show how random stochastic variation in species interactions may result in relative stability at a higher community level. In other systems the reverse may take place, in which deterministic interactions result in unpredictable chaotic dynamics. Analysing the processes and dynamics at these different scales has led to an increasing mechanistic understanding of the variation in ecological communities in space and time. Unambiguous identification of cause and effect relations from this work is of the greatest importance, as many ecosystems in the world are not amenable to experimentation. This work should form the scientific basis for identifying the threats to ecosystems and defining proper conservation and mitigation measures.

1 Introduction: The Fascinating Complexity of Ecosystems

One central problem in ecology is understanding the distribution of species and individuals over the landscape. Species are organised in ecological communities of producers (generally plants) and consumers (herbivores and predators) that change across the landscape. Climatic factors and soil and water conditions may change already over short distances and vary with altitude vs latitude. Adaptations determine the distribution of species over gradients. Beautiful nature documentaries

H. de Kroon (✉) · E. Jongejans
Faculty of Science, Departments of Plant Ecology and Animal Ecology,
Institute for Water and Wetland Research, Radboud University, Nijmegen, The Netherlands
e-mail: h.dekroon@science.ru.nl

© The Author(s) 2016
K. Landsman and E. van Wolde (eds.), *The Challenge of Chance*,
The Frontiers Collection, DOI 10.1007/978-3-319-26300-7_11

often focus on these amazing characteristics of species by which they are able to cope with the challenges of their often extreme environment.

An important goal in ecology as a scientific discipline is understanding the driving forces, or underlying mechanisms responsible for differences in distribution of species in their natural habitats. However, how much mechanistic understanding is possible in ecosystems in which chance processes play a prominent role? For example, long-distance migration of plants is subject to the coincidental combination of a rare event like a heavy storm taking place at exactly the right time and place carrying ripe seeds to another location with exactly the right conditions for establishment. Such events are hardly tractable in the field. How much does chance affect distributions of individuals and interactions between them, and how much do actual ecological and evolutionary processes contribute? The question is important not only for the progress of ecology as a scientific discipline, but also for understanding the impact of disturbances (such as global climate change) and formulating appropriate interventions to mitigate such disturbances.

Illustrated with two examples, we argue that coincidence, variation and causality are scale-dependent. With scale we imply the extent of time and space (McGill 2010), but also the hierarchical structure of ecosystems, in which individuals of the same species are grouped within populations, populations of different species are grouped within structured ecological communities, which in turn interact with abiotic conditions regarding climate, soil and water within the landscape. Patterns expressed at one scale are driven by causal processes at a smaller underlying scale. Vice versa, random processes at a lower scale sum up to measurable variation at a higher scale. As a result, rare events at a lower scale can be predicted at a higher scale, e.g. under which climatic conditions new soybean rusts from South America can be expected in North America (Isard et al. 2011).

In the first example we give an overview on current theory explaining the maintenance of species diversity, with emphasis on hyper-diverse communities such as tropical forests. The complexity is daunting. Such communities exist of hundreds, sometimes thousands of species, each with their own characteristics, ecological relationships with other species and responses to environmental conditions. What are the stabilizing forces preventing species from extinction? How important are stabilizing forces preserving these communities relative to chance effects?

In the second example we investigate trends of populations of species over time, as they are influenced by deterministic and stochastic factors. Studying such trends is of great importance for the conservation of species and the prediction of the impact of environmental stress factors. We will see that in the currently fragmented landscapes all over the world, populations are ruled by chance events affecting the extinction of small populations, as well as rare long-distant dispersal events. How can we gain control over this stochasticity, in order to understand and predict how environmental factors influence the viability of populations? Answering this question very much depends on the spatial scale at which we are studying processes, from a very local patch of suitable habitat where a limited number of individuals survive and reproduce, to a region (such as an entire country) harbouring numerous of these small populations that together form a predictive trend.

2 Example 1: Explaining the Maintenance of Species Diversity

2.1 Coexistence Theory: Species Differ in Niches

One of the most long-lasting questions in ecology is to explain how so many organisms can coexist in a community. Hyper-diverse communities (Box 1) are tantalizing examples challenging a long-standing paradigm in ecology. The 'competitive exclusion principle', formulated by the Russian biologist Georgy Gauss in the 1930s and based on laboratory experiments with *Paramecium* (unicellular ciliated protozoa), states that two species can only stably coexist if they differentiate in their fundamental requirements such as their food source (their 'niches'). Early on, the competitive exclusion principle received theoretical support from population models (Lotka 1920; Volterra 1928). The Lotka-Volterra equations describing the competition between two species and defining the conditions for competitive exclusion or stable coexistence can be considered the $E = mc^2$ of community ecology. They still form the cornerstone of modern coexistence theory (Chesson 2000).

BOX 1: the dazzling number of species that coexist in natural plant communities

Plant communities can harbour very high numbers of species in a given area. Communities differ in composition and complexity. Why are some communities more species-rich than others? Why are the tropical forests overwhelmingly species-rich and why are these levels of biodiversity not reached in the temperate or boreal forests?

The differences are enormous. Current estimates suggest the minimum number of tropical tree species in the world between 40,000 and 53,000 (Slik et al. 2015). The number of tree species described globally for temperate forest is only 1166 (Latham and Ricklefs 1993). Also at smaller scale, tropical forests can contain an astonishing number of species. For example, a single hectare (approximately one baseball field or two soccer fields) can support hundreds of species of trees (record: 942 species of trees per hectare in Amazonian Ecuador; Wilson et al. 2012). An area of the size of a fraction of the Radboud University campus would thus harbour approximately the same the number of species as the entire temperate forest region in the world including Europe, Asia and North-America (4.2 million km^2). How did this diversity arise, and how is the diversity maintained?

Extensively managed, relatively nutrient-poor grasslands all over the world are another example of extreme plant species richness, albeit at a smaller scale (Wilson et al. 2012). Per m^2 such communities can have dozens of species of higher plants (record: 89 species of vascular plants m^{-2} for a mountain grassland in Argentina). How is it possible that such communities are maintained, without a few superior species starting to dominate and drive competitively inferior species to extinction?

In its essence, coexistence theory states that different species in a community can stably coexist if a species gains a competitive advantage over the resident community when that species becomes rare. Consequently, if for whatever reason a species gets low in numbers, its population will bounce back resulting in coexistence. Such frequency-dependent population dynamics is only possible if species differ in their requirements to complete their life cycle, i.e. differ in their niches. Niche differences can arise from many different characteristics, with food source as the most obvious one. Differences in reproduction (the 'regeneration niche', i.e. requirements for nesting in birds, micro-climatic conditions for seedlings to establish) and natural enemies (herbivores and diseases) also constitute niche axes. What is crucial is that these differences in requirements result in differences in survival and reproductive schemes between species. Consequently, if a species becomes rare in the community, its species-specific niche 'opens up', resulting in positive population growth rates and recovery. For all species combined, niche differences are a necessary stabilising force.

2.2 Natural Enemies as Niche-Axes: The Janzen-Connell Hypothesis

This theory sparked a quest for important niche axes, particularly for plants for which niche differences are hard to conceive because plants all have essentially the same nutritional requirements. In the early nineteen-seventies, Daniel Janzen and Joseph Connell invoked natural enemies in explaining the high tropical tree diversity (Condit 1995; Connell 1971; Janzen 1970). Co-evolution between the feeding adaptations of herbivores and the defence mechanisms of plants has led to sophisticated adaptations resulting in numerous specific plant-herbivore relationships. In what is now known as the Janzen-Connell hypothesis, they argued that each plant species accumulates its own specific community of natural enemies, which is more detrimental to this particular plant species than to other species in the community. Consequently, offspring of a tree has relatively lower chances for establishment close to the parent tree than at further distance where other tree species are growing. A given species therefore cannot stand its local ground forever, but, Janzen and Connell hypothesized, if this is a reciprocal process applying to all species in the forest it will lead to stable coexistence of large numbers of tree species. Nearly fifty years after its conception, the Janzen-Connell hypothesis has only gained in importance in community ecology (Comita et al. 2014). Attention has shifted from aboveground herbivores to belowground enemies (root feeding larvae, worms and insects, and particularly soil pathogenic micro-organisms including bacteria, fungi and other unicellular organisms) (Mangan et al. 2010). Janzen-Connell effects are also considered an important driving force in species-rich grasslands (Bever et al. 2012; de Kroon et al. 2012; Petermann et al. 2008).

But how can the Janzen-Connell hypothesis result in stable coexistence of many species? The reason is that Janzen-Connell effects balance competition between trees of the same species relative to competition between trees of different species. If a tree species becomes dominant it will be at disadvantage relative to other species in the community. Conversely, when a species gets rare in the community, its seedlings may easily find suitable areas for growth and the species will gain in competitive ability and abundance. Such frequency-dependent responses are consistent with the general theory of species coexistence (Chesson 2000).

2.3 Coexistence Through Intransitive Competition and Rock-Paper-Scissor Games

The frequency-dependent population dynamics expected from Janzen-Connell effects have been compared to intransitive competitive networks. Intransitive competition implies that competitive abilities of different species cannot be ranked along a hierarchy in which a single species gains competitive dominance (Buss 1980; Gilpin 1975). An example of an intransitive competitive network is when species A is superior to species B, and B is superior to C, but C is superior to species A (A > B; B > C; C > A). Models of spatial distributions of individuals and populations suggest that intransitive competitive relationships result in coexisting populations (Laird and Schamp 2006). They contrast with transitive or hierarchical competition, as is predicted from competition for essential resources. Transitive competition implies that when species A is competitively superior to species B, and B to species C, species A will also win in competition with species C (A > B; B > C; => A > C). Indeed, if species compete for a limited soil resource, it is inconceivable that a superior competitor A that wins in competition with a species B will lose in competition with species C that is itself competitively inferior to species B (de Kroon et al. 2012; Lankau 2010). However, species-specific belowground interactions may result in species interactions consistent with intransitive competitive networks (Lankau et al. 2011).

Intransitive competition is also referred to as a rock-paper-scissor game. This concept has been developed as an example of game theory (Nowak and Sigmund 2004; Weitzel and Rosenkranz, this volume), and it is easily conceivable because you play it with your kids. Rock wins from scissor, scissor wins from paper, paper wins from rock, there is not a single winner. If each player makes one of the three choices completely at random, independently from what the other players chose or have chosen, and is therefore unable to predict any of the other players, all players will end with a similar proportion of runs won. The trick is to predict a pattern of choice with your competitors, which is never completely random but is for instance based on previous choices.

2.4 Tests with Bacterial Communities: Rock-Paper-Scissor Dynamics Is not Enough for Stable Coexistence

How can a rock–paper–scissor game based on unpredictable interactions among the players result in stable (i.e. predictable) coexistence of the players themselves? It should be realized that in ecology the play is implemented somewhat differently from human politics and economics. In a sizable ecological community the number of players (i.e. individuals) is almost infinite, and it is assumed that individuals of a given species share a common strategy (i.e. they behave either as rock, paper, or scissor and do not change). So the game is played among species, but over numerous of individuals interacting with each other, and the unpredictability lies in the random encounters of individuals of different species at one place and time. Does intransitivity in competitive relationships among species indeed result in stable coexistence?

With life spans of hundreds of years, this question is hard to investigate empirically for tropical trees. Bacteria, however, with well-defined characteristics and a short generation time, have shown to be interesting model systems for testing questions of species coexistence (Hol and Dekker 2014; Kerr et al. 2002). And the answer is no, intransitivity all by itself does not necessarily lead to coexistence. Kerr et al. (2002) carried out a compelling test with the model bacteria *Escherichia coli*. Three strains that together constitute an intransitive competitive network were grown in mixtures. One strain produces the toxin colicin (colicinogenic cells C), to which other strains are either sensitive (sensitive cells S) or immune (resistant cells R). Colicin production, and to a lesser degree immunity, is costly to the cells and compromises the growth rates of the C and R strains. As a result, this C- S- R system satisfies the rock–paper–scissor relationship because C can displace S (because C kills S), R can displace C (because R has a growth-rate advantage) and S can displace R (because S has a growth-rate advantage) (Kerr et al. 2002). When all three strains were mixed together in liquid medium in a flask and shaken, maximizing interactions among the bacteria, the S strain was rapidly driven to extinction by C, and subsequently C was outcompeted by R due to the higher growth rate of the latter.

Why do the strains fail to show coexistence, although the conditions of intransitivity are met? Theoretical models predict that if the competitive relationships are transitive, but the dynamics lead to different strengths of interaction, fluctuations may appear and one strategy may eventually win (Nowak and Sigmund 2004). In the case of *E. coli*, the toxin produced by the C strain is immediately lethal to the S strain, but the growth advantage of the S strain over the other two strains results in slower replacement. Consequently, when the communities interact 'globally' in a shaken flask, S is eradicated quickly and the intransitive network collapses.

Interestingly, Kerr et al. (2002) showed that when interactions among the three types of bacteria were not global but local, at the surface of a petri dish filled with agar, coexistence did occur. Here encounters were no longer random because the strains occurred in patches (clumps) and interacted at the borders where patches of different strains met. Pictures of the petri dishes over time show that strains were chasing each other, as predicted by the rock–paper–scissor relationships, resulting in a pattern of clumps that is changing all the time. Kerr et al. (2002) concluded that "balanced chasing in a spatially structured, non-hierarchical community may result in the maintenance of diversity". Spatial structure where individuals with similar strategy clump and limited dispersal may give much better chances for the maintenance of diversity than well-mixed populations (Nowak and Sigmund 2004).

2.5 Global Stability in Hyper-Diverse Plant Communities Consistent with Local Rock-Paper-Scissor Dynamics

To what extent is this coexistence mechanism also to be expected for hyper-diverse communities of tropical forest or grassland? An increasing number of studies have shown that competitive relationships between plant species are not hierarchical but intransitive (de Kroon et al. 2012; Soliveres et al. 2015). Dynamics are obviously orders of magnitude slower than in bacterial communities but, interestingly, long-term observations have revealed patchy dynamics of grassland species that are reminiscent of those of the bacterial patches in petri dishes. In grasslands at the slopes of the Krkonoše mountains in the Czech Republic, species form patches that change position all the time because individuals die and are replaced by other species at one location while they appear at locations nearby as a result of clonal expansion or germination (Herben et al. 1993a). The replacements of species are largely random and to some degree intransitive and thus resemble the "balanced chasing" described above (Herben et al. 1997; Herben et al. 1993b). The consequence is a very stable community as a whole, while paradoxically numerous replacements take place at a local scale. Indeed, a 10 × 10 m area of these grasslands would look very much the same with the same species co-occurring year after year, but if one could make a movie of the area over decades the species would be seen to move around like ants in an ants nest. For tropical forest, the Janzen-Connell hypothesis predicts very similar spatial dynamics. Although there is a huge number of trees in a tropical forest, a particular tree will interact most with its direct neighbours, while dispersal distances are limited in most cases. However, replacements are even slower than in grasslands. The oldest forest dynamics plot at Barro Colorado Island in Panama (where all trees of all species over 50 ha are mapped; Condit 1995) was laid out in the early 1980s and is still way too young for a demonstration of such dynamics.

2.6 Global Stability Through Neutral Dynamics if Species Are Demographically Equal

The growth rate differences between the bacterial strains of *E. coli* in the study of Kerr et al. (2002) hinge upon an important element in current coexistence theory, i.e. fitness differences between species (Chesson 2000). As explained above, the key stabilizing force in communities are the niche differences, the fundamental requirements between species affecting their population growth rates. Because of these differences, individuals of the same species have stronger competitive interactions than individuals of different species. In other words, species limit their own growth more than they limit the growth of other species, i.e. intraspecific competition is stronger than interspecific competition. The degree to which intra- and interspecific competition coefficients must be different for stable coexistence to occur depends on the average fitness differences between species (Adler et al. 2007; Chesson and Kuang 2008). Fitness in this context refers to the relative degree of adaptedness of that species to the conditions of the habitat in the absence of niche differences (Chesson 2011). Species with higher fitness develop higher population growth rates and will win the competition. Niche differences balance the fitness differences stabilizing the dynamics and providing the conditions for coexistence.

Coexistence will also be promoted not only if niche differences are larger, but also if fitness differences are smaller. Spatial structure is one way to reduce the effects of fitness differences, as the examples of *E. coli* and grasslands illustrate. Indeed, competitive replacement may be slowed down considerably if competitors are growing in patches, with interspecific interactions taking place at the border. In such cases, most of the interactions in the community are between members of the same species, rather than between members of different species, favoring the weaker competitor (Stoll and Prati 2001). Spatial structure and limited dispersal, both prominent in most ecosystems, are thus forces that equalize fitness differences between species and promote coexistence.

The most radical and influential idea with respect to the maintenance of species diversity has been the formulation of 'neutral theory' (Hubbell 2001). Neutral theory essentially states that all species are demographically equal, i.e. that fitness differences do not exist. Species may differ in numerous traits and resource requirements, but do not translate into a net difference in population growth rates between the species under prevailing habitat conditions. In this theory, there are no niche differences and there is no stable coexistence, but there is an opportunity for long-time co-occurrence. Individuals do compete for limited resources but competitive strengths are similar for all individuals, irrespective of species identity, resulting in replacements driven by chance. All these random replacements add up to neutral dynamics in which populations of different species are maintained if the community is of sufficient size. Population numbers do fluctuate as a result of stochastic (e.g. climatic) influences and are not buffered against extinction. Particularly in communities of limited size, such as in fragmented habitats, populations have a chance of going extinct under neutral dynamics due to demographic stochasticity (as further explained in de second example below).

When published in 2001, Hubbell's book was a provocation to the many community ecologists studying niche axes in their communities. Fifteen years later, neutral theory has been shown to predict community characteristics surprisingly well (Rosindell et al. 2011). It has been accepted as an inherent element in community theory and not only in the tropical forest that formed its inspiration. Indeed, also in species-rich grasslands much of the competitive interactions appear largely equivalent among species (Law et al. 1997), resulting in random replacements at the local scale, and near neutral dynamics at the larger scale of the community, despite the fact that we know that these species differ in their ecological requirements.

2.7 Coexistence Mechanisms May Result in Unpredictable Dynamics

While consensus is now emerging about how neutral and niche processes together govern community dynamics, we should realize that they do not necessarily result in overall community stability. Ground-breaking mathematical theory developed in the 1970's by Robert May, showed that simple differential equations with density-dependent feedback could result in very complex non-linear dynamics of the system with chaotic fluctuations that are by definition unpredictable (May 1976; Weitzel and Rosenkranz, this volume). Work of Jef Huisman and co-workers has demonstrated that such dynamics bear relevance for the coexistence of many species of plankton in aquatic ecosystems. Huisman and Weissing (1999) showed theoretically that a well-parameterised competition model, describing the competition for limiting resources (such as nitrogen, phosphorus, silicon, light and inorganic carbon) gave rise to coexistence of many different species of plankton. The number of species coexisting was much more than expected on the basis of their differences in resource requirements as predicted by the competitive exclusion principle (i.e. their niche differences alone). The model predicted that the species displaced each other in a cyclic fashion, giving rise to oscillations and chaotic dynamics, reminiscent of the non-linear dynamics described by May (1976). Despite sometimes major fluctuations in species numbers, when a species became dominant, other species at low numbers bounced back, though at different rates. Later empirical work eloquently demonstrated that these predictions may actually occur in reality (Benincà et al. 2008). In a laboratory setup with a plankton community in tanks many different species coexisted for a period up to 2300 days, covering a couple of hundred generations. They did so while showing population size fluctuations over several orders of magnitude that were essentially unpredictable, yet leading to overall persistence of the community. Note that the conditions were stable and there was no spatial structure within the tanks, all plankton species interacted with random encounters (as in the flasks of Kerr et al. 2002). The chaotic fluctuations were attributed to different species interactions in the planktonic food web, giving rise to different periodicities in the ups and downs of the various populations. It is important to realize that these dynamics arise from inherent deterministic relationships, i.e. from competition and predation process (Huisman and Weissing 1999).

Recently, Benincà et al. (2015) demonstrated for the first time that near-chaotic dynamics may occur in the world outside. In an intertidal ecosystem in New Zealand, cyclic replacements occur of barnacles colonizing bare rock, brown alga overgrowing barnacles, mussels settling on barnacles and algae, giving rise to bare rock as the mussels eventually detach. The cyclic fluctuations of the populations in this community become irregular through the seasonality of the system and the time needed for the establishment of each of the species. Interestingly, the cyclic replacement is reminiscent of rock-paper-scissor interactions (Benincà et al. 2015) but the dynamics do not resemble those of the *E. coli* strains described above (Kerr et al. 2002). The reason might be that with sessile stages but global dispersal of recruits the intertidal community is neither global (leading to the dominance of a single species in the well-mixed flasks), nor completely spatially structured (resulting in balanced chasing at patch edges and overall stability).

2.8   Conclusion: The Interplay Between Scale-Dependent Predictable and Unpredictable Patterns in Community Dynamics

We have seen three archetypes of long-term persistence of complex communities with many different species, all operating in a very different way. In the case of niche differences between species (as in the case of rock–paper–scissor games), stabilizing forces may be strong and promote stability. Such communities are likely spatio-temporally structured with many predictable species replacements at a local scale. Neutral theory confronts us with the situation that numerous random interactions at a local scale sum up to stochastic dynamics at a global scale. However, in a relatively stable environment, competitive equivalence among species, as assumed in neutral theory, can lead to overall predictable community patterns (long-term co-occurrence). Finally, the plankton example shows the reverse, where predictable species replacements result in chaotic dynamics of a community which is as unpredictable as the weather (Benincà et al. 2008).

These contrasts feed the uneasy relationship that ecology has with deterministic versus stochastic processes underlying the structure of ecological communities (Bjørnstad 2015; Chase and Myers 2011; Vellend et al. 2014). Is any fundamental process really stochastic, or does it always have an underlying deterministic origin? These examples indicate that the scale of processes must be considered, with many different possibilities. Truly stochastic (i.e. unpredictable) dynamics may find their origin in underlying deterministic processes, while stochastic interactions at a local scale may give rise to relatively stable (and hence predictable) dynamics at a global scale. Each scale calls for its own methodologies describing processes and dynamics (Vellend et al. 2014). It is important to understand the interplay of deterministic vs. stochastic processes with the scales of organisation, as it affects the nature of causality in ecology. We will see this now in our second example.

3 Example 2: Understanding Species Population Trends

3.1 Species Survive in Metapopulations with a High Incidence of Chance Effects

In the current fragmented landscape, in almost all regions of the world, species are distributed in discrete populations. At some point in time every single population started with a colonization event of an area where the species did not occur at that moment, and after a while (which may take days, years, or centuries) every population will go extinct when the last individual has died or left the area. The Finnish ecologist Illka Hanski coined the collection of discrete populations in the landscape a metapopulation (Hanski 1998). The success of a species is defined by its metapopulation dynamics, determined by the processes of immigration and extinction. The classical example, and the one where metapopulations were first described, is the Baltic Sea with numerous islands for the coast of Finland inhabited by butterflies that form small populations on the islands, connected by dispersal of the butterflies between the islands. Hanski et al. (1994) demonstrated how the islands are colonized and vacated by the butterflies, leading to continuous changes in island occupation, but with remarkable stability of the metapopulation of butterflies in the archipelago.

The success or decline of species is described by the fluctuations in the metapopulation as a whole. These global fluctuations are the accumulation of numerous extinction and colonisation events at the local scale. As favourable habitat may be small (as is the case for many of the islands in the Finnish archipelago) chance effects play a large role. Any local population can be subject to accidental hazards such as a fire or storm leading to local extinction. The local unpredictable variation is referred to as environmental stochasticity (Lande 1993). In addition, demographic stochasticity exists whereby small populations can simply go extinct due to chance effects (Lande 1993). The smaller the population, the bigger the chance that all individuals leave the local habitat, die, or fail to reproduce, partly due to difficulties of finding mates and/or inbreeding depression. Colonisation events, whereby unoccupied habitat is discovered by animals from elsewhere, also have a high element of chance (unless dispersers actively search for empty habitat).

3.2   Farmland Birds: Understanding Population Trends

An urgent question nowadays is to what extent climate change, pollution, or changes in land use form a threat to species of plants and animals (Bowler et al. 2015). And if so, can negative effects be mitigated? But how can we understand these threats if species survive in metapopulations, where the dynamics are the cumulative effects of numerous events where chance plays a major role? How can we control this unpredictable variation, identify and quantify causes of decline, and suggest measures to counteract the threats? As in our previous example, we deal with local processes scaling up to patterns at larger scales, and stochastic local dynamics leading to global stability.

We illustrate these questions with an example of the status of farmland birds in the Netherlands and its association with neonicotinoid insecticides in the environment (Hallmann et al. 2014). Farmland bird species in the Western landscape have been decimated over the last hundred years as a result of agricultural intensification, increased fertilisation and pesticide use. Many bird populations are now confined to small suitable habitat patches like hedgerows, or along water bodies, often consisting of only few bird territories. Bird territories in the Netherlands are counted in a standardized way by thousands of volunteers under auspices of Sovon, the Dutch Centre for Field Ornithology. Bringing this information together we know that, over recent decades, some bird species in the Netherlands show signs of recovery, albeit not to the same extent throughout the country. Zooming in, we typically see a patchwork across the Netherlands with local areas in which bird populations increase, interspersed with areas with negative population trends. Investigating 15 insectivorous farmland bird species, Hallmann et al. (2014) demonstrated that these differences correlated with the local concentration of imidacloprid in the surface water. Imidacloprid is the most widely use neonicotinoid, a group of insecticides introduced in the mid-nineties. Neonicotinoids specifically target the nervous system of insects and are therefore highly lethal to invertebrates and much less so to vertebrates like humans or birds. Applied as seed coating or by spraying, major quantities of insecticides are not taken up by the crop to be protected but wash out in the soils and accumulate in the waterways. It is in these environments that larval stages of insects grow up, which form the bulk food of the bird species investigated. Hallmann et al. therefore hypothesised that local populations of these bird species, relying on insects particularly in the breeding season, are in decline due to food shortage.

Hallmann et al.'s study is essentially correlative, showing that local bird population trends are more likely to be negative when local imidacloprid concentrations in the surface water are higher. The case for imidacloprid as a cause of bird decline was reinforced in two ways. First, local bird trends over the last ten years were also correlated to local changes in land use that were known to affect bird populations, including changes in levels of nitrogen use, and changes in areas of maize, winter cereals, fallow land, greenhouses and alike. In this analysis imidacloprid stood out as by far the best explanatory variable for local bird population trends. Second, the correlation between imidacloprid and bird trends was much weaker and non-significant when bird trends in the same areas were considered before the introduction of the compound, suggesting that the correlation was not due to some unknown explanatory variable already present before the introduction of neonicotinoids.

3.3 Mastering Chance Effects at Local Scale to Explain Global Trends

Nevertheless, the Hallmann et al. (2014) study remains correlative, raising the pertinent question: Are neonicotinoids the causal factor for the trends in bird decline? In the worldwide press attention that the study received, this question came up repeatedly. In a response to Hallman et al., the Dutch Minister of the Environment expressed her concern about the effects of neonicotinoids on the environment, but also said that changes in legislation could not be based on a correlative study only. In the strict sense, demonstrating causality would require experiments (Dively et al. 2015; Godfray et al. 2014; Rundlof et al. 2015) but for birds this would entail field trials at such a large spatial scale and over such a long time that they are impossible to conduct. We therefore must find other ways to demonstrate causality. This is very difficult because, as explained above, mechanisms of colonisation, growth and extinction of the populations operate at a small spatial scale, where chance processes and other local factors may prevail. Indeed, even the interpretation of correlations at the global (metapopulation) scale in the context of local effects can lead to apparent contradictions of the kind we also see in epidemiology.

Emerging global trends such as the bird trends in relation to imidacloprid may not necessarily be seen locally everywhere. At the scale of the Netherlands, the trend was quite strong: bird populations declined with a rate of 3.5 % per year where local imidacloprid concentrations exceeded 20 ng/l in the surface water. While this trend was highly significant, part of the variation in bird trends remained unexplained, and appeared as noise around the correlation. Some of this variation may be due to chance effects related to environmental and stochastic stochasticity. But, how then can a population escape the hazards of neonicotinoids at a local scale? This remains to be investigated, but as an example the following situation can easily be envisioned. Imagine an agricultural field adjacent to the dunes, which are important nature reserves for birds in the Netherlands. If the insecticide pollutes the local soil and waterways and deteriorates local insect populations, birds in the vicinity may be little affected as they can forage in the dune area with its own hydrology, not affected by the pollution nearby. Such a population may be healthy and increasing, while the overall analysis predicts a declining population at this location with a high imidacloprid load. However, given enough data at the scale of the Netherlands, the 'noise' of such local situations will not mask a general correlation between bird trends and insecticide concentrations.

Throughout ecology such unexplained variation is rather common: ecosystems are influenced by a very large number of factors at the same time. Ecologists therefore strive to quantify more and more of the environmental factors that influence e.g. bird trends, or at least try to determine what the most important explanatory factors are. But even when we understand a fair amount of what drives local populations (50 % of variation explained is certainly a glass well filled for ecologists), the mechanisms (and chance events) of dispersal between areas remain even more elusive. Recent attempts to model dispersal focus more and more on all aspects of dispersal: what local conditions lead to the initiation of dispersal, how far does an individual travel through a landscape, what makes him/her stop, and what is the impact at the destination? Combining spatial population models and mechanistic 'gravity' models of dispersal (reviewed in Jongejans et al. 2015) might therefore be a way forward in linking local processes and global trends, although unexplained variation will remain (due to chance but also due to unmeasured factors).

3.4   Understanding Causality: A Comparison with Epidemiology

Ecology is not the only field of research that struggles with the reconciliation of processes and patterns that are apparent (or not) at different scales. Similar difficulties in relating global trends to local effects appear in epidemiology. Epidemiological studies investigate large groups of people and identify the factors that may explain the differences in health. Well-known examples include how smoking is related to human mortality (Banks et al. 2015; Thun et al. 2013), and how obesity (and diet) is related to an increased chance of diseases and premature death (references in Würtz et al. 2014). However, as we all know, 'local' exceptions to these convincing 'global' trends exist. Many people are acquainted with a person reaching old age in relatively good health while smoking like a chimney. We can ask a similar question as with the bird example: how can a person escape the hazard of smoking? Detailed investigation of the medical condition and the habits of such a person could perhaps give indications. If unsuccessful we consider the health of the person as a happy coincidence, but this would not dismiss the hazards of smoking in general. Still, the global trends between smoking and mortality remain essentially a correlation.

As with our bird study, proper controlled experiments on humans regarding effects of unhealthy diet or smoking are considered unethical and cannot be done. Epidemiology has long recognized the weakness of the correlative nature of its investigations for identifying the biological and behavioural causes of disease (Galea et al. 2010; Hill 1965; March and Susser 2006). Still, it is possible to discover cause-effect relationships from purely observational data (Pearl and Verma

1991). Current methodologies attempt to solve this problem by capturing the complexity of the many risk factors for human health in complex systems modelling (Galea et al. 2010). Another way forward is advanced statistics on large datasets together with targeted measurements. Human metabolic profiles are important health indicators and an important question is to what extent they are related to Body Mass Index (BMI) or to a genetic disposition for adiposity, even for people in the non-obese range. By using a Mendelian randomisation framework, Würtz et al. (2014) have recently shown how health indicators can be causally related to BMI, by incorporating a gene score for predisposition to elevated BMI. This statistical method is designed to infer causality in observational studies while taking possible confounding effects into account.

Similar to the situation in epidemiology, observational studies in ecology are often the only field instruments to gauge the 'health' of populations of species of conservation interest. Statistical and modelling techniques are only beginning to be applied to master the explained against the unexplained variation, and to quantify causation taking into account the many confounding factors operating at different scales. Further work in this direction is required to convince public and decision makers that effects are real and require appropriate action. The history of the implementation of smoke restrictions indicates that this is not an easy trajectory.

4 Epilogue

Chance is pervasive in ecological systems. However, chance events never come alone. They may have a solid deterministic origin or they may scale up to predictable variation. In many cases stochasticity and determinism are closely intertwined through the different scales of biological organisation. The scale-dependency of cause and effect has an uneasy relationship with the scale-dependency of stochasticity and determinism. We should take this relationship into account when defining causality, as well as in what can be considered as scientific proof. Methods are to be developed to quantify causal relationships and distinguish them from random effects and confounding factors.

Acknowledgements This essay was inspired by the excellent work of our graduate students Caspar Hallmann, Marloes Hendriks, Janneke Ravenek and Marco Visser. We are grateful to our collaborators Ruud Foppen, Tomas Herben, Liesje Mommer, Helene Muller-Landau, Chris van Turnhout, Wim van der Putten and Joe Wright for enlightening discussion over the years.

Open Access This chapter is distributed under the terms of the Creative Commons Attribution-Noncommercial 2.5 License (http://creativecommons.org/licenses/by-nc/2.5/) which permits any noncommercial use, distribution, and reproduction in any medium, provided the original author(s) and source are credited. The images or other third party material in this chapter are included in the work's Creative Commons license, unless indicated otherwise in the credit line; if such material is not included in the work's Creative Commons license and the respective action is not permitted by statutory regulation, users will need to obtain permission from the license holder to duplicate, adapt or reproduce the material.

References

Adler, P. B., HilleRisLambers, J., & Levine, J. M. (2007). A niche for neutrality. *Ecology Letters, 10*, 95–104.

Banks, E., Joshy, G., Weber, M. F., Liu, B., Grenfell, R., Egger, S., et al. (2015). Tobacco smoking and all-cause mortality in a large Australian cohort study: Findings from a mature epidemic with current low smoking prevalence. *BMC Medicine, 13*, 281.

Benincà, E., Ballantine, B., Ellner, S. P., & Huisman, J. (2015). Species fluctuations sustained by a cyclic succession at the edge of chaos. *Proceedings of the National Academy of Sciences of the United States of America, 112*, 6389–6394.

Benincà, E., Huisman, J., Heerkloss, R., Jöhnk, K. D., Branco, P., Van Nes, E. H., et al. (2008). Chaos in a long-term experiment with a plankton community. *Nature, 451*, 822–825.

Bever, J. D., Platt, T. G., & Morton, E. R. (2012). Microbial population and community dynamics on plant roots and their feedbacks on plant communities. *Annual Review of Microbiology, 66*, 265–283.

Bjørnstad, O. N. (2015). Nonlinearity and chaos in ecological dynamics revisited. *Proceedings of the National Academy of Sciences of the United States of America, 112*, 6252–6253.

Bowler, D., Haase, P., Kröncke, I., Tackenberg, O., Bauer, H., Brendel, C., et al. (2015). A cross-taxon analysis of the impact of climate change on abundance trends in central Europe. *Biological Conservation, 187*, 41–50.

Buss, L. W. (1980). Competitive intransitivity and size-frequency distributions of interacting populations. *Proceedings of the National Academy of Sciences of the United States of America-Biological Sciences, 77*, 5355–5359.

Chase, J. M., & Myers, J. A. (2011). Disentangling the importance of ecological niches from stochastic processes across scales. *Philosophical Transactions of the Royal Society B: Biological Sciences, 366*, 2351–2363.

Chesson, P. (2000). Mechanisms of maintenance of species diversity. *Annual Review of Ecology and Systematics, 31*, 343–366.

Chesson, P. (2011). Ecological niches and diversity maintenance. In I. Pavlinov (Ed.), *Research in Biodiversity—Models and Applications*. Rijeka: In Tech.

Chesson, P., & Kuang, J. J. (2008). The interaction between predation and competition. *Nature, 456*, 235–238.

Comita, L. S., Queenborough, S. A., Murphy, S. J., Eck, J. L., Xu, K., Krishnadas, M., et al. (2014). Testing predictions of the Janzen-Connell hypothesis: A meta-analysis of experimental evidence for distance- and density-dependent seed and seedling survival. *Journal of Ecology, 102*, 845–856.

Condit, R. (1995). Research in large, long-term tropical forest plots. *Trends in Ecology & Evolution, 10*, 18–22.

Connell, J. H. (1971). On the role of natural enemies in preventing competitive exclusion in some marine animals and in rain forest trees. In P. J. den Boer & G. R. Gradwell (Eds.), *Dynamics of populations* (pp. 298–312). Wageningen: Center for Agricultural Publishing and Documentation.

de Kroon, H., Hendriks, M., van Ruijven, J., Ravenek, J., Padilla, F. M., Jongejans, E., et al. (2012). Root responses to nutrients and soil biota: Drivers of species coexistence and ecosystem productivity. *Journal of Ecology, 100*, 6–15.

Dively, G. P., Embrey, M. S., Kamel, A., Hawthorne, D. J., & Pettis, J. S. (2015). Assessment of chronic sublethal effects of imidacloprid on honey bee colony health. *PLoS One, 10*, e0118748.

Galea, S., Riddle, M., & Kaplan, G. A. (2010). Causal thinking and complex system approaches in epidemiology. *International Journal of Epidemiology, 39*, 97–106.

Gilpin, M. E. (1975). Limit cycles in competition communities. *American Naturalist, 109*, 51–60.

Godfray, H. C. J., Blacquiere, T., Field, L. M., Hails, R. S., Petrokofsky, G., Potts, S. G., et al. (2014). A restatement of the natural science evidence base concerning neonicotinoid insecticides and insect pollinators. *Proceedings of the Royal Society of London B: Biological Sciences, 281*(1786), 20140558.

Hallmann, C. A., Foppen, R. P. B., van Turnhout, C. A. M., de Kroon, H., & Jongejans, E. (2014). Declines in insectivorous birds are associated with high neonicotinoid concentrations. *Nature, 511*, 341–343.

Hanski, I. (1998). Metapopulation dynamics. *Nature, 396*, 41–49.

Hanski, I., Kuussaari, M., & Nieminen, M. (1994). Metapopulation structure and migration in the butterfly *Melitaea cinxia*. *Ecology, 75*, 747–762.

Herben, T., Krahulec, F., Hadincova, V., & Kovarova, M. (1993a). Small-scale spatial dynamics of plant species in a grassland community over six years. *Journal of Vegetation Science, 4*, 171–178.

Herben, T., Krahulec, F., Hadincová, V., Pechacková, S., & Kovárová, M. (1997). Fine-scale spatio-temporal patterns in a mountain grassland: Do species replace each other in a regular fashion? *Journal of Vegetation Science, 8*, 217–224.

Herben, T., Krahulec, F., Hadincová, V., & Skálová, H. (1993b). Small-scale variability as a mechanism for large-scale stability in mountain grasslands. *Journal of Vegetation Science, 4*, 163–170.

Hill, A. B. (1965). Environment and disease—association or causation? *Proceedings of the Royal Society of Medicine-London, 58*, 295–300.

Hol, F. J., & Dekker, C. (2014). Zooming into see the bigger picture: Microfluidic and nanofabrication tools to study bacteria. *Science, 346*, 1251821.

Hubbell, S. P. (2001). *The unified neutral theory of biodiversity and biogeography*. Princeton: Princeton University Press.

Huisman, J., & Weissing, F. J. (1999). Biodiversity of plankton by species oscillations and chaos. *Nature, 402*, 407–410.

Isard, S., Barnes, C., Hambleton, S., Ariatti, A., Russo, J., Tenuta, A., et al. (2011). Predicting soybean rust incursions into the North American continental interior using crop monitoring, spore trapping, and aerobiological modeling. *Plant Disease, 95*, 1346–1357.

Janzen, D. H. (1970). Herbivores and the number of tree species in tropical forests. *American Naturalist, 104*, 501–528.

Jongejans, E., Skarpaas, O., Ferrari, M. J., Long, E. S., Dauer, J. T., Schwarz, C. M., et al. (2015). A unifying gravity framework for dispersal. *Theoretical Ecology, 8*, 207–223.

Kerr, B., Riley, M. A., Feldman, M. W., & Bohannan, B. J. M. (2002). Local dispersal promotes biodiversity in a real-life game of rock-paper-scissors. *Nature, 418*, 171–174.

Laird, R. A., & Schamp, B. S. (2006). Competitive intransitivity promotes species coexistence. *American Naturalist, 168*, 182–193.

Lande, R. (1993). Risks of population extinction from demographic and environmental stochasticity and random catastrophes. *American Naturalist, 142*, 911–927.

Lankau, R. A. (2010). Intraspecific variation in allelochemistry determines an invasive species' impact on soil microbial communities. *Oecologia, 165*, 453–463.

Lankau, R. A., Wheeler, E., Bennett, A. E., & Strauss, S. Y. (2011). Plant-soil feedbacks contribute to an intransitive competitive network that promotes both genetic and species diversity. *Journal of Ecology, 99*, 176–185.

Latham, R. E., & Ricklefs, R. E. (1993). Global patterns of tree species richness in moist forests— Energy-diversity theory does not account for variation in species richness. *Oikos, 67*, 325–333.

Law, R., Herben, T., & Dieckmann, U. (1997). Non manipulative estimates of competition coefficients in a montane grassland community. *Journal of Ecology, 85*, 505–517.

Lotka, A. J. (1920). Analytical note on certain rhythmic relations in organic systems. *Proceedings of the National Academy of Sciences of the United States of America, 6*, 410–415.

Mangan, S. A., Schnitzer, S. A., Herre, E. A., Mack, K. M. L., Valencia, M. C., Sanchez, E. I., et al. (2010). Negative plant–soil feedback predicts tree-species relative abundance in a tropical forest. *Nature, 466*, 752–755.

March, D., & Susser, E. (2006). The eco- in eco-epidemiology. *International Journal of Epidemiology, 35*, 1379–1383.

May, R. M. (1976). Simple mathematical models with very complicated dynamics. *Nature, 261*, 459–467.

McGill, B. J. (2010). Matters of scale. *Science, 328*, 575–576.

Nowak, M. A., & Sigmund, K. (2004). Evolutionary dynamics of biological games. *Science, 303*, 793–799.

Pearl, J., & Verma, T. (1991). A theory of inferred causation. In J. Allen, R. Fikes & E. Sandewall (Eds.), *Knowledge Representation and Reasoning: Proceedings of the Second International Conference*. San Mateo, CA: Morgan Kaufmann.

Petermann, J. S., Fergus, A. J. F., Turnbull, L. A., & Schmid, B. (2008). Janzen-Connell effects are widespread and strong enough to maintain diversity in grasslands. *Ecology, 89*, 2399–2406.

Rosindell, J., Hubbell, S. P., & Etienne, R. S. (2011). The unified neutral theory of biodiversity and biogeography at age ten. *Trends in Ecology & Evolution, 26*, 340–348.

Rundlof, M., Andersson, G. K., Bommarco, R., Fries, I., Hederstrom, V., Herbertsson, L., et al. (2015). Seed coating with a neonicotinoid insecticide negatively affects wild bees. *Nature, 521*, 77–80.

Slik, J. F., Arroyo-Rodríguez, V., Aiba, S.-I., Alvarez-Loayza, P., Alves, L. F., Ashton, P., et al. (2015). An estimate of the number of tropical tree species. *Proceedings of the National Academy of Sciences of the United States of America*, 201423147.

Soliveres, S., Maestre, F. T., Ulrich, W., Manning, P., Boch, S., Bowker, M. A., et al. (2015). Intransitive competition is widespread in plant communities and maintains their species richness. *Ecology Letters, 18*(8), 790–808.

Stoll, P., & Prati, D. (2001). Intraspecific aggregation alters competitive interactions in experimental plant communities. *Ecology, 82*, 319–327.

Thun, M. J., Carter, B. D., Feskanich, D., Freedman, N. D., Prentice, R., Lopez, A. D., et al. (2013). 50-year trends in smoking-related mortality in the United States. *New England Journal of Medicine, 368*, 351–364.

Vellend, M., Srivastava, D. S., Anderson, K. M., Brown, C. D., Jankowski, J. E., Kleynhans, E. J., et al. (2014). Assessing the relative importance of neutral stochasticity in ecological communities. *Oikos, 123*, 1420–1430.

Volterra, V. (1928). Variations and fluctuations of the number of individuals in animal species living together. *Journal du Conseil. Conseil Permanent International pour l'Exploration de la Mer, 3*, 3–51.

Wilson, J. B., Peet, R. K., Dengler, J., & Pärtel, M. (2012). Plant species richness: The world records. *Journal of Vegetation Science, 23*, 796–802.

Würtz, P., Wang, Q., Kangas, A. J., Richmond, R. C., Skarp, J., Tiainen, M., et al. (2014). Metabolic signatures of adiposity in young adults: Mendelian randomization analysis and effects of weight change. *PLoS Medicine, 11*, e1001765.

The Size of History: Coincidence, Counterfactuality and Questions of Scale in History

Olivier Hekster

Abstract Historians try to interpret the past by analysing patterns in human behaviour in earlier periods of time. In some ways, that excludes 'coincidence' as a mode of interpretation. Most historians view coincidences as closely related events that lack causal relationship. That type of coincidence does not fit into a historical narrative, because historians tend to focus on causality, action, and consequence. This is noticeably linked to questions of historical scale: the choice for the scale of a specific narrative decides whether certain events are coincidental to the history which is being described, or causal factors within that history. This relation between historical coincidence and the scale of writing history is at the centre of this contribution. It focuses on different trends in writing history, and analyses the possibilities to use 'coincidence' as an interpretative tool in each of them. In doing so, this article discusses counterfactual historical analysis ('what if history'), determinist views of history and their relation to speculative philosophy of history, 'cliodynamics' and 'big history'. It ultimately argues for historical accounts that pay attention to both the large processes that are likely to lead to certain trajectories, and the enormous number of micro-causes that triggered the events as they happened. Coincidence might fall outside of the analysis of (macro-) historians who are looking for a comprehensive view of historical processes, but could still play a proper role in thinking about historical trajectories.

1 Introduction: Coincidence and Comparisons

Historians try to interpret the past. They do so by analysing patterns in human behaviour in earlier periods of time. In some ways, that excludes 'coincidence' as an interpretative tool. The word 'coincidence', after all, derives from the Latin *cum incidere*, which means 'happening together'. Most historians take this to imply that coincidences are events that seem closely related but lack a causal relationship. They

O. Hekster (✉)
Faculty of Arts, Radboud University, Nijmegen, The Netherlands
e-mail: o.hekster@let.ru.nl

© The Author(s) 2016
K. Landsman and E. van Wolde (eds.), *The Challenge of Chance*,
The Frontiers Collection, DOI 10.1007/978-3-319-26300-7_12

215

just happen to occur at (roughly) the same time, or in a similar mode. This makes them, at first sight, less suitable as a historical explanatory notion.[1] This rather negative meaning of the word 'coincidence' in a historical context is best exemplified by the among American historians infamous list of 'creepy coincidences' between Abraham Lincoln and John F. Kennedy. The coincidences in question range from 'the names Lincoln and Kennedy both contain seven letters' to 'both presidents were shot in the head on a Friday before a major holiday'. The challenge to professional historians is then to explain this set of coincidences. Since the sort of similarities assembled on the list can be drawn between any historical figures, these 'coincidences' only amass to 'pseudo-historical demonstrations of data-massaging' (Kern and Brown 2001, p. 534). What lacks is any attempt to explain the (historical) significance of these similarities, meaning that they do not usefully contribute to modes of interpreting past events.

This is not to say that analysing similarities in itself is something that is condemned by modern historians. In fact, comparing and contrasting individuals or societies which seem similar in their structural set-out but have only limited inter-relation is one of the methodological starting points of comparative history. By comparing given individuals, institutes or areas, core aspects of specific phenomena can become clearer. But it is tacitly acknowledged that the historical relevance of these similarities needs to be explained, in terms of cause and effect, for them to be useful as a historical term. This can be explicitly contrasted with 'coincidental' similarities, as is clear from a recent overview of the progress of historical scholarship by J.H. Elliott. Elliott reflects on his earlier comparative study of the two seventeenth-century statesmen cardinal Richelieu and the Count Duke of Olivares, in which he noted several similarities between the two, one of the more trivial being that both men were third sons of noble fathers, finding employment in the service of a monarch. 'But', he observes, 'this simple fact suggests one of the problems inherent in comparative history. Are we dealing here with coincidence, or does the similarity point to some wider consideration that is worthy of note?' (Elliott 2012, p. 180; Elliott 1991). Coincidences in themselves, clearly, are not deemed noteworthy.

Both Olivares and Richelieu received an early education for an ecclesiastical career. The death of Olivares' older brother meant that Olivares became head of the family and needed to marry. Unwillingness of Richelieu's older brother Alphonse to become bishop meant that Richelieu would make a career in the church. Elliott describes the developments in their respective families as 'an initial coincidence' but one that 'went on to create a number of similarities', which he does consider 'of obvious significance', such as the influence of neo-Stoic philosophy on their actions as statesmen. He contrasts these to 'other coincidences' that 'may lead to a dead end', although even such 'chance resemblances' can help sharpening the mind (Elliott 2012, p. 180–1). There is clear differentiation here between (if no real definition of) what is deemed relevant for explaining actions and events, and what is

---

[1]For the development leading to the exclusive use of *cum incidere* as coincidence, see Vogt (2011), pp. 43–66 followed by a discussion of the Greek *Tyche*.

deemed merely anecdotal, and without explanatory power. Something is either 'a mere coincidence' or, more often encountered within historical scholarship, 'not a coincidence', and therefore relevant.

Coincidence, in this reading, becomes something that does not fit into a historical narrative, mainly because historians, as a very recent *Manifesto* formulated: 'focus on the question of *how*: Who did the changing, and how can we be sure they were the agents? These analytics of causality, action, and consequence make them specialists in noticing the change around us' (Guldi and Armitage 2014, p. 14). Notions that happened to happen but (the agency of which) cannot be explained fall outside of such a view of historical analysis. If (human) agency is a core interest, the question of coincidence becomes linked to discussions about the randomness of human behaviour or about free will (see the articles of Weitzel and Rosenkranz, Thijssen and Loy, and Van Elk, Friston and Bekkering in this volume), neither of which cohere easily with the attention (or competence) of most historians.

There is also an aspect of scale involved. Whether a historian interprets something as coincidental to, or a central focus of causality of, events depends somewhat on the historical scope of his or her analysis. The education of historical figures, for example, may explain their historical actions. Few biographers would formulate their protagonist's education as 'mere coincidence', but for a historian who is interested in larger historical trajectories, the education that a single individual happened to have had, and which may explain specific actions that were part of a wider chain of events, has much less explanatory force. In such a reading the definition of historical coincidence becomes almost a matter of taste. To be more precise: the choice for the scale of a specific historical narrative decides whether certain events are coincidental to the history which is being described, or causal factors within that history. This relation between historical coincidence and the scale of writing history is at the focus of this contribution.

2 Contingency, Causality and Counterfactuality

As must be clear from the above, there is certainly some space for coincidence as a factor in historical explanation. There is a number of semi-synonyms that feature somewhat more frequently in modern historiography, among which 'accident', 'singular event', 'chance' and 'contingent circumstances' are especially noticeable. The term 'luck' also features, but mainly as a (rather self-effacing) explanation for success in the career trajectory of a particular historian, or as part of cautionary reflections on the ways in which processes which are difficult to influence or trace have influenced historical discoveries. (DuPont Chandler 2004; McClellan III 2005; Cushing 1992). Luck, it seems, can be allowed to have influenced individual historians, but is not explicitly acknowledged as a factor in the historical processes that these historians investigate.

The situation seems different when 'contingency' is mentioned. There is a long-standing strand of historical research that deals with 'turning points', and at some of these turning points, in a famous dictum by the great British historian George M. Trevelyan 'history failed to turn'.[2] The notion that events may have '*failed to turn* as expected is to admit the role of contingency, the element of chance —randomness' (Post 2009). This has ultimately led to a flourishing (though not uncontested) strand of historical analysis which is known as 'virtual history' or 'counterfactual history'. The question posed by historians in this form of research is 'what would have happened if events at a certain turning point would have gone marginally differently'. Even before explicit attention to such counterfactuality, this had been an implicit mode of reasoning within scholarly historical works, going back at least to the Roman historian Livy's analysis of how a war between the armies of Alexander the Great and Rome would have played out (*Ab Urbe Condita* 9.17–19; Morello 2002). Perhaps the most famous example of modern implicit historical use of counterfactuality is the statement by the German historian Eduard Meyer in his *Geschichte des Altertums* that if the Persians had beaten the Greeks (especially the Athenians) at the battle of Marathon in 490 BC European culture would have developed entirely differently:

> Das Endergebniss wäre schliesslich doch gewesen, dass eine Kirche und ein durchge-bildetes theologisches System dem griechischen Leben und Denken ihr Joch aufgelegt und jede freiere Regung in Fesseln geschlagen hätte, dass auch die neue griechische Kultur so gut wie die orientalischen ein theologisch-religiöses Gepräge erhalten hätte.[3]

This line of arguing was recognised as 'counterfactual' by Max Weber, who started his career as an ancient historian. Shortly after the publication of Meyer's work, he noted how the argument that the development of European history would have shifted dramatically with a different outcome of the battle at Marathon rested on a series of assumptions. Meyer argued that a Persian victory would have blocked the preconditions for Athenian supremacy, and with it (still according to Meyer) the development of democracy and rationality. Weber showed the analogies that stood at the basis of the argument (Persian behaviour elsewhere), and showed the methodological pitfalls within such a line of argument. Meyer' conclusions were less at stake than his methods, as is clear from the title of Weber's 1905 essay that has become justly famous: 'Objective possibility and adequate causation in historical explanation'. In it, Weber questions

> how the attribution of a concrete result to a single 'cause' is … feasible … given that … it is always an infinity of causal factors that brought about the single 'event'.[4]

---

[2]Trevelyan (1918), p. 79. The quote is often contributed to Taylor (1945), p. 71, who describes Germany in the contexts of the revolutions in 1848: 'German history reached its turning-point and failed to turn'.

[3]Meyer (1901), 446: 'The end result would ultimately have been that a Church and a well-developed theological system had brought the Greek life and thought under their yoke, and placed chains on any aspiration to freedom; that the new Greek culture, much like oriental culture, would have had a theological-religious character'.

[4]Weber (1905) with the comments by Ringer (2002). Cf. also Huizinga (1937), p. 137.

The answer is counterfactual reasoning, by which, Weber argued, it becomes possible to conjecturally sort and rank different possible causes. The historian can take various events out of the equation, and then conjecture what the new historical trajectory would be. This makes it possible to see which potential causes would have brought about which effect. Causality and intention are at issue here. The crucial Weberian terms are 'objective probability' and 'adequate causation':

> Where an actual result was brought about by a complex of antecedent conditions that made it 'objectively probable', the 'cause' is termed 'adequate' in relation to the 'effect'. Where a causal factor contributed to a historically interesting aspect of an outcome without being 'adequate' in this sense, it may be considered its 'accidental cause' (Ringer 2002, pp. 165–6).

The 'accidental cause' in the Weberian sense is of course not the same as coincidence, but his line of arguing highlights an awareness that developments in history could have gone differently, and that often events that seemed less directly relevant may have had wide-ranging consequences. Noticeably, in Weber's argument a single concrete cause leads to one concrete effect, making it particularly important for him to distinguish between causal and coincidental relations.

There is, in fact, an element of the counterfactual in almost all assumptions of causality. If we argue that one event causes a second event, we also argue that the second event would not have taken place (or at least not in that way) without the first. In that sense, there is little difference between 'true' causes and coincidences, as the coincidence still caused later events to happen. It is here that scale becomes an issue. Historians accept events as a 'true' cause if they fit the size of their narrative interest. If they are too 'small', they are often waylaid as coincidental, if they are too 'big' they become background context. Whether a cause is deemed of the right size depends on what the historian is trying to (re)construct. Within that reconstruction, there are protagonists, turning points and contexts. The scale of the historical trajectory decides which events are too small (coincidental) or too big (background). In that sense, a historical coincidence is a cause (in the counterfactual sense) which is without explanatory force within the chosen narrative framework.

This is an important background to so-called 'what if' history, which tends to focus on causes that seem too small for the historical trajectory which is being analysed—and therefore seem to suggest that coincidence plays a major part in history. Such attention for (single) events that could have gone either way has had numerous advocates in the 20th century. One of the earliest examples of what ultimately grew to be a proper genre of counterfactional history is a collection of essays by professional historians called *If It Had Happened Otherwise*, which best claim to fame is the inclusion of an essay by Winston Churchill on what would have happened if Lee had won the Battle of Gettysburg.[5] For much of the remainder of the 20th century these essays were mainly influential on writers of fiction, lying at the base of the genre of 'alternate history/reality'.

---

[5]Squire (1931). Cf. Hacker and Chamberlain (1981), with an overview of early examples of counterfactual history, and now Gallagher (2011).

Yet, some historians also engaged with this line of thinking, and recognised that the ultimate consequence of it would be that 'the course of the world's history depends on accidents'. Notably, the historian and classicist J.B. Bury devoted an essay on contingency, starting with the importance of the shape of Cleopatra's nose. If this nose had not been so attractive, the argument went, Mark Antony would not have been so distracted as to lose the pivotal battle at Actium (31 BC) against the later emperor Augustus. Of course, the 'shape of Cleopatra's nose was rightly conditioned by the causal sequence of her heredity', but since this causal link had nothing to do with causal links determining the politics of Ancient Rome, the 'collision' of these two unrelated sequences, in Bury's view, boiled down to chance: 'the valuable collision of two or more independent chains of causes'. He refines this definition by differentiating between 'pure' and 'mixed' contingencies:

> If Napoleon at an early stage in his career had been killed by a meteorite, that would have been the purest of pure contingencies ... The meteorite was completely disinterested in his death.[6]

This emphasis on the role of chance on historical processes led to what has been coined the 'accidental view of history', which opposed notion of historical inevitability (Berenson 1952, 88). It was embraced by, amongst others, the philosopher Isaiah Berlin. In his essay on 'Historical inevitability', he argued strongly for recognition of the role of accidents in history, which he placed in opposition to the (in his thoughts) determinist philosophies of history of Marx and Hegel, to which we will return later in this essay. According to Berlin, explaining human behaviour in terms of causality denied free will, and falsely absolved historians of the task to morally evaluate historical actors: 'to assess degrees of their responsibility, to attribute this or that consequence to their free decision, to set them up as examples or deterrents, to seek to derive lessons from their lives'. Recognising that there are accidents that influence the course of events, and that only these, *'force majeure*—being unavoidable—are necessarily outside the category of responsibility and consequently beyond the bounds of criticism'.[7]

This view of history, and of the role of the historian, was famously opposed by the historian and diplomat E.H. Carr, who took issue with 'Cleopatra's Nose' in his influential *What is History*. Carr explicitly argued against those who 'pointed out the absurdity of failing to recognise the role of accident in history', singling out Berlin whom he blames for talking nonsense (conceding that he did so 'in an engaging and attractive way') and for flogging 'this very dead horse back into a semblance of life'. Berlin's (and Karl Popper's) opposition to a deterministic outlook in history, with their emphasis on chance, was in Carr's view little more than 'a parlour game with the might-have-beens of history' (Carr 1961, pp. 92–99). The role of the historian is to recognise patterns of historical significance:

---

[6]Bury (1916/1930), pp. 60 (course of history), 61 (chance), 62 (heredity), 67 (Napoleon).
[7]Berlin (1954/2012), pp. 119–190, citing Berenson (p. 119), discussing the importance of moral judgement (pp. 140–142), and discussing *force majeure* (p. 146).

Just as from the infinite ocean of facts the historian selects those which are significant for his purpose, so from the multiplicity of sequences of cause and effect he extracts those, and only those, which are historically significant; and the standard of historical significance is his ability to fit them into his pattern of rational explanation and interpretation. Other sequences of cause and effect have to be rejected as accidental, not because the relation between cause and effect is different, but because the sequence itself is irrelevant. The historian can do nothing with it; it is not amenable to rational interpretation, and has no meaning either for the past or the present (Carr 1961, p. 105).

For a long time, Carr's criticisms made 'what-if-history' suspect for serious scholars. But counterfactual constructions of historical developments have blossomed as an academic branch of writing history since the publication of Ferguson (1997). Ferguson, like Bury and especially Berlin before him, ultimately tries to take distance from deterministic views of historical processes (Marxism prime amongst them) and aims to show how a limited set of changes, in which contingent factors would have changed crucial outcomes, would have resulted in an almost unrecognisable historical trajectory; a 'virtual history' for the period 1646–1996.

There have, almost inevitably, been counter-reactions, most noticeably by Richard Evans (2014), written at least partly to explain his 'initial, somewhat allergic reaction' (p. xvi) to counterfactuals. Evans argues amongst other that the impression that history could have developed in a radically different way by a combination of minor moments of chance will challenge interest in what really happened, and 'allows historians to rewrite history according to their present-day political purposes and prejudices' (p. 63). Moreover, Evans argues, counterfactual thinking does not, in fact, function well

as a vehicle for overcoming "determinism", in the sense of the prioritization of larger historical forces over smaller, personal, chance, and contingent events and circumstances (Evans 2014, p. 104).

By selecting a finite number of 'alternative outcomes that were at least plausible', counterfactual studies, according to Evans, highlight the importance of forces beyond individual control. Noticeably, Evans' criticism of these 'altered pasts' does not focus on the existence and importance of chance events as such within the functioning of history, but on the assumption that one minor alteration in the historical timeline would cascade towards major changes over a prolonged period of time. In Evans' view, history is not necessarily fixed, but historians should not have the liberty to adapt it as they see fit (Evans 2014, p. 46).

Criticism on counterfactual history is not quite the same as criticism on counterfactual thinking within historical analyses, which is much less disputed. Indeed, the above-mentioned *History Manifesto* sees historical expertise in counterfactual thinking as an important asset for answering questions of sustainability, and possibly other problems that are acute in modern society (Guldi and Armitage 2014, pp. 31–4, with Booth 2003 and Thompson 2010). Such a use of counterfactual history partly closes the gap between counterfactual history and counterfactual analysis. The latter aims to draw clear distinctions between causal and coincidental relations, but does not suggest alternative trajectories. It also tries to apply counterfactual analysis on events that are of the 'right' scale.

'What if' history clearly does suggest alternative trajectories, but the suggestion in the *Manifesto* seems to lean more towards using counterfactuality as a mode of drawing distinctions. There is, in any case, a burgeoning of recent literature discussing the importance of counterfactual history, in which the plausibility of the selection of historical changes is placed at the fore. If historical alternatives are grounded in probability they are deemed to be 'good reasoning'. If not, they are not (Bunzl 2004, p. 845). Ultimately, coincidence in one form of the other lies at the basis of these 'imagined' historical developments. Something in the past needed to have gone differently. But the emphasis seems less on the importance of that 'coincidence' than on the plausibility of the counterfactual consideration: how likely is it that something would have gone differently. History may not be predetermined, but randomness is still kept in check. Coincidence has its role, but is not used as an interpretive tool.

3 Coincidence and the Construction of a Clear Course of History

As stated above, at least some of the historians promoting counterfactual thinking, and emphasising the importance of contingency in an 'accidental view of history', did so in reaction to a determinist perspective. One of the 'determinist' views that leaves least scope for chance is the one that sees historical events as the result of divine providence. As formulated in the late-nineteenth century: 'History, when written rightly, is but a record of Providence; and he who would read history rightly, must read it with his eyes constantly on God' (Read 1862, p. 4). Very few historians, none taken seriously, would now present an academic historical analysis in such stark terms. Yet, there is still a noticeable absence for coincidence as a real factor in interpreting history among the various approaches that look at the larger schemes of history, ranging from speculative philosophy of history to macro-historical explanations and long-term causes.

Speculative philosophy of history is in a way similar to Read's notion of 'a record of Providence'. The ultimate trajectory is considered a given, and the process is then interpreted in historical terms. Thus, Hegelian dialectics assume that history follows a specific trajectory, through 'thesis' (historical developments) and the resulting 'antithesis' (historical reaction) to synthesis until history fulfils itself. In that sense: 'world history exhibits nothing other than the plan of providence' (Hegel 1807/1977). Hegel, as opposed to Read, might still be able to accept that chance occurrences were part of history, and will have influenced the lives of historical individuals over the course of time. These occurrences, however, were not significant in the greater scheme of things. They do not fit the size of the framework in which he places his historical narrative. Hegel's large-scale path of history is inevitable, and though it allows for coincidences to have happened, coincidence as a concept is excluded as a relevant mode of interpretation.

In fact, all forms of history run some risk of taking a determinist point of view in that too often an historical argument is geared towards the known outcome of a historical process. In the famous dictum by Schlegel: 'Der Historiker ist ein rückwärts gekehrter Prophet'. In its extreme form, this way of working leads to so-called 'Whig history' in which the present is seen as the inevitable (and desirable) outcome of historical progress. Such a presupposition results in loosing track of roads not taken or dead ends.[8] The risk of falling in that methodological trap seems to be larger when the scope of the historical argument is wider. Applied at a cosmic scale, this appears to be at the basis of the so-called 'Fine Tuning Argument', debunked by Landsman in his contribution to this volume.

There have of course been (and still are) nuanced, and influential, interpretations of history by historians that focus on the structuring factors of human behaviour through the ages. Best known, and a school of thought to which the likes of Popper and Berlin reacted, is Marxism. Interestingly enough, Marxism is not prominent in current discussions on determinism in analytical philosophy. It does, however, feature with near inevitability in debates regarding macro-historical scholarship (Adcock 2007, p. 351–2). Following Popper's criticisms, Marxism (much like the 'speculative philosophies' of the likes of Hegel or Spengler) is sometime described as a form of 'historicism', which is then presented as an inflexible method of predicting the 'future course of human history'. The term 'historicism' is, however, more commonly associated with the notions of the famous nineteenth-century historians Ranke and Humboldt, in which the historical notions are in continuous flux. Historical ideas and historical developments define the nature of institutions or states and are the historicists object of study (Ankersmit 1995, p. 143; Vogt 2011, pp. 367–92 and 495–502). To avoid confusion, this chapter will avoid the massive debates about the different sorts of historicism, and formulate the argument in a context of determinism and macro-history.

It is important to note the difference in the deterministic aspects of Marxist scholarship, and the absolute 'hand of god', and slightly less absolute Hegelian dialectics, with which this section opened. Marxist views of history, in a simplified form, assert that there is an inevitable sequence within societies ('social formations') which depends on inter linkage between 'modes of productions' (economic systems) and internal conflicts ('contradictions') (Burke 1992, pp. 141–4). This certainly implies economic determinism; the notion that economic causes determine human events, or at least have a primary role in explaining outcomes. Such mono-causal explanations of historical events are looked at with suspicion by most modern historians, which partly explains the bogey-man role Marxism has come to play for those arguing against determinist points of view. Does Marxist theory also imply historical determinism? Much, as always, depends on definition:

> When unpacked more fully this belief [historical determinism] is usually interpreted as a substantive claim that the aggregate course of major historical events traces a structured process of successive changes unfolding inevitably through the events of the past, present, and onto the future (Adcock 2007, pp. 352–4).

---

[8]F. V. Schlegel, *Athenaeum* 1798.2, 20: 'The historian is a prophet looking backwards'. Whig history: Mayr (1990). The term 'Whig history' was coined by Butterfield (1931). A recent contribution on coincidence in history, discussed further below, stresses the importance for historians to keep an open mind regarding the course history did *not* take: Nijhuis (2003).

This excludes the possibility that coincidence influences the ultimate—inevitable —course of history.

Marxism's certainty that civilization will progress from primitive communism to the rise of private property and the development of an aristocracy, and then through feudalism and capitalism to true socialism/communism fits that definition of historical inevitability. Yet Marxism leaves scope for (temporary) developments that do not fit its pattern of history, such as the 'refeudilization' of Spain and Italy, and for individual actions or exogenous explanations to decide the pace if not the course of history. Coincidence, in these contexts, is possible and may influence the course of history. Within a smaller scale of analysis, 'coincidence' is deemed acceptable. But it does not influence the outcome of the large-scale historical process. As an explanatory notion it is irrelevant for the clear course of historical developments that are at issue in Marxist (historical) thinking (Burke 1992, pp. 141–2).

Is the exclusion of coincidence as a 'historical tool' from such large views of history inevitable? Elsewhere in this volume, De Kroon and Jongejans set out how, when looking at the complexities of ecosystems, the scale of organisation is a crucial factor in distinguishing between developments that appear unpredictable or determined. Possibly, the same applies to the study of historical 'systems'. Popper's and Berlin's sustained attack on Hegel and Marx, combined with the economic determinism in Marxist theory and its resulting mono-causality, has made the mere mention of historical determinism suspect. Yet, already half a century ago Ernst Nagel countered many of Berlin's objections to determinism as such. The central philosophical premise of determinism—every human event is the effect of antecedent causes—does not, he held, necessarily imply a claim about inevitably structured patterns in history as a whole (Nagel 1966). In fact, historical narratives can easily be construed as a sequence of events, in which specific actions were the trigger for subsequent events. Tracing a 'series of causal mechanics' can form the basis of relevant historical explanations (Hedström and Swedberg 1998). With relative frequency, recent attempts at macro-historical explanations of the past have been criticised 'for displaying unjustified determinism'. Yet macro-history (or even determinism) is not in itself incompatible with explanations that include causal complexity or choice. Like we saw in this contribution so far, and was made clear in the contribution of De Kroon and Jongejans, much depends upon the level at which we are analysing (or narrating) the historical system:

> Truly stochastic (i.e. unpredictable) dynamics may find their origin in underlying deterministic processes, while stochastic interactions at a local scale may give rise to relatively stable (and hence predictable) dynamics at a global scale.

Even contingency, it seems, can be included in a macro-historical approach, if contingency effectively means contingent upon causes that cannot be explained by current theory (Adcock 2007, p. 347; pp. 354–5). So it seems that there is theoretically nothing that blocks a macro-historical approach that incorporates contingency, if not perhaps coincidence, as an interpretive framework.

Indeed, the *History Manifesto*, which, as we have already seen, advocated the use of counterfactual thinking among historians, also makes a bold statement for the return of 'grand sweeps of history', emphasising the importance of larger historical models—Braudel's *longue durée*—for modern society. Yet it warns that the timescale should still be such 'that historians can do what they what they do best: comparing different kinds of data side by side'. Only then can multiple causality be stressed, showing that 'the reality of natural laws' nor 'the predominance of pattern' is ultimately decisive. Individuals can choose (Guldi and Armitage 2014, pp. 52–60). Historical events may be decided by such complex combinations of causes that 'contingency' should be incorporated in an historical analysis. Still, looking at historical developments in the longest-term perspective raises the risks of losing sight of contingency. As famously set out by A.J.P. Taylor: 'every road accident is caused, in the last resort, by the invention of the internal combustion engine and by men's desire to get from one place to another... But a motorist, charged with dangerous driving, would be ill-advised if he pleaded the existence of motor cars as his sole defence' (Taylor 1961, pp. 102–3, paraphrased by Guldi and Armitage 2014, p. 57).

4   Coincidence, Big History and Accidental Cause

Scale, it seems, is a relevant if not necessarily decisive factor in the extent to which coincidence can be included in an analysis of historical processes. At the same time, we have seen the *Manifesto*'s plea to reclaim the telling of large historical narratives as a skill for professional historians. That attempt has much to do with the influence of discussions about human history by non-historians. Prime among them is Jared Diamond. Diamond (2005) is mentioned by Guldi and Armitage (2014), 57 as 'a gripping account of the fates of societies stricken by plague, mixing archaeological evidence with the history of species extinction and ethnic deracination', though it is simultaneously noted that the work lacks the engagement with detail that characterises the works of some recent historians. More influential than Diamond (2005) is Diamond (1997).[9] It is a key example of a successful account of long-term historical processes. But it is not written by a historian, nor does it refer to writings by historians. The time-scale, also, is wholly different. Where the largest scope of a macro-historical analysis by a historian focuses on 3,000 years, Diamond's 'big history' covers more than 13,000 years.[10]

At the core of Diamond's book lies the theory that the main reason for Eurasian dominance in history has to do with environmental advantages. Physical geography

---

[9]This book has amassed prizes (amongst which the Pulitzer and the Aventis Prize), sold in enormous numbers, and was filmed by National Geographic. It has become so well-known that Mitt Romney attempted to use the book to support one of his claims in the 2012 American presidential campaign, which led to a published reaction by Diamond in the New York Times, followed by a host of tweets and articles: http://www.nytimes.com/2012/08/02/opinion/mitt-romneys-search-for-simple-answers.html, visited on 20.3.2015.

[10]The term 'Big history' was coined by Christian (2004).

underlies historical developments. In Diamond's own words, European historical dominance resulted from

> accidents of geography and biogeography—in particular to the continent's different areas, axes, and suites of wild plant and animal species. This is, the different historical trajectories of Africa and Europe stem ultimately from differences in real estate (Diamond 1997, p. 401).[11]

The book is well written and an extremely rewarding read, but its very strong reading of historical processes and near-negation of human agency has led to continuous critical reactions. One returning feature in that criticism is Diamond's exclusion of cultural autonomy. The environmental historian J. R. McNeill formulated his objections to this emphatically, in an exchange of views with Diamond in the *New York Review of Books* (May 15, 1997)

> Much more powerfully than any other species, [humans] change the environment around us; and have done so ever since our ancestors began to control fire and to use tools. Learned behavior, channeled along innumerable different paths by divergent cultures, is what allows us to do so. Human beings do indeed often "approach limits imposed by environmental constraints" only to find a way to overcome and escape those constraints, as the history of technology repeatedly illustrates.[12]

A second, related, point of criticism is that of the book's apparently linked geographic and historical determinism:

> At its worst, it develops an argument about human inequality based on a deterministic logic that reduces social relations such as poverty, state violence, and persistent social domination, to inexorable outcomes of geography and environments (Correia 2013, p. 1).

One can discuss the merits of and problems with *Guns, Germs and Steel* at length. For the purposes of this article, however, it is especially interesting to compare the outlook of history in the book to that of Marxism as discussed above, and note the reactions by professional historians. Marx's economic determinism fed a form of historical determinism, leading to critical reactions by historians both because of the mono-causal explanation of the course of history, and because of the absence of human agency. Replace 'economic' with 'geographic' and the outline of argument and reaction is the same for *Guns, Germs and Steel*, again an approach to history written by a non-historian. The comparison goes awry at several levels, but both theories deny coincidence the power to influence the outcome of historical trajectories. Diamond's dialogue with McNeill is telling:

---

[11]Note how, in some ways, Diamond seems to work from a 'first cause', the existence of which is not further explained. The underlying geographies of the world are, in Diamond's own words: 'accidents'. This is, however, a very different mode of using coincidence as an instrument in historical analysis than the one this chapter is interested in.

[12]http://www.nybooks.com/articles/archives/1997/jun/26/guns-germs-and-steel/, visited on 20.3. 2015.

> Yet the emergence of such [literate] societies in Eurasia was no accident. It had long antecedents with clear environmental causes ... over the hundreds of generations of post-Ice Age human history, and over a large continent's thousands of societies, cultural differences become sifted to approach limits imposed by environmental constraints.[13]

Clear courses of history seem to leave little room for coincidence as an interpretative tool—and even less so if human agency is excluded.

5 Coincidence and Cliodynamics

Not all large-scale narratives of human history exclude the importance of agency or coincidence. Yet there is an inherent tension between the drive towards explaining long-term dynamical processes in history and attention to individual human actions. This comes clearly forward in the recently developed school of thought called Cliodynamics, consisting mainly of sociologists with an historical interest. Its agenda is to recognise 'laws of history':

> *Cliodynamics* (from *Clio*, the muse of history, and *dynamics*, the study of temporally varying processes) is the new transdisciplinary area of research at the intersection of historical macrosociology, economic history/cliometrics, mathematical modeling of long-term social processes, and the construction and analysis of historical databases ... ultimately the aim is to discover general principles that explain the functioning and dynamics of actual historical societies.[14]

Unsurprisingly, 'coincidence' does not feature in 'Cliodynamic' scholarship, into which some of the recent scholarship of Jared Diamond might also be included (Diamond and Robinson 2010, with Thomas 2010). What Cliodynamics aims at are general principles, explaining how historical societies developed and functioned. Binding the various scholarship within this new subdiscipline of historical sciences is (1) attention for the general principles that explain the dynamics of societies, that lead to (2) models, often formulated in mathematical terms, which are the confronted with (3) empirical content. The data from historical societies are used to develop general patterns, and test the accuracy of assumption from the models. Data from (other) historical societies can then be used to test the model predictions. The basic assumption is that history can be modelled, and that these models can help us 'predict' what happened in individual societies (Turchin 2011).

At first sight, attempts to discover 'laws of history' would seem to side with the above-discussed systems of deterministic history. The much more specific attention to modelling in this new approach to large-scale historical processes, however, allows randomness to be systematically incorporated into thinking about these historical processes. In brief, the main purpose of Cliodynamic scholarship is to

---

[13]http://www.nybooks.com/articles/archives/1997/jun/26/guns-germs-and-steel/, visited on 20.3. 2015.

[14]http://cliodynamics.info/.

develop a highly sophisticated model to describe historical societies. The model includes a range of (competing) general principles that are 'translated' into systematic regularities, which can then be described as 'structural events'. Some of these events are predictable and can be modelled. Some events cannot. Which is where randomness comes in.

In the terms just described, randomness is a modelling device. There are developments and actions that are excluded from the model, in order to make events understandable. The reason to exclude them can be because the actions *cannot* be predicted (because, for instance, they follow from the human free will), or because the predictable causes behind the actions are unknown to the scholar designing the model, or, finally, because these causes are known but too complex for the model. The absence of these actions from the model does not deny that both structural and non-structural forces are continuously operating within societies and should be included in a perfect model. But describing the unknowable/unknown/too complex aspects of historical processes as 'random' allows the model to function, and gives the scholar possibilities to differentiate between structural forces and factors that are now described as random. Almost all authors working within this framework define human agency as 'random' (Goldstone 2003; Skocpol 2003).

The underlying assumption is one of complex causality, in which there is differentiation between structural causes and triggering events. In terms of the model, these triggering events might be 'random', but that does not deny them their place of importance. These triggering events are, however, difficult to predict. The underlying 'macro-causes' amplify these triggering notions ('micro-causes'), in which agency is crucial. Agency is essential, but not really what this field of study is particularly interested in. Instead, the structural causes underlying the effects of the triggering events are focus of attention. Randomness is accepted, but the interest is in the grand 'mechanisms' of societal change (Turchin 2006). Coincidence is not really denied, but it is certainly pushed to the background. There seems to be little attention to the role of coincidence in history when non-historians pose the questions—and it is certainly not used as a mode of interpretation.

6 Coincidence as an Interpretative Tool?

Might there be a mode to include coincidence as an interpretative tool in historical research, in light of the above brief overview? It seems clear that *if* such a notion can be used, it should give sufficient attention to so-called micro-causes, in which human agency is of great importance. Individuals have freedom to take specific actions. They of course do so within socio-biological frameworks (see the contributions by Weitzel and Rosenkranz and Van Elk, Friston and Bekkering) and within the context of a larger historical frameworks, which is what macro-history focuses on. However, as Ton Nijhuis recognised in a stimulating recent contribution on exactly this topic—coincidence in history—the action chosen out of a range of possible courses of action has influence on historical developments, and needs to be taken seriously. History was not a closed trajectory *before* it happened, and the outcome that we happened to arrive at is not necessarily more worthwhile because this is the course events took. Nijhuis draws the analogy between studying history and walking in an unknown landscape. If you only look at the trodden path, you fail to see the various avenues that you can take, and others could have taken (Nijhuis 2003, p. 63). In a macro-historical perspective, developments are homogenous and continuous. In reality, there are heterogeneous and discontinuous serial events, that only the analysis of micro-history can make visible. In this context both Isaiah Berlin and the well-known Italian historian Carlo Ginzburg have used Tolstoy's *War and Peace* as an example. Tolstoy assumed that 'a historical phenomenon can become comprehensible only by reconstructing the activities of *all* the persons who participated in it' (Ginzburg et al. 1993, p. 24; Berlin 1978). In the intersection between private actions and the public world, history develops. For a systematic recognition of the role of coincidence in history, our analysis needs to be at the right scale. It needs to value human agency, and the stochastic element this brings into historical developments, whilst at the same time identifying the underlying processes which may be much more predictable.

The ideal historical account should take both into consideration, going back and forth between the large processes that are likely to lead to certain trajectories, and the enormous number of micro-causes that triggered the events as they happened. In that context, one might be able to usefully employ the famous thought experiment of the evolutionary biologist Stephen Jay Gould:

> I call this experiment "replaying life's tape": You press the rewind button and, making sure you thoroughly erase everything that has actually happened, go back to any place in the past ... Then let the tape run again and see if the repetition looks at all like the original.[15]

Gould's work has been intensely discussed (and occasionally maligned), but the metaphor of replaying the tape might be a way of coming to grips with the relation between coincidence as an interpretive tool and discussion on the size of history. How likely do we deem certain historical events if they were to be played out again? To what extent does that depend on the scale of the historian's narrative? How often would Antony's distraction with Cleopatra's nose lead him to lose the battle at Actium if we were to reply the events of 2 September 31 BC? Would it be in the region of 2 out of 10.000 times (a contingent chain of events) or 9998 out of 10.000 (quasi-determined)? And might the result be different depending on our perspective? If we were to replay the whole year 31 BC, or the longer period of 44–31 BC, would that decrease the likelihood that this chain of events were to happen again? This could be a way to start recognising the possible role of coincidence in various 'triggering events'. Coincidence might fall outside of the interpretative toolbox of the sociologists and macro-historians who are looking for a comprehensive view of historical processes, but could still play a proper role in thinking

---

[15]Gould (1989), 48, with the discussion in Vogt, *Kontingenz und Zufall*, 231–234. The possibilities of using Gould's thoughts for my argument, and their possible repercussions in this context, were kindly pointed out to me by Robert-Jan Wille.

about historical trajectories. Not 'what if' history exactly, but rather history with an open mind.

Acknowledgment I am grateful to Remieg Aerts, Chiel van den Akker, Jelle Goeman, Klaas Landsman, Pieter Muysken and Robert-Jan Wille for comments on earlier versions of this paper, and to Peter Turchin for being willing to explain and discuss some of the underlying notions of Cliodynamics. None of them necessarily agree with the articles as it stands, but they have been extremely generous with their ideas, and massively improved the argument that I am trying to make.

Open Access This chapter is distributed under the terms of the Creative Commons Attribution-Noncommercial 2.5 License (http://creativecommons.org/licenses/by-nc/2.5/) which permits any noncommercial use, distribution, and reproduction in any medium, provided the original author(s) and source are credited. The images or other third party material in this chapter are included in the work's Creative Commons license, unless indicated otherwise in the credit line; if such material is not included in the work's Creative Commons license and the respective action is not permitted by statutory regulation, users will need to obtain permission from the license holder to duplicate, adapt or reproduce the material.

References

Adcock, R. (2007). Who's afraid of determinism? The ambivalence of macro-historical enquiry. *Journal of the Philosophy of History, 1*, 346–364.

Ankersmit, F. (1995). Historicism: An attempt at synthesis. *History and Theory, 34*, 143–161.

Berenson, B. (1952). *Rumour and Reflection: 1941: 1944*. London: Constable.

Berlin, I. (1954/2012). *Historical inevitability* (OUP Oxford) (pp. 119–190) (Reprinted in *The proper study of mankind: An anthology of essays*). Springer: Random House 2012.

Berlin, I. (1978). The hedgehog and the fox: An essay on Tolstoy's view of history. In: H. Hardy & A. Kelly (Eds.), *Russian Thinkers* (pp. 22–81). London: Weidenfeld & Nicolson.

Booth, C. (2003). Does history matter in strategy? The possibilities and problems of counterfactual analysis. *Management Decision, 41*, 96–104.

Bunzl, M. (2004). Counterfactual history: A user's guide. *The American Historical Review, 109*, 845–858.

Burke, P. (1992). *History and social theory*. Ithaca: Cornell University Press.

Bury, J. B. (1916/1930). 'Cleopatra's nose', RPA Annual (pp. 16–23) (Reprinted in J. B. Bury, *Selected Essays*, pp. 60–69). Cambridge: Cambridge University Press 1930.

Butterfield, H. (1931) *The whig interpretation of history*. London: Bell.

Carr, E. H. (1961). *What is history*. London: Penguin.

Christian, D. (2004). *Maps of time. An introduction to big history*. Berkeley: University of California Press.

Correia, D. (2013). F**k Jared diamond. *Capitalism Nature Socialism, 24*, 1–6.

Cushing, J. T. (1992). Historical contingency and theory selection in science. *PSA: Proceedings of the Biennial Meeting of the Philosophy of Science Association*, 446–457.

Diamond, J. (1997). *Guns, germs and steel. The fates of human societies*. New York: W.W. Norton.

Diamond, J. (2005). *Collapse: How societies choose to fail or succeed*. New York: Viking Press.

Diamond, J., & Robinson, J. A. (Eds.). (2010). *Natural experiments of history*. Cambridge, MA: Belknap Press.

DuPont Chandler, A. (2004). Luck and the shaping of a historian's professional education. *Massachusetts Historical Review, 6*, 1–20.

Elliott, J. H. (1991). *Richelieu and Olivares*. Cambridge: Cambridge University Press.

Elliott, J. H. (2012). *History in the making*. London: Yale University Press.

Evans, R. (2014). *Altered pasts: Counterfactuals in history*. Brown: Little.

Ferguson, N. (1997). *Virtual history: Alternatives and counterfactuals*. London: Picador.

Gallagher, C. (2011). What would Napoleon do? Historical, fictional, and counterfactual characters. *New Literary History, 42*, 323–325.

Ginzburg, C., Tedeschi, J., & Tedeschi, A. C. (1993). Microhistory: Two or three things that I know about it. *Critical Inquiry, 20*, 10–35.

Goldstone, J. A. (2003) Comparative historical analysis and knowledge accumulation in the study of revolutions. In J. Mahoney & D. Rueschmeyer (Eds.), *Comparative historical analysis in the social sciences* (pp. 41–90). Cambridge: Cambridge University Press.

Gould, S. J. (1989). *Wonderful life: The burgess shale and the nature of history*. New York: W.W Norton.

Guldi, J., & Armitage, D. (2014). *The History Manifesto*. Cambridge: Cambridge University Press.

Hacker, B., & Chamberlain, G. B. (1981). Pasts that might have been: An annotated bibliography of alternative history. *Extrapolation, 22*, 334–368.

Hedström, P., & Swedberg, R. (Eds.). (1998). *Social mechanisms: An analytical approach to social theory, studies in rationality and social change*. Cambridge: Cambridge University Press.

Hegel, G. F. (1807/1977). Phenomenology of spirit (A. V. Miller Trans. The original *Phänomenologie des Geistes*, 1807). Oxford: Oxford University Press 1977.

Huizinga, J. (1937). De wetenschap der geschiedenis: Groei van de historische wetenschap sedert het begin der negentiende eeuw. In *Geschiedwetenschap/hedendaagse cultuur* (pp. 104–172). Haarlem: H.D. Tjeenk Willink & zoon.

Kern, K., & Brown, K. (2001). Using the *list of creepy coincidences* as an educational opportunity. *The History Teacher, 34*, 531–536.

Mayr, E. (1990). When is historiography Whiggish? *Journal of the History of Ideas, 51*, 301–309.

McClellan, J. E., III (2005). Accident, luck and serendipity in historical research. *Proceedings of the American Philological Society, 149*, 1–21.

Meyer, E. (1901). *Geschichte des Altertums. Dritter Band: Das Perserreich und die Griechen. Erste Hälfte: Bis zu den Friedensschlüssen von 448 und 446 v. Chr.* (Stuttgart).

Morello, R. (2002). Livy's Alexander digression (9.17-19): Counterfactuals and apologetics. *Journal of Roman Studies, 92*, 62–85.

Nagel, E. (1966). Determinism in history. In W. Dray (Ed.), *Philosophical analysis and history*. New York: Harper & Row.

Nijhuis, T. (2003). Geschiedenis, toeval en contigentie. In S. Haakma & E. Lemmens (Eds.), *Toeval* (PP. 49–72). Utrecht: Studium Generale.

Post, R. C. (2009). Chance and contingency: Putting Mel Kranzberg in context. *Technology and Culture, 50*, 839–872.

Read, H. (1862). *The hand of god in history: Or, divine providence historically illustrated in the extension & establishment of christianity*. Glasgow: W. Collins.

Ringer, F. (2002). Max weber on causal analysis interpretation and comparison. *History and Theory, 41*, 163–178.

Skocpol, T. (2003). Doubly engaged social science: The promise of comparative historical analysis. In J. Mahoney & D. Rueschmeyer (Eds.), *Comparative historical analysis in the social sciences* (pp. 407–428). Cambridge: Cambridge University Press.

Squire, J. C. (1931). *If it had happened otherwise*. Greens: Longmans.

Taylor, A. J. P. (1945). *The course of German history, a survey of the development of German history since 1815*. London: Hamish Hamilton.

Taylor, A. J. P. (1961). *The origins of the Second World War*. London: Hamish Hamilton.

Thomas, C. E. (2010). 'Test in Time', *Cliodynamics*, 1 (2010). Retrieved from http://escholarship. org/uc/item/368956r6.

Thompson, W. (2010). The lead economy sequence in world politics (from Sung China to the United States): Selected counterfactuals. *Journal of Globalization Studies, 1*, 6–28.

Trevelyan, G. M. (1918). From Waterloo to Marne. *Quartely Review, 229*, 73–90.

Turchin, P. (2006). *War and peace and war: The life cycles of imperial nations.* NY: Pi Press.

Turchin (2011). Social tipping points and trend reversals: A historical approach. Presented at the *Tipping Point Workshop* May 20–22. Switzerland: Mount PilatusLucerne.

Vogt, P. (2011). *Kontingenz und Zufall. Eine Ideen- und Begriffsgeschichte* Berlin: Akademie Verlag.

Weber, M. (1905). Objective possibility and adequate causation in historical explanation. In M. Weber, E. Shils & H. Finch (Eds.), *The methodology of the social sciences* (pp. 164–88). Glencoe: Free Press 1949 (original 1905).

Accidental Harm Under (Roman) Civil Law

Corjo Jansen

Abstract A leading idea under Roman private law and nearly all European legal systems is that an owner has to bear the risk of an accidental loss (*casus*). An accident is a circumstance for which a third party cannot be blamed (*culpa* or fault). A person suffering damage from an accident had to bear that damage himself. This idea has been subject to attack throughout history. Every once in a while, it is said that 'bad luck must be righted' ('*pech moet weg*'). This position has not become the prevailing viewpoint among lawyers. Although it does not seem very realistic, 'bad luck must be righted' did form the basis of social security policies of the Netherlands and some other western countries after World War II: social security 'from womb to tomb'. The scope of social security benefits has been reduced in many countries in the last decades of the twentieth century, because the costs were no longer affordable. The idea that a owner has to bear the risk of *casus* has withstood the test of time quite well. That accidental harm must be borne by the one suffering it, is legally and morally justifiable.

1 Introduction

While engaging in activities, a person may suffer damage himself or, even worse, cause damage to someone else. An example for each situation is very easy to provide for. The first situation involves damage brought on by one's self: An individual slips and falls after a heavy rain shower and breaks a leg. The second situation involves damage brought on by another person: A bicyclist knocks down a pedestrian, who winds up with a broken arm. The moment at which the damage arises is often the point in time when the law comes into the picture. A significant task of the law is to formulate rules to specify in which cases damage must be

C. Jansen (✉)
Legal History and Civil Law, Business and Law Research Centre, Radboud University, Nijmegen, The Netherlands
e-mail: c.j.h.jansen@jur.ru.nl

© The Author(s) 2016
K. Landsman and E. van Wolde (eds.), *The Challenge of Chance*, The Frontiers Collection, DOI 10.1007/978-3-319-26300-7_13

233

compensated, for a person's freedom of action is not unlimited. The question arises whether someone who has suffered damage has to bear the loss himself or whether another person should bear liability for this.

A leading idea under the law is that the loss should lie where it falls.[1] This notion reflects the old adage under Roman law, *'casum sentit dominus'*. An owner has to bear the risk of an accidental loss or an accidental deterioration which has resulted in harm to him.[2]

This viewpoint has been criticised by, for example, the renowned German professor of Roman law, B. Windscheid (1817–1892): *"Unbrauchbar und in dieser Allgemeinheit unrichtig (…)."*[3] Another prominent German scholar, H. Dernburg (1829–1907), was equally disapproving of the *casum sentit dominus* rule, stating: *"Ihre Unhaltbarkeit wird (…) nicht leicht bestritten."*[4] And yet, under nearly all European legal systems, this rule appears to be authoritative. The legislator in Austria even laid it down in that country's civil code (§ 1311 *ABGB* of 1811): *"Der bloße Zufall trifft denjenigen, in dessen Vermögen oder Person er sich ereignet."*[5]

The Roman law adage *'casum sentit dominus'* is still important to day. The civil law systems of the Continent stem from Roman law. This law has become the intellectual ground for a largely homogenous legal culture on the Continent, based on the reception of Roman legal rules and principles. The two most important civil codes influenced by Roman law are the *Code civil* (*Cc*) of France (1804) and the *Bürgerliches Gesetzbuch* (*BGB*) of Germany (1900). Spain, Portugal, Belgium, the Netherlands, the former colonies of France and nearly the whole of South America have adopted the French *Cc*. Greece and Japan adopted the German *BGB*. Contrariwise to the Continent, England developed an unique legal tradition. This tradition is called *Common Law*. Nearly 40 % of all the people live in common law countries, like the United States of America, Australia, New Zealand, India, Pakistan and the other former colonies of England. Also the common law tradition has accepted the idea that loss should lie where it falls. That's why nearly all legal systems in the world know the adage *'casum sentit dominus'*. In this article, the focus is on the Roman law tradition.[6]

An issue which crops up under Roman law (and thus under Continental civil law) is the meaning which needs to be given to the concept of *'casus'* ('accident'). Roman-law sources suggest two meanings here. In the first instance, these sources teach, an 'accident' refers to some natural phenomenon ('act of God'), such as a

---

[1]Hartlief (1997); Sieburgh (2000); Markesinis and Deakin (2003), p. 42.

[2]Code of Justinian (C.) 4,24,9; Digests (D.) 50,17,23 (*casus a nullo praestantur*). See Wacke (1984), p. 271; Zimmermann (1996), pp. 154, 162, and 281; Sieburgh (2000), p. 5. Also referred to as *'res perit domino'*.

[3]Windscheid and Kipp (1900), § 264, fn. 5: 'Useless and in this indefinite way wrong'.

[4]Dernburg (1903), p. 123, Note 3. For other examples, see Ranieri (2009), pp. 569–572.

[5]Wacke (1984), p. 670. *Cf.* Article 1105 *Código Civil* (Spanish Civil Code). In the Netherlands, see the Flood Damage Act 1953, Article 8:543 Dutch Civil Code and Article 8:1004(2) Dutch Civil Code.

[6]Zweigert and Kötz (1998), pp. 68–69, 218 *et seq.*, 298; Zimmermann (2011), 27 *et seq.*

lightning strike, flood, earthquake or compact impact.[7] Further, an 'accident' may involve some misfortune, an unfortunate confluence of circumstances or '*allgemeine Lebensrisiko*'.

To figure out the place '*casus*' occupies under (Roman) private law, let's revisit my two earlier examples. If a person slips and falls after a heavy rain shower and breaks his leg, that person must bear the risk of his loss. The fact that the person slipped and fell is a risk which everyone runs in daily life. Another party cannot be made to pay for the damage. The situation might be different if an individual slips on a banana peel which was deliberately put on the floor. There is a very good chance in that case that someone else will be liable for the loss suffered. In the example where a pedestrian breaks his arm because of the bicyclist's actions, the question likewise arises whether the person suffering damage can hold someone else liable or not. The major reason for having another person foot the bill for the injury in the last two examples is that the injury can be traced back to the fault (*culpa*) of the persons causing the damage. The bicyclist bears blame for the pedestrian's broken arm, just as the person placing the banana peel on the floor can be blamed for the broken leg (liability without fault is also possible under the law, by the way, but I won't get into that here). "Fault is the basic element of the law of torts."[8] In these examples, 'accident' stands in contrast to fault (in the sense of blameworthiness).[9] Unlike with 'fault', the damage is not attributable to a person in the case of an 'accident'. By law, an accident is a circumstance for which a third party cannot be blamed or which is not attributable to someone.

Are there other meanings given to 'accident' which are relevant in (Roman) private law and which modify or expand the definition? In his inaugural lecture, De Mul, professor of philosophy in Rotterdam, argued that, under private law, the term 'accident' mainly signifies something being determined by fate and is often associated with such concepts as 'unforeseen' or 'unforeseeable'.[10] I won't get into the first part of De Mul's definition. As to the second part, I note the following. For lawyers, concepts like 'unforeseen' and 'unforeseeable' are primarily factors in determining whether certain harm came about because of someone's fault. There is fault when what could have been foreseen by a diligent man was not foreseen. Fault entails someone's having acted differently than he should have, given the circumstances of the case.[11] If that is so, he can be blamed for the harm. The meaning

---

[7] I won't get into the tricky distinction between *casus* and *vis major* (force majeure or 'scourge of God': D. 19,2,26,6). There is some overlap between the two terms. Coing (1989), p. 462, and Deroussin (2007), pp. 592–593.

[8] Owen (1995), pp. 201, 208–209, 223 *et seq.*; Atiyah (1997), p. 3; Zimmermann (1996), p. 1034; Markesinis and Deakin (2003), 41 *et seq. Cf.* D. 50,17,23.

[9] Bruins (1906), pp. 71–72; Rümelin (1896), p. 17; Von Bar (1999), Nos. 318-322, 485; Sieburgh (2000), 14 *et seq.*

[10] De Mul (1994), pp. 8–12, mentions two other basic connotations of 'accident': fortuitous (incidental or inessential) and contingent. See also Rümelin (1896), pp. 8–9 and 18.

[11] *Cf.* D. 9,2,31; Zimmermann (1996), pp. 1007–1009; Hartkamp and Sieburgh 6-IV* (2011), No. 100.

propounded by De Mul leaves the above definition of 'accident' intact. In Rümelin's words: *"[Im Gegensatz zur Schuld spielt Zufall, das nicht Vorhergesehene und erfahrungsgemäss nicht Vorhersehbare] die Rolle eines Grenzbegriffs für irgend welche Verantwortlichkeit."*[12] Moreover, unintended consequences of people's conduct are often referred to as 'accidental'.[13] They, too, may be taken into account when assessing whether a certain action may be imputed to someone or not. Again, modifying the definition of 'accident' (in contrast to 'fault') does not seem necessary.

A brief digression about the sources of the obligation helps us to analyse the function of *casus* (accident) and *culpa* (fault) under (Roman) private law and the legal consequences which ensue from this. With regard to each Roman-law or other source, the notion of *casus* or *culpa* is, to a certain degree, developed differently.

2   Sources of Obligation Under Roman and Modern Law

An obligation to compensate another person's damage is not obvious. In his Institutes, the Roman lawyer Gaius (110–185 AD) distinguished two categories from which such an obligation might arise. "Every obligation arises either from contract or from wrongdoing."[14] The distinction between these two sources of obligation has withstood the test of time. It can also be seen in modern legal systems. Obligations to pay compensation do not, however, only arise from contract or wrongdoing. Gaius recognised this as well. In the second book passed down in his name, *Res cottidianae sive aurea*, *Golden Words*, Gaius added a third category as a source of obligation: "Obligations arise either from contract or from wrongdoing or, by some special right, from various types of causes."[15] What were examples of such another cause? Gaius talked about several of these in the third book of his *Golden Words*. One example was managing someone else's interest, *negotiorum gestio* ('agency without a mandate'). If someone looked after an absent person's affairs without having been instructed to do so, the two individuals became connected to each other and could litigate against one another based on the notion of management of another's affairs without authorisation. These legal actions did not arise from contract or wrongdoing. After all, the party managing the absent party's affairs had not concluded a contract with the absent party beforehand. Managing a person's affairs without a mandate was not a form of wrongdoing, either.[16]

---

[12]Rümelin (1896), p. 53.

[13]Rümelin (1896), p. 18; De Mul (1994), p. 12.

[14]Institutes of Gaius (Gai. Inst.) 3,88: (…): *omnis enim obligatio vel ex contractu nascitur vel ex delicto.*

[15]D. 44,7,1,pr.: *Obligationes aut ex contractu nascuntur aut ex maleficio aut proprio quodam iure ex variis causarum figuris.*

[16]D. 44,7,5,pr.

The emperor Justinian (527–565) adopted the distinction from the *Golden Words* in his Institutes and changed it a little: "A further sub-classification results in four categories: specifically, obligations arising from contract, as if they were a contract (quasi-contract), or from delict, as if they were a delict (quasi-delict)."[17] Many European legislators built upon Justinian's views in their legal codes. This also held true for the Dutch Civil Code of 1838. Under Article 1269 of this Code, all obligations resulted from either contract or the law [de wet]. Pursuant to Article 1388 Dutch Civil Code 1838, the latter category could in turn be sub-divided into obligations emanating "from the law alone or from the law as a result of human conduct". Obligations ensuing from the law due to people's conduct arose from "a lawful or an unlawful act" (Article 1389 Dutch Civil Code 1838). The provisions in Article 1269 Dutch Civil Code 1838 were criticised in the literature. In laying the foundations for the current Dutch Civil Code (1992), E.M. Meijers (1880–1954), professor of civil law in Leyde, tried to restore Gaius's definition. Meijers's original wording of Article 6:1 Dutch Civil Code stated that obligations arose from contract, tort or other juridical facts if these ensued from the law. He wanted to prevent obligations from rising outside the law based on the principle of reasonableness and fairness ('good faith') or based on unwritten law or custom. That simply produced legal uncertainty and legal inequality.[18] Legislators after him looked to the Dutch Supreme Court's decision in *Quint v Te Poel*[19] to formulate the sources of obligation under the new Dutch Civil Code. According to Article 6:1 Dutch Civil Code, obligations can only arise from the law. The words 'from the law', the Dutch Supreme Court explained, do not in any way mean that each obligation has to be expressly provided for by the law. In situations not provided for by law, the solution must be accepted "which fits into the statutory legal system and is in line with rules already laid down for similar situations." Still, Justinian's ancient Roman classification system lies behind the sparse wording of Article 6:1 Dutch Civil Code.

I will successively discuss the role of *casus* under Roman tort law, Roman contract law and Roman law concerning *negotiorum gestio*.[20]

---

[17]Institutes of Justinian (Inst.) 3,13,2: *Sequens divisio in quattuor species diducitur: aut enim ex contractu sunt aut quasi ex contractu aut ex maleficio aut quasi ex maleficio.* See also Inst. 4,5,pr.

[18]Asser, Hartkamp and Sieburgh 6-I* (2008), No. 47 *et seq.*

[19]Dutch Supreme Court (Hoge Raad), 30 January 1959, *Nederlandse Jurisprudentie (NJ)* 1959, 548.

[20]*Casus* does not come into play with quasi-delicts.

3   Accidental Harm Under Roman Tort Law and Subsequent Criticism

The sources are very clear about the fact that, under Roman tort law, a person suffering damage from an accident had to bear that damage himself. Gaius contended that a party causing damage without fault or not intentionally, but instead by accident (*casu*), should remain 'unpunished'.[21] The lawyer Alfenus (who lived in the first century before Christ) responded similarly when asked whether the owner of a slave who had broken his leg during a game after being pushed hard could litigate pursuant to the *lex Aquilia*, a statute reforming the law on wrongful damage to property by the Roman assembly of the *plebs* (the *comitia centuriata*) from 286 or 287 BC, named after the person who proposed it, the tribune *Aquilius*. Litigation was not possible, "[because] the unfortunate event needed to be deemed the result of an accident (*casu*) rather than fault."[22] This position has been subject to attack throughout history. The natural law scholar Chr. Thomasius (1655–1728) regarded 'you must not injure your neighbour' (*alterum non laedere*), a perspective which could also be found in the sources,[23] as the guiding principle of tort law, from which he concluded that every damage ought to be compensated, even if it was caused by accident. "It is not only equitable but even just that I should make good damage done by accident." If, for instance, a person accidentally dropped someone's crystal glass, that person was, Thomasius felt, liable for the damage. It was his curiosity, not the owner's curiosity, which made the glass fall.[24] Thomasius's position did not reflect the view of Roman lawyers. It has also not been embraced very much by today's lawyers. Every once in a while, it is said that 'bad luck must be righted' [pech moet weg], even if one can only blame one's self for it, but this is not the prevailing attitude. Lord Steyn, a justice of the current Supreme Court of the United Kingdom, articulated this feeling well:

> "But we do not live in Utopia: we live in a practical world where the tort system imposes limits to the classes of claims that rank for consideration as well as to the heads of recoverable damages. This results, of course, in imperfect justice but it is by and large the best that the common law can do."[25]

---

[21]Gai. Inst. 3,211: *Itaque inpunitus est qui sine culpa et dolo malo, casu quodam damnum committit.*

[22]D. 9,2,52,4: *Respondi non posse, cum casu magis quam culpa videretur factum.*

[23]D. 1,1,10,1.

[24]Thomasius (2000), Sec. IV; Jansen (2009), 231 et *seq.*; Atiyah (1997), pp. 178–179. To his mind, the assumption that every type of loss ought to be compensated makes the compensatory system unaffordable.

[25][1998] 3 *WLR* 1539B-C (per Lord Steyn).

4   Accidental Harm Under Roman Contract Law

A contract usually involves two people: a debtor and a creditor. They must act with due care towards each other. The degree of care they need to exercise depends on the type of contract and the object to which the contract pertains (a horse, for instance, requires different care than a slave or painting). If debtors or creditors breach their duties of care, they will be liable for the ensuing damage. Here's an example. A person lends his bicycle to his neighbour ('the contract of *commodatum*'). The bicycle is stolen from the neighbour. Is the neighbour liable for this theft? This question cannot easily be answered. The crux of the issue is the scope of the neighbour's duty of care. Generally speaking, if a person does not exercise the requisite duty of care, he is at fault and is liable. *Commodatum* consisted of a gratuitous loan of a corporeal thing (mostly movables). The party borrowing the bicycle does not owe any money to the other party. Furthermore, lending out the bicycle is in the borrower's interest. Hence, Roman lawyers believed that the party who borrowed the bicycle, the neighbour in my example, had a very weighty duty of care (*'custodia'*, or a duty of safekeeping). Because of this duty, the neighbour was liable to the lender of the bicycle in the event of theft. The borrower, who was responsible for *custodia*, had to therefore compensate the lender's damage, even though, subjectively, the borrower bore no blame.

This basic principle regarding *commodatum* did not automatically apply to other contracts. Some agreements merely entailed liability for intentional misconduct (*dolus*), others, for intentional misconduct (*dolus*) and fault (*culpa*), and still others, for intentional misconduct (*dolus*), fault (*culpa*) and *custodia*. The contract of *mutuum* received special treatment. Such a contract consisted of transferring ownership of a quantity of fungible goods (such as money or grain) to another party, who undertook to return an equal quantity of goods of the same sort. The most prominent example of a *mutuum* was the moneylending contract. The borrower became the owner of the goods. Consequently, the borrower bore the risk of destruction of the objects, even if this occurred by accident (*casu fortuito*), due to, say, fire, collapse, shipwreck or an attack by bandits or enemies.[26] This principle was consistent with the rule that owners bear the risk of an accidental loss.

With other contracts, the required level of due care varied, but was never absolute. To quote the applicable Roman legislative text verbatim, "no one need bear responsibility for accidents and deaths occurring to living beings which are not attributable to anyone's fault, escapes by slaves usually left unguarded, or robberies, riots, fires, floods or attacks by bandits."[27] Such occurrences corresponded in large measure to 'acts of God' under English law. A borrower had to act with the utmost due care, for example. As we have seen, this meant, too, that the borrower was liable to the lender for theft. The borrower only avoided liability for events

---

[26] Inst. 3,14,2 and D. 44,7,1,4. See Wallinga (2009), p. 225 *et seq.*

[27] D. 50,17,23: (….) *casus mortesque, quae sine culpa accidunt, (…) a nullo praestantur.*

which nobody could defend against (*casus non praestat*), such as attacks by bandits, enemies or pirates, fire and so on.[28] As Schulz put it:

> "[A borrower] was absolutely liable for certain typical accidents which were regarded as avoidable by properly watching and guarding the borrowed thing, and on the other hand he was not liable for other typical accidents which were invariably regarded as not avoidable by the exercise of care."[29]

A depositee who had offered to take possession of someone else's property was liable for intentional misconduct, negligence and *custodia*, but not for fortuitous events (*casus fortuitos*).[30] A man who could show that he had lost his bookkeeping records on account of a shipwreck, collapse, fire or similar accident (*alio simili casu*) was not accountable to the banker from whom he had borrowed money.[31] Likewise, losses incurred by accident (*casu*) were not chargeable to the balances which slaves had to pay their masters.[32] In contrast, a mandatary could not seek reimbursement of his costs from the mandator if the mandatary had been robbed by bandits or had lost property during a shipwreck. These events were attributable to accident (*magis casibusquam*) rather than the mandate.[33] If *casus* was involved, the party who had suffered damage thus bore this damage himself. There was no reason to shift the risk onto somebody else's shoulders.

5 Accidental Harm in the Case of Negotiorum Gestio

The idea of the management of another's affairs is a peculiarity under the law, since furnishing unsolicited help to another person is a precarious undertaking. Such conduct is readily seen as an undesirable interference or as a curtailment of someone's freedom. Under Roman law, no one was obliged to help another person. Nevertheless, there was a strong notion that citizens *should* help their fellow citizens in times of distress, by, for example, giving advice, providing a loan or voluntarily managing someone's interests without a mandate to do so. Schulz described 'management of another's affairs' as a "quite original genuinely Roman creation without parallels in the laws of other peoples not dependant on Roman Law."[34]

---

[28]D. 13,6,18 pr. and D. 44,7,1,4.

[29]Schulz 1969, p. 515.

[30]D. 16,3,1,35. Zimmermann (1996), pp. 208–209. Such wide-ranging liability for a custodian is rather exceptional. It can be seen, too, in French law, but not in German law. Normally, a custodian is liable for *dolus*.

[31]D. 2,13,6,9.

[32]D. 40,5,41,7.

[33]D. 17,1,26,6; Zimmermann (1996), pp. 430–431.

[34]Schulz (1969), p. 624; Kortmann (2005), p. 99 *et seq.*; Jansen (2014), p. 43 *et seq.*

The special nature of *negotiorum gestio* was also apparent in the manager's scope of liability; if an individual took care of the interests of an absent person who was unaware of this, the manager became liable for both wilful misconduct and negligence. A leading Roman lawyer maintained, however, that the party looking after the affairs even "had to answer for accident (*casus*), for example if in the name of the absent principal he transacts business the principal did not usually do." It goes without saying that such liability would be very extensive. Presumably, this far-reaching liability was prompted by the deep-seated aversion to representing someone's interests against the will of that person.[35]

6   Other Meanings of 'Casus' in the Roman Sources

'*Casus*' does not just mean 'accident' in the Roman-law sources. Sometimes, the word signifies 'misfortune', 'fate', 'adversity' or 'setback'. In these instances, it refers—just like in the case of accident—to an event resulting in damage which cannot be traced back to another party's fault. One text, for instance, categorises trees being uprooted or trees being blown over because of a storm as '*casus*'.[36] A Roman lawyer used the word '*casus*' in a similar sense, when he noted that it is neither decent nor natural to speculate about the misfortune or setback which a free man has suffered.[37]

'*Casus*' sometimes relates to destiny, say, to the fact that a person is deaf or blind.[38] These meanings of 'accident' suggest something of the incomprehensibility or arbitrariness of life and the vicissitudes of fate (see De Mul's definition given in this article's introduction). 'Accident' here pertains not to human conduct, but to divine or similar intervention. It sets forth the limit of what lies within a person's control.[39]

Finally, the Roman-law sources seem to imply that '*casus*' also means 'independent of a person's will'. The law made it possible for giving rise of a legal consequence to hinge on a condition. Someone decided that the legal consequences would only arise if or until an uncertain future event took place (such as ownership not being transferred until the entire purchase price was paid). A slave, for example,

---

[35]D. 3,5,10; Zimmermann (1996), pp. 446–447. See also § 678 *BGB* (German Civil Code), which adopted this solution. The Dutch Supreme Court has embraced this position as well: Supreme Court, 19 April 1996, *NJ* 1997, 24.

[36]D. 7,1,12,pr.

[37]D. 45,1,83,5. See also D. 4,6,1,pr.

[38]D. 3,1,1,3 and 5. See also D. 4,4,11,5.

[39]Eijsbouts (1989), pp. 2, 16, 19–20. See also the definition of 'treasure' in Article 642(2) Dutch Civil Code 1838 (Article 716 *Code Civil* (French Civil Code): a 'treasure' had to have been discovered by pure chance (*le pur effet du hasard*) (Eijsbouts (1989), pp. 6–7). In the current Article 5:13 Dutch Civil Code (a translation of § 984 *BGB*), chance is no longer an element of the definition of treasure.

might have been set free under a certain condition. This might have consisted of a fact, an action or one or another fortuitous circumstance (*casu*).[40]

7 Brief Interlude: 'Casus' Under Roman Criminal Law

Unlike in modern criminal law, '*casus*' (by accident) is mentioned in the text of several Roman-law criminal provisions. The term had to do with the state of mind with which a crime was committed: with premeditation (*proposito*); in the heat of the moment (*impetu*) or by accident (*casu*, when, for example, a spear thrown at a wild animal during a hunt killed a man).[41] The state of mind was relevant in determining the severity of the punishment. In the case of the more serious crimes, ascertaining whether these had been committed with premeditation or by accident (*casus*) was crucial, said the lawyer Ulpianus (who died in 223). For all crimes, this distinction had to result in either a just punishment or reduced punishment.[42] Hence, according to the emperor Hadrian (117–138), the punishment for an individual who had committed manslaughter accidentally (*casu*) rather than intentionally (*magis quam voluntate*) during a scuffle was moderated.[43] In modern criminal-law systems, intent, premeditation and negligence are the subjective elements of a crime. These days, *casus* is a factor in determining the degree of guilt which can be ascribed to accused individuals when they have engaged in potentially criminal conduct.

Accident always plays a role in criminal law to some extent. Whether certain punishable conduct must be characterised as an assault or as manslaughter depends on the consequence which ensues. Often, whether someone dies or is merely seriously injured as a result of a sharp blow to the face is a matter of accident. Potentially punishable conduct may likewise be nipped in the bud purely by accident; consider, for example, the case of a heavy rain shower which extinguishes a deliberately set fire.[44]

---

[40]D. 40,5,33,1.

[41]D. 48,19,11,2.

[42]D. 48,19,5,2. See also D. 47,9,9.

[43]D. 48,8,1,3.

[44]Whether such an arson attempt constitutes a crime will depend on the circumstances of the case. See Dutch Supreme Court, 19 March 1934, *NJ* 1934, p. 350 (the Eindhoven arson judgment) and Dutch Supreme Court, 19 September 1977, *NJ* 1978, 126.

8   Accidental Harm Under Modern Private Law

The viewpoints found in Roman law have—as stated above—often remained valuable for modern civil law. To the extent they still apply, the scope of their application must be determined. As we have seen, in Roman law, *casus* (in the sense of 'accident') played a role in contract law, the management of another's affairs and tort law. The force majeure doctrine ('non-attributable non-performance') has greatly diminished the role of accident in the area of contract law. Briefly stated, a situation of force majeure exists if the debtor, for reasons beyond his control, cannot fulfil his obligations. The failure of performance does not result from his fault and is not at his risk. Notwithstanding this development in modern private law, especially in the Dutch and German civil law, the *ABGB* (Austrian Civil Code) and *Código Civil* (the Spanish Civil Code, 1889) still include —consistent with the Roman-law tradition—*casus* (accident) in addition to force majeure as a circumstance which frees debtors from their obligations (§1447 *ABGB* and Article 1105 *Código Civil* respectively). French courts still take accidental elements (*l'aléa*) into account as well.[45]

Almost all Continental legal systems look to the fault principle when a contract is not performed or not in a timely or proper manner: If a debtor cannot be blamed for the failure to perform, the debtor is not liable for damages. This principle was expanded in the Dutch Civil Code of 1992. Debtors can claim force majeure if performance is hindered for reasons for which they do not bear any fault and for which they do not bear the risk.[46] The standard concerning the debtor's conduct is objective insofar as the debtor must have acted as a prudent debtor would have acted in the given situation (*cf.* Article 6:27 Dutch Civil Code and § 276 *BGB*). If the debtor has violated this objective standard, it must be examined whether the debtor can personally be blamed for this. If such personal blameworthiness (fault) is lacking, the question becomes whether the debtor is liable based on the law, juridical act or 'generally accepted principles'/common opinion (see Article 6:75 Dutch Civil Code).[47] A person may invoke force majeure, for instance, if his life or liberty is threatened. Under generally accepted standards, the person is not liable then. For a comprehensive comparative–law analysis of fault and wilful or intentional misconduct in determining whether a debtor has breached his obligations towards the other party, I refer to Ranieri.[48]

Casus seems to play a larger role in modern tort law than in modern contract law. That is certainly true for the Netherlands. Whether a court must assume liability based on tort in a specific case will involve a weighing of the two viewpoints mentioned earlier which dominate this area of the law: 'the loss should lie where it

---

[45]Ranieri (2009), pp. 579–580, 584; Deroussin (2007), p. 594.
[46]Parlementaire Geschiedenis Boek 6 (1981), pp. 263–264; Hartkamp and Sieburgh 6-I* (2008), No. 343; Ranieri (2009), p. 572.
[47]Hartkamp and Sieburgh 6-I* (2008), No. 344.
[48]Ranieri (2009), pp. 572–650.

falls' and 'you must not harm your neighbour'. The general consensus in Continental legal systems is that the maxim 'the loss should lie where it falls' must prevail if the damage arose through an accident, such as a failed harvest, flood or lightning strike. Even if the actions by a person resulting in damage actually come down to nothing more than an unfortunate confluence of circumstances, the party suffering that damage must bear that damage himself.

> I mention a few examples. An old woman who wanted to get into a bus stepped back to allow someone to go in front of her. In stepping back, she bumped up against another old woman, who fell and broke her hip. A hiker in a forest kicked a branch. This branch lashed the eye of the hiker behind him. The hiker who got hit with the branch lost an eye. Two sisters were moving house. One of them lost her grip on a cabinet while going down the staircase. As a result, one sister's arm became wedged between the cabinet and the wall. The arm had to be amputated.[49]

In each of these situations, the conduct leading to harm cannot be said to have been improper or unlawful. What's more, the people in question can hardly be blamed for their conduct. The injury was related to an everyday risk, to the fact that we live and participate in society. These incidents are sometimes termed 'common or garden accidents'. The ensuing damage ought to remain where it fell. The risk did not exceed the general risk of damage which an individual runs in daily life. Further, the nature of the activity was not so dangerous that precautionary measures needed to be taken.[50] Of course, the situation changes if a person's traits and abilities should have kept that person from participating in certain activities. "Just as it is ethically acceptable for people to claim personal credit for conduct which is partly a product of their good luck in having a certain personality and certain capacities, so people must accept responsibility for conduct which is partly a product of bad luck in having a certain personality and certain capacities."[51]

9   Concluding Observations

Law and *casus* go back a long way. Accident was and is mainly important in answering the question whether certain damage must be borne by the party suffering the damage (the 'owner') or whether another party can be made to pay for this. This other party may be a natural person (the one causing the damage) or a legal entity (in particular, an insurance company). All of the Continental legal systems assume that damage arising by accident remains the responsibility of the party suffering the damage. For liability for the damage to be passed on to another

---

[49]Dutch Supreme Court, 11 December 1987, *NJ* 1988, 393; Dutch Supreme Court, 9 December 1994, *NJ* 1996, 403; and Dutch Supreme Court, 12 May 2000, *NJ* 2001, 300.

[50]Von Bar (1999), Nos. 319–320; Sieburgh (2000), 12 *et seq.*; Hartkamp and Sieburgh 6-IV* (2011), No. 20. Contrary view: Van Dam (2000), No. 808. See also Rümelin (1896), p. 12 *et seq.*

[51]Cane (1997), p. 51.

party, the rule is and was that the damage must have been the fault of or attributable to that other party. The Roman-law adage *'casus sentit dominus'* has withstood the test of time quite well. That accidental harm must be borne by the one suffering it is also morally justifiable. Owen expressed this as follows: "To the extent that risks of harm from action may be deemed a necessary part of 'proper' choices of action in an uncertain world, and hence 'reasonable' according to some fair standard, they should be viewed as 'background risks' of life for victims to protect against and bear."[52]

Obviously, accident could be excluded as much as possible under private law. 'Bad luck must be righted' [*pech moet weg*] could be taken as the point of departure. Although this starting principle does not seem very realistic, it did form the basis of Dutch social security policy after World War II. The starving, humiliated and exhausted Dutch population expected a future in which socio–economic security was guaranteed for everyone and was no longer left to chance. This dream had to be realised by seeking high employment and an extensive system of social security and social welfare benefits. The Van Rhijn Commission (established on 7 April 1943)[53] was the *auctor intellectualis* of this philosophy in the Netherlands. It articulated the following legal basis for a complete system of social security benefits encompassing the entire population:

> "The community, organised in the form of the State, is responsible for the social security and protection against want of all of its members, provided that those members themselves do what is reasonable to furnish such social security and protection against want."[54]

The Van Rhijn Commission gained inspiration from overseas. Reports and plans in the United States and Great Britain served as models for the Dutch proposals. Winston Churchill (1874–1965) was instrumental here. Together with the American president Franklin Delano Roosevelt (1882–1945), he was the originator of the Atlantic Charter of 19 August 1941. That charter set out all sorts of freedoms, including 'freedom from want'. Such protection against want was intended to accomplish the following: "to bring about the fullest collaboration between all nations in the economic field, with the object of securing for all improved labour standards, economic advancement, and social security." The ideal here reflected the famous principle derived from Churchill, namely, 'social security from womb to tomb'.[55] This welfare state ideal was taken too far by some, who subscribed to the 'bad-luck-must-be-righted' notion and who wanted to shift any damage which a citizen might suffer primarily to the State (through social security, state funds, state

---

[52]Owen (1995), pp. 226–227.

[53]A.A. van Rhijn (1892–1986) was a secretary-general in several government departments (1933–1940), Minister of Agriculture and Fisheries (1940–1941) and Secretary-General at Social Affairs (1945–1950). His commission was responsible for providing a preliminary overview regarding the principles and main features of social security in the Netherlands.

[54]Social Security Report (1945) II, p. 10.

[55]Jansen and Loonstra (2013), pp. 269–270. A variation on this saying is 'from the cradle to the grave'.

pensions and the like). This goal entails many risks, as "[p]eople who have grown up believing that the state would look always after them, no matter what misfortunes should strike, are now driven to find someone to sue, when they discover that the state will not and cannot deliver on this expectation."[56]

The scope of social security benefits has been reduced recently in countries such as the Netherlands. The costs were no longer affordable. The 'damage' was shifted too much to the community, so that the pressure on private insurance and tort law grew.[57] Atiyah therefore argued that a no-fault system of liability should be developed for accidents and the personal harm ensuing from these and that a system of private and group insurance ought to be implemented for other cases of personal harm.[58] In principle, insurance, which provides the right to a benefit if a particular contingent event occurs, is not concerned about the cause of the event. Whether the party entitled to the benefit was at fault is irrelevant.[59] Atiyah's proposal, however, has hardly generated any response. The part of *casus* under modern private law has been anything but played out. The idea that the owner has to bear the consequences of 'accidents' is still very much alive in all Continental legal systems.

Open Access This chapter is distributed under the terms of the Creative Commons Attribution-Noncommercial 2.5 License (http://creativecommons.org/licenses/by-nc/2.5/) which permits any noncommercial use, distribution, and reproduction in any medium, provided the original author(s) and source are credited. The images or other third party material in this chapter are included in the work's Creative Commons license, unless indicated otherwise in the credit line; if such material is not included in the work's Creative Commons license and the respective action is not permitted by statutory regulation, users will need to obtain permission from the license holder to duplicate, adapt or reproduce the material.

References

Atiyah, P. S. (1997). *The damages lottery*. Oxford.
Bruins, G. W. J. (1906). *Een onderzoek naar den rechtsgrond der schadevergoeding*, diss. 's-Gravenhage.
Cane, P. (1997). *The anatomy of tort law*. Oxford.
Coing, H. (1989). *Europäisches Privatrecht 1800 bis 1914*, II (19. Jahrhundert). München.
De Mul, J. (1994). *Toeval*, inaugural lecture. Rotterdam.
Dernburg, H. (1903). *Pandekten*, 7ᵉ Auflage. Berlijn.
Deroussin, D. (2007). *Histoire du droit des obligations*. Parijs.
Eijsbouts, W. T. (1989). *Recht en toeval. Premissen van 'het beleid' in het licht van de feiten*, diss. Amsterdam.
For the Institutes of Gaius (Gai. Inst.), the Institutes of Justinian (Inst.), the Digests (D.) and the Code of Justinian (C.), the following were utilised: J. E. Spruit & K. Bongenaar, *De Instituten van Gaius*, Second Edition, Zutphen 1994; J. E. Spruit, R. Feenstra & K. E. M. Bongenaar

---

[56]Atiyah (1997), p. 176.

[57]Hartlief (1997), p. 72. I won't get into the role of strict liability.

[58]Atiyah (1997), p. 173 *et seq.*

[59]Hartlief (1997), p. 28.

(red.), *Corpus Iuris Civilis. Tekst en Vertaling*, I, Zutphen/'s-Gravenhage 1993; J. E. Spruit, R. Feenstra & K. E. M. Bongenaar (red.), *Corpus Iuris Civilis. Tekst en Vertaling*, II–VI, Zutphen/ 's-Gravenhage 1994–2001; J. E. Spruit, J. M. J. Chorus & L. de Ligt (red.), *Corpus Iuris Civilis. Tekst en Vertaling*, VII–IX, Amsterdam 2005–2010.

Hartkamp, A. S., & Sieburgh, C. H. (2008). *Mr. C. Asser's Handleiding tot de beoefening van het Nederlandse Burgerlijk Recht, Verbintenissenrecht, De verbintenis in het algemeen*, 6-I*, 13ᵉ druk. Deventer.

Hartkamp, A. S., & Sieburgh, C. H. (2011). *Mr. C. Assers Handleiding tot de beoefening van het Nederlands Burgerlijk Recht, Verbintenissenrecht, De verbintenis uit de wet*, 6-IV*, 13ᵉ druk. Deventer.

Hartlief, T. (1997). *Ieder draagt zijn eigen schade*. Deventer.

Jansen, C. J. H. (2009). 'De lex Aquilia en de moderne onrechtmatige daad'. In K. J. Krzeminski en M.C.A. van den Nieuwenhuijzen (red.), *De Digesten en de receptie van het Romeinse recht in het Nederlandse privaatrecht*. Nijmegen, 231 sqq.

Jansen, C. J. H. (2014). 'Enkele historisch-vergelijkende beschouwingen over de grondslagen van de zaakwaarneming'. In L. van den Berge e.a. (red.), *Historische Wortels van het Recht*. Nijmegen, 41 sqq.

Jansen, C. J. H., & Loonstra, C. J. (2013). 'De personele werkingssfeer van de sociale verzekeringswetten 1900–1960'. In C. J. H. Jansen en C. J. Loonstra, *Opstellen over de historische ontwikkeling van het arbeidsrecht*. Den Haag, 261 sqq.

Kortmann, J. (2005). *Altruism in private law: Liability for nonfeasance and negotiorum gestio*. Oxford: Oxford University Press.

Markesinis, B., & Deakin, S. (2003). *Markesinis and Deakin's Tort Law* (5th ed.). Oxford: Oxford University Press.

Owen, D. G. (1995). 'Philosophical foundations of fault in Tort law'. In D. G. Owen (Ed.), *Philosophical Foundations of Tort Law*. Oxford.

Ranieri, F. (2009). *Europäisches Obligationenrecht. Ein Handbuch mit Texten und Materialien*, 3ᵉ Auflage. Wien/New York.

Rümelin, M. (1896). *Der Zufall im Recht*. Freiburg/Leipzig.

Schulz, F. (1969). *Classical roman law*. Oxford.

Sieburgh, C. H. (2000). *Toerekening van een onrechtmatige daad*, diss. Deventer.

Sociale Zekerheid, I–II. (1945). 's-Gravenhage.

Thomasius, C. (2000). *Larva Legis Aquiliae* [etc.] (M. Hewett, Ed., Trans.). Oxford/Portland: Oregon.

Van Dam, C. C. (2000). *Aansprakelijkheidsrecht. Een grensoverschrijdend handboek*, Den Haag.

Van Zeben, C. J. e.a. (red.). (1981). *Parlementaire Geschiedenis van het Nieuwe Burgerlijk Wetboek*, Boek 6 (Algemeen gedeelte van het verbintenissenrecht). Deventer.

Von Bar, C. (1999). *Gemeineuropäisches Deliktsrecht*, II. München.

Wacke, A. (1984). Gefahrerhöhung als Besitzerverschulden. Zur Risikoverteilung bei Rückgabepflichten im Spannungsfeld der Zurechnungsprinzipien *casum sentit dominus, fur semper in mora* und *versari in re illicita*. In G. Baumgärtel (u.A.) (Hrsg.) *Festschrift für Heinz Hübner*. Berlin/New York, 269 sqq.

Wallinga, T. (2009). Opzet, schuld en toeval. In K. J. Krzeminski en M. C. A. van den Nieuwenhuijzen (red.), *De Digesten en de receptie van het Romeinse recht in het Nederlandse privaatrecht*. Nijmegen, 215 sqq.

Windscheid, B. (1900). *Lehrbuch des Pandektenrechts*, II. Band, 8. Auflage, bearbeitet von Th. Kipp. Frankfurt.

Zimmermann, R. (1996). *The Law of Obligations. Roman Foundations of the Civilian Tradition*. South Africa/Deventer-Boston.

Zimmermann, R. (2011). Roman law and the Harmonization of Private Law in Europe. In A. Hartkamp et al. (Eds.), *Towards a European civil code* (4th ed.). Alphen aan den Rijn/Nijmegen, 27 sqq.

Zweigert, K., & Kötz, H. (1998). *Introduction to Comparative law* (T. Weir, Trans.). Oxford.

Taming Chaos. Chance and Variability in the Language Sciences

Roeland van Hout and Pieter Muysken

Abstract This paper focuses on chance and variability in language, and how the language sciences have dealt with that variability. After describing four types of variability found: (a) Inter-species variability, (b) Inter-language variability, (c) Variability in the linguistic signal within a given language, and (d) Inter-individual variability, the paper discusses the work of two pioneers who have tried to deal with this variability: Joseph H. Greenberg and William Labov. These near-contemporaries have tried to grapple with variability of types (b) and (c), as two separate enterprises. Thus these researchers have tried to separate pure chance or randomness from meaningful variability in two different ways, and in doing so have tried to tame the chaos. For them indeed the mission of linguistics as a discipline is to eliminate chance as much as possible, as the target of any scientific enterprise by definition is to isolate, separate or exclude what cannot be explained or understood. Nonetheless, chance and variability are key elements in language, and a proper understanding of language will take these as the point of departure. What does it mean to say that chance is an inherent property of human language? The paper outlines the beginning of answer to this question.

1 Introduction

The publication of Ferdinand de Saussure's *Cours de linguistique générale* a hundred years ago, in 1916, heralded the beginning of modern linguistics. Since then the field has unfolded and developed into many directions.

Among the achievements of this past century is the discovery of the incredible variability in human language. At the same time this variability continues to present

R. van Hout · P. Muysken (✉)
Faculty of Arts, Radboud University, Nijmegen, The Netherlands
e-mail: p.muysken@let.ru.nl

© The Author(s) 2016
K. Landsman and E. van Wolde (eds.), *The Challenge of Chance*,
The Frontiers Collection, DOI 10.1007/978-3-319-26300-7_14

a set of fundamental puzzles that need to be solved to find the key in explaining and understanding variability as an inherent property of human language. Variability can be found at all levels of language and language use. We may distinguish four types of variability.

(1) *Inter-species variability*: The communication system of humans differs in many ways from that of other species, in the channels used (speech, sign, gesture, body posture), the structure of the code used, and the purposes of communication. Nonetheless, there are also specific features shared to various degrees between human and non-human communication: vocal learning, imitation, structure, exchange patterns, that need to be taken into account.[1]

(2) *Inter-language variability*: The 7000 languages currently identified (a small subset of the languages that have existed over the last 100,000 years or so) vary enormously among each other. Their words and sounds differ, as well as the distinctions they encode, and their grammatical patterns. This is often referred to as the curse of Babylon. A special place is reserved for the many signed languages of the deaf, which differ considerably among each other, but also of course from spoken languages.

(3) *Variability in the linguistic signal within a given language*: Every utterance is unique in its physical properties given shape by the human articulators, which partly reflects aspects of the setting in which it is uttered (formal/informal, for instance), features of the interlocutors (gender, class, education, ethnicity, etc.), and other factors to be identified. The sounds in speaking are complex, with an overwhelming set of details. This is one of the mean reasons why automatic speech recognition is so hard.

(4) *Inter-individual variability*: Despite recent approaches emphasizing the homogeneity within languages, speakers differ on many levels, which allow us to recognize an individual through her or his speech signal. Speakers differ in their linguistic abilities and verbosity, in their communicative styles, in their timbre and voice quality, etc., but also in the perceptual systems they have built up. The same physical or acoustical signal may be perceived differently, not only in segmenting the signal but also in its social evaluation.

[1]The relation between these four levels is the subject of systematic exploration in one of the teams operating in the NWO research consortium Language in Interaction (2013–2023), involving researchers from Nijmegen and Leiden. Pieter Muysken's contribution to this paper is funded through the Language in Interaction grant.

2 The Field

Given these types of variability, there have been two main reactions in the linguistic research community in the recent past.

One important school of thought, *generative linguistics*, was inspired by the towering figure of Avram Noam **Chomsky** (1928-). Chomsky, professor of linguistics at the Massachusetts Institute of Technology for most of his career, simply ignored the variability in natural language. In his work the universal, cognitive principles underlying our formal knowledge of grammar were the target of investigation, rather than the variable and transient actual usage. What underlying abstract patterns play a role in determining the well-formedness (grammaticality) of sentences (viewed as strings of words), and how do we derive the meaning of these strings?

In Chomsky's work, only Type 2 variability was deemed to be of interest, as it was meant to be reduced to a universal, finite set of principles and parameters underlying all human languages. Type 3 and Type 4 variability were considered to either only noise (fine mud grains floating in the water, irrelevant for a hydraulic engineer) or outside the domain of linguistics (being part of psychology or the study of human development). Type 1 variability was assumed to be beside the point, given the uniqueness of the human language faculty.

Other researchers, however, have tried to separate pure chance or randomness from meaningful variability in other ways, and in doing so have tried to tame the chaos. It could be said that for them indeed the mission of linguistics as a discipline is to eliminate chance as much as possible, as the target of any scientific enterprise by definition is to isolate, separate or exclude what cannot be explained or understood. On the other hand, chance or randomness can be made part of a theory on language and language use. The concept that seems to be most relevant in the latter approach is inherent variability, meaning that language is per definition heterogeneous, in its very foundations. This concept does not define however what the role of chance is. Chance in linguistics thus has no special definition, but it is tackled nevertheless from various angles.

A researcher famous for attempting to tackle Type 2 variability is Joseph H. **Greenberg** (1915–2001). Greenberg was an anthropologist and linguist who spent most of his career as a scholar at Stanford. He started out with a study of the influence of Islam on the Hausa in Africa but soon turned to languages. He first attempted to classify all the languages in the world in large groupings (language macro-families). These were generally accepted for Africa, but which met with skepticism for the Pacific and the New World. More important for our concerns, however, is his attempt to find language universals, based on correlations between structural traits, and thus coming to grips with Type 2 variability. For this purpose he created a database with systematic data on around 30 languages from all over the world. Current data bases are much larger, cf. the often cited WALS database (Dryer & Haspelmath 2013). In his work, Greenberg built on earlier studies which had proposed specific 'language types', and therefore this approach is called *linguistic typology*.

The scholar best known for attempting to come to grips with variability of Type 3 is William **Labov** (1928-). Labov was initially trained and employed as an industrial chemist, but soon started using new techniques to record the English spoken around him on the United States East Coast, almost like an engineer (Labov 1972). Initially based at Columbia, but later moving on to the University of Pennsylvania, he has pursued a life-long career in trying to capture Type 3 variation in speech, both theoretically and empirically. How can we systematically study the variability found in everyday language use, and how can we model it in a way that does justice both to the nature of language itself and to the embedding of language in social systems? Why do some people in New York pronounce the /r/ in 'fourth floor', while others leave it out, and what does this tell us about the variable nature of the sound system of New York English? Labov's approach is referred to as *variationist linguistics*.

While the research programs initiated by Greenberg and the one associated with Labov differ in many respects, they share the crucial strategy of attempting to tame the chaos in their data by going to higher levels of aggregation, following the strategy pioneered by Durkheim (1897) in his work on suicide. It is only at the aggregate level of the whole population that we can understand suicide behavior, since we cannot ask individuals afterwards why they did it. While Durkheim's concrete findings have been criticized both from the perspective of Simpson's Paradox[2] and from that of the Ecological Fallacy, the strategy of moving from seemingly chaotic and accidental behavior at the level of separate individuals ('tokens') to general patterns at the level of aggregated groups ('types') has been very successful in many sciences. For Greenberg, the aggregated group was the population of human languages as a whole, for Labov it is the speech community (like the inhabitants of a village, a city, an island, or even a region or country; again the problem of the level of aggregation pops up).

Following in the footsteps of Greenberg and Labov, in this paper we will focus on variability types 2 and 3, reflecting our own expertise.[3] Thus, we will first explore different parts of the language sciences: the chance and variability in the constitution of languages (type 2 variability), and then chance variability in production (type 3 variability) and perception (taking in type 4 variability). Finally we will combine these two perspectives and briefly discuss the consequences for language change. We will focus here on the interaction between biological systems

---

[2]Simpson's paradox, is a paradox in statistics: a trend which appears in different groups of data disappears or reverses when these groups are combined in the sample.

[3]Pieter Muysken is a specialist in inter-language variability and language contact, and Roeland van Hout has worked in the area of variation studies and statistics. Type 4 variability is being studied at the Max Planck Institute for Psycholinguistics in Nijmegen in a group led by Antje Meyer. The work on Type 1 variability is progressing rapidly, but has not yet reached even an interim level of conclusiveness.

and social constructs. The biological systems involved are constrained but open, flexible and adaptive to all kinds of circumstances and they are made up by our articulators, our ears, but also our brains (and even our bodies). The speech they produce must be communicative but transferable and learnable at the same time, to serve the emergence and establishment of communicative networks and social groups.

3 Linguistic Typology: Chance and Variability in the Constitution of Languages

Languages vary in almost infinite ways: their sounds, their words, the order of the words in the sentence, the distinctions encoded. How can we reconcile that variability with the fact that languages also show unity? While there are other dimensions to variability, as noted (cf, our four types of variability), we will focus here on inter-language variability.

3.1 L'arbitraire du signe

The most striking variability no doubt is that in the words of the different languages. Thus the favorite four-legged creature that is being loved and fed in many Western households is called *Hund* in German, *chien* in French, and *perro* in Spanish. In many languages in the Bolivian Amazon it is called *paku* (but the creatures there are not nearly as pampered). Form to meaning mappings are in fact coincidental, as pointed out by Saussure: *l'arbitraire du signe*, the arbitrariness of the sign. There is nothing inherent in dogs that gets them these different names. Is it pure chance only?

Not completely. A good place to start is historical linguistics. It has been known for a long time that words in different languages may or may not be related. The following words are all related:

pater	Latin
padre	Spanish
Vater	German
father	English
vader	Dutch

Indeed, they all go back to a reconstructed Indo-European form *pH$_2$tér 'father' (the subscript on H refers to a particular sound combination). Forms and meanings are passed not only from one generation to another, but also from one language to another, when new languages split off from their predecessors. Variability comes in, but somehow the origin remains visible or deducible, constraining the role of chance by chains of inheritance.

However, there are other factors as well. The fact that a number of languages in Bolivia share the word *paku* is due to word spread or borrowing in language contact. The word went from one language to the other, possibly as the practice spread of keeping dogs as a domestic animal (used for hunting mostly). Thus there is a number of words which have an extremely wide distribution in the languages of the world, such as the words for 'coffee' and 'tea', or quite recently, 'tsunami'.

Besides **inheritance** and **contact**, sometimes the presence of a word has a more intrinsic explanation. Consider the following:

mamma	Dutch, English, Italian, ...	(Europe)
mama	Quechua	(South America)
mama	Lingala, Luo, Swahili	(Africa)
mama	Mandarin Chinese	(E. Asia)

Even though there are striking correspondences here, we assume that these words are not historically related, but that their similarity is due to properties of the vocal tract. Opening the mouth widely to give room to outgoing air produces an a-like sound. Closing it, to stop the air, gives a m-like sound. In combination with a repeating syllable, *ma-ma* is the result. Babies often will have mama as one of their first words, because it is easy to pronounce. Its frequent occurrence is to be explained by ease of pronunciation rather than random developments (Jakobson 1960).

Some intrinsic explanations are referred to as **motivation**. The workings of chance are undone or constrained by factors having to do with the way language is processed, produced and learned. Motivations come in many forms, and are often more quantitative and statistical rather than qualitative and absolute. While there are various ways in which motivation plays a role in the lexicon, its role in the rest of the language system is much more obvious.

One special such type of motivation comes from sound symbolism. A striking example is the *kiki—bouba* effect described by Ramachandran & Hubbard (2001), building on much earlier work by Köhler (1929). Sharp, pointed objects are often referred to as *kiki*, by speakers of very different languages, smooth, rounded objects as *bouba*, when forced to make the choice in a matching experiment.

Within a language, a particular sound combination may be associated with particular sets of meanings. Examples from English include words starting with "sl" to mark frictionless motion:

slide   slick   sled   slip   slither   slosh

However, there is even a much larger class of words with negative or pejorative meanings, some of them related to the previous set.

slab	sleepy	slough
slack	sleet	slovenly
slang	slime	slow
slant	slipshod	sludge
slap	slit	slug
slash	slobber	sluggard
slate.	slog	slum
slattern	slope	slump
slaver	sloppy	slur
slay	sloth	slut
sleek	slouch	sly

Given the mixture of explainable and accidental/occasional forms and structures, a main question in language science is how to detect the mechanisms or processes that connect and perhaps partially explain the heterogeneity or variability by investigating preferential aspects or patternings and how these are related to inherent properties of a language.

While the diversity of human languages and the specific forms they take appear accidental and governed by chance, the chance factor is constrained by all kinds of processes and external factors. Is the consistency between the characteristics of several languages occasional or are there preferential aspects or patterns? Motivation can be **external**, in terms of iconicity, or **internal**, in terms of systemic harmony. We will first give a few examples of **external motivation** through **iconicity**, which makes patterns of variation less accidental.

3.2   Iconicity

Iconic motivation can be defined as pressure from the similarity or analogy between a sign or linguistic structure and its meaning. To give a simple example, when I say: 'I went to buy a book and had an ice cream,' normally I want to indicate that buying the book preceded eating the ice cream. The temporal sequence in the utterance mirrors the temporal sequence of events portrayed. This is temporal iconicity (Givón 1985).

Similarly, there is quantity iconicity. If I say *druk druk* 'busy busy' in Dutch in response to the question 'how are you doing?', I mean to say that I am more than just busy. Reduplication can be iconic in this way, but need not be; in many West African languages reduplicating a predicate makes it into an adjective or noun.

Another set of phenomena linked to quantity iconicity can be illustrated with the following two sets of English prepositions:

of	without
to	until
by	during
in	in spite of
at	because of

On the whole, the prepositions on the left are much shorter than those on the right. They are also much more basic (and often grammatical) in their meaning.

On the whole, short words may have more basic meanings than longer words. This effect is fairly general. Consider some Quechua case endings or postpositions (Muysken 2008):

-pa/-q 'genitive, of'	*-manta* 'ablative, from'
-ta 'accusative'	*-kama* 'until'
-man 'dative, to'	*-rayku* 'because of'
-pi 'locative, in'	*-hina* 'like'

Again we find a correlation between length and meaning complexity.

Sound symbolism may bring about iconicity as well. High front vowels (notably/i/) are associated with small sizes, and low back vowels like /ɑ/ and /ɔ/, with large sizes. Think of French *petit* 'small' (with /i/) and *grand* 'large'(with /ɑ/). There are exceptions, but this may well be a trend when we would study a whole range of languages.

There is also intonational iconicity. In a great many languages, a question has a rising, higher fundamental pitch than a statement. Ohala (1997) links this to the acoustic frequency code, and claims there is possibly a cross-species association of high acoustic frequency with small sizes and low acoustic frequency with large sizes.

3.3 Dependencies

Internal motivation is a complicated issue as well, and subject to much debate, a debate that centers around the concept of dependencies. How does property X of a language system depend on, or how is it predicted by, property Y? There are all kinds of dependencies that have been proposed, with various degrees of success. Indeed, some people would claim that finding and accounting for these dependencies is the key mission of linguistics as a discipline.

To take a simple example, consider a five vowel system such as the one of Spanish:

This is highly symmetrical (for every front vowel there is a back vowel, and vice versa), and fully occupies the 'vowel space'. Notice also that it contains an uneven number of vowels, with a single /a/ at the bottom.

Contrast this with a (non-existent) system like:

This system is not at all symmetrical, and further more does not exploit the 'high' vowels /i/ and /u/ in the vowel space.

The following table, from Schwartz et al. (1997, p. 244), shows the distribution of vowel systems in a data set of 189 languages from different parts of the world. The odd-numbered symmetrical systems are marked in bold italic, and constitute 144 of the total set of 189 languages. The non-symmetrical language are by far more rare (41 versus 148), often being left asymmetrical (more front than back vowels). In the front the vowel space is simply larger than in the back.

	Number of languages		
Number of vowels	Symmetrical	Left	Right
3	*17*	1	0
4	0	14	4
5	*97*	1	0
6	3	12	4
7	*23*	0	0
8	0	3	2
9	*7*	0	0
10	1	0	0
Total	148	31	10

The symmetries in the vowel system can be viewed as a case of structural dependency: the presence of /o/ in Spanish 'depends on' or is 'predicted by' the presence at the same level of /e/, and thus not a pure accident, even though the fact that Spanish has a five vowel system is in itself accidental. Related languages such as Portuguese and French have more complicated vowel systems.

Similar symmetries are found in the consonants. Consider the stops of Cuzco Quechua, which includes a regular, an aspirated (pronounced with aspiration), and an ejective (pronounced with a sudden burst of air) series:

Regular	p	t	č	k	q
Aspirated	p^h	t^h	$č^h$	k^h	q^h
Ejective	p'	t'	č'	k'	q'

This system is highly symmetrical: for each regular stop there is an aspirated and an ejective stop. Another Quechua variety, Ecuadorian Quechua, has a slightly simpler system, which is likewise symmetrical:

Regular	p	t	č	k
Aspirated	p^h	t^h	$č^h$	k^h

The overall presence of aspirated and ejective consonants in these varieties of Quechua may be an accident (which has a historical explanation through influence from a neighboring language, Aymara), but the fact that they come in a series or sets can be viewed as a result of a dependency, and hence not as accidental. Various theories have been proposed to explain sound symmetries, but this need not concern us here.

The dependencies that are found in the languages of the world are the object of research in language typology, the research program started by Greenberg. The team of Frans Plank at the University of Konstanz has created a data base containing no less than 2029 of statements about such dependencies.

A typical example (#1 in fact in the list compiled in Konstanz), based on Greenberg (1963), would be:

IF adpositions precede their noun phrases (i.e., they are prepositions), THEN head nouns almost always precede their attributive nouns (genitives or possessor (poss) phrases).

This would predict a dependency such as:

'**In** the house' (preposition) > 'The house **of Mary**' (poss)[4]

Dependency #2 is the complement of #1:

IF adpositions follow their noun phrases (i.e. they are postpositions), THEN head nouns almost always follow their attributive nouns (genitives).

---

[4]Notice that in English we also have 'Mary's house', which illustrates the problems in making general statements about a language, of the type *Language X has Property Y*.

Thus we find in Quechua the following examples:
wasi-pi [house-in] (postposition) > ***Mariya-q*** *wasi-n* [Mary-poss house-her]
These statements of dependencies are generally statistical in nature: there are always some exceptions to the general pattern.

Much research has been done on trying to explain these dependencies in terms of processing constraints, but many questions remain in this general area, including the question to which extent such dependencies are truly universal, or lineage-specific, as argued by Dunn et al. (2011)? Also: why are some dependencies (almost) without exceptions, and others more a tendency than an absolute?

Typological patterns and dependencies are the result of inheritance and contact, but at the same time of internal motivation and external social factors, unfolding in time and space. We can use chance to model this enormous variability, admitting that our understanding is incomplete and our models are too global to catch the complexity of languages. The alternative is to give room to the concept of chance/probability, by including it as an inherent property of the language system or to put it somewhere on the interface between language and the social, epigenetic factors in which language and language use are embedded.

Summarizing and taking a very broad perspective, we can say that linguists have discovered a number of things in the typological paradigm initiated by Greenberg:

(a) There is much more variability than had been imagined. Many putative 'universals of language' turn out to have counterexamples somewhere among the 7000 languages in the world (Evans and Levinson 2009).
(b) Many universals hold only for a large group of languages.
(c) There are regularly exceptional pattern, some of which can be classified as 'rarissima' (Wohlgemut and Cysouw 2010).
(d) There are 'local optima', i.e., correlations and dependencies which hold between features in specific language groups. Some of these are lineage specific, i.e., limited to specific languages.
(e) Many majority solutions may be due to a functional explanation or constraint. Examples: (i) The almost universal noun/verb distinction may be linked to the need of humans to be able to refer to both objects and actions/activities. (ii) The almost universal ordering ... subject ... object ... (but not ... object ... subject ...) may be linked to the facts that subjects are often the topic and topics occur early in the sentence.

Could some functional constraints be indeed absolute, and hold for all languages because they are wired into the human brain as a result of evolution?

4 Variationist Linguistics: Chance in Production and Perception

The variability in the acoustic signal is enormous. No two speech sounds are the same, because of varying physical circumstances, differences between vocal tracts and the complexity of producing sounds. Nevertheless, in concrete interactions speakers and listeners interact smoothly and understanding seems to proceed in a self-evident way. Speakers seem to abstract from concrete sounds, handling language on the level of words and utterances. On that level however, the problem of variability reoccurs. Speaking implies making choices, continuously, between constructions, between words and even between pronunciations. To what degree is variation free and what are the constraints? Chance plays a role in the many decision processes involved in speaking, but to estimate its role we have to explain as much as possible the role of all sources of variation involved in the process of communication, i.e. in using language. Substantial parts of the variability is reducible to (a) priming by the communicative context, (b) intention of the participants in the process of communication (including 'free will'), (c) language internal constraints (properties relating the various linguistic elements; internal motivation), and (d) external constraints that characterize the speech community involved as a whole (community profile) and its individual speakers (their social profiles). These factors are rooted in the way we speak (production) and the way we perceive and understand (perception).

Leaving out word final t

The complexity can be illustrated by a simple phenomenon, **t-deletion**, which may have different sources:

1. the distinction between *nie* and *niet* (not) and *da* and *dat* (that), which are different small words, stored in the lexicon of many southern speakers in the Dutch language area;
2. phonetic reduction in consonant clusters at the end of words: *herfst* vs *herfs* (autumn), resulting in the absence of the word final plosive sound in the speech produced, a phenomenon that is present in many other languages;
3. phonetic reduction may be restricted by the morphological status of a word; in Nijmegen reduction occurs less in past participles than in nouns (*feest* (party, noun) en *gefeest* (partied, past particple, verb *feesten*); the same constraint has been found for American English (Guy 1980).
4. morphological analogy may lead to the deletion of the/t/in first person present tense in words ending in consonant clusters: *ik vin* vs. *ik vind* (I think);

> 5. this analogy wrongly applies to specific irregular past forms: *ik moes* vs. *ik moest* (I must).
>
> All these sources of variation are active in speakers from the town of Nijmegen, for instance.

The differences between speakers can partly be explained by using a mixture of factors, from internal and also external origin, but we cannot, despite advanced statistical modeling, predict what happens at the level of the individual occurrence. The predictions are fairly correct only on higher levels of aggregation. Predictions are sometimes fairly successful in explaining inter-individual variability by taking into account the social profiles of speakers, including social background characteristics such as age, gender and educational background. It means that speech is indexical for social characteristics of the speaker: the speech signal carries social meaning. Young people are marked by other speech features than older speakers. Parts of the variability keeps out of touch however, as unexplained error, perhaps based on pure probability.

Even if much language behavior is probabilistically determined, certain behavior lies closer to our consciousness threshold, implying that it is more under our control (avoid using *zij* (them, subject) instead of *hun* (them, the object form). The problematic relationship between consciousness and variability is a classical problem in studies of language variation and we have to investigate the type of relationships between them by using the scale [unconscious/probabilistic] .... [conscious/categorical], to ascertain that variability is not the outcome of insufficient cognitive control or interfering cognitive mechanisms.

Another approach is the distinction between active control on the level of the speaker ('agency') and a passive, more computationally oriented approach where 'control' is being carried out by 'variable constraints'. Speakers have possibilities of cognitive control over their speech and language behaviour. The impact of control can clearly be observed on higher levels of aggregation, where decisions are being taken and which can be successful. In French, there is active policy to resist word borrowings and to use native words. In English, the old counting order of one-and-twenty, five-and-ninety has been replaced by the order of going from larger to smaller numbers (twenty-one, ninety-five). The numerals between 10 and 20 were kept out from this revision. It links cognitive control and social forces.

The role of social forces can be illustrated as well by the course of sound changes in language. The Dutch vowel system is currently undergoing several related sound changes. The tense mid vowels [e:,ø:,o:] tend to become realized as diphthongs [ei,øy,ou]. The diphthongs [ɛi,œy,ɔu] are beginning to lower (referred to as 'Polder Dutch'; Stroop 1998, van Heuven, van Bezooijen & Edelman 2005; Jacobi 2009), causing e.g. *kijk* 'look' to sound more like [kaik] rather than [kɛik].

Change means that variation may lead to a change in the speech or language of a community. Again, different views can be proposed whether sound change originates in production or in perception. Is it the speaker, who realizes speech forms differently because of structural/systemic constraints—for example, pronouncing the Dutch verb *kijken* (to watch) with a novel vowel [ai] rather than conservative [ɛi] to preserve distinctiveness from *keken* (past tense *kijken*), whose vowel [eː] is changing into [ei] (Stroop 1998; Jacobi 2009)—or articulatory constraints (e.g. Ohala 1983; Browman & Goldstein 1989; Zsiga 1997)? Is it perhaps the listener, who may misperceive speech forms (e.g. Ohala 1981; Blevins 2004)? Or is it because the novel speech form is positively evaluated, leading to the desire to sound like and imitate the other speaker (Giles 1973; Gussenhoven 2000; Pierrehumbert 2001; Bybee 2002)? In sum, in sound change at least three different perspectives play a role: production, perception and evaluation, and the complex interplay between these three perspectives helps to define the process of selecting new variants in the language community (Yu 2013). It makes clear that we have to add the social embedding of patterns of language variation to understand what is going on in a language.

The aim of variationist linguistics is to explain patterns of variation as much as possible by maximizing the sources of variation involved in language use: the properties of the vocal tract and the ears (both being originally biological sources), social forces (the environment, the social group) and cognitive processes (the brains).

5   Chance: Conundrum or Inherent Property?

Now that we have described the ways in which both language typology and variationist linguistics have attempted to come to grips with accidental aspects of language behavior, we can try to understand where they intersect.

First of all, there is no principled difference between variation between (studied by Greenberg) and variation within (studied by Labov) languages. We can give an example from syntax and one from phonology.

> In syntax, we often find, as was discovered by Greenberg, that the position of the verb at the end of the sentence (called SOV) correlates with that of possessor (poss) phrases before the noun, as in the following Quechua example:
>
> Mariya wasi-ta **riku-n** [Mary house-object see-s] ⇔ **Mariya-q** wasi-n [Mary-poss house-her]
>
> Likewise, a verb in the middle of the sentence often correlates with a possessor phrase after the noun, as in Spanish:
>
> Maria **ve** la casa [Mary sees the house] ⇔ la casa **de Maria** [the house of Mary]
>
> This patterns holds at the level of a large language sample. However, Luján et al. (1984) have shown that it also holds with the bilingual Quechua/Spanish

> speaking community of Cuzco: those Spanish varieties more influenced by
> Quechua show the Quechua word order both in verb placement and in pos-
> sessor placement, leading to patterns such as:
>
> La casa Maria **ve** [the house Mary sees] ⟺ **de Maria** la casa [of Mary the
> house]

Thus syntactic variation between languages may also occur within a single language community, and there is no reason why it should be different.

We also find instances in pronunciation where the same variation patterns occur at the community and at the global level. The rhotic consonant/r/comprises a large class of sounds. Most language have a rhotic consonant (about 75 % of the world's languages) (Maddieson 1984). The most common rhotic is the alveolar trill (with the tongue tip), occurring in about half of the languages of the world (Maddieson 1984), but many other variants are found, the uvular trill being one of the infrequent ones (but being the standard pronunciation in French and German). Ladefoged and Maddieson (1996, p. 235) point out that all different forms of rhotics in the languages of the world occur as well in the various dialects of English. The same is true for German (Wiese 2011).

In a study of the/r/in the Dutch language area Sebregts (2015) distinguished 20 different rhotic forms. He did not study dialects but standard Dutch as spoken by ordinary speakers. The different forms he found are grouped in six variant types in the figure below, where their distribution is given for ten towns (with a sample of about 40 speakers per town), six in the Netherlands (n) (upper part of the map) and four in Flanders (f) (lower part of the map). The bars represent the six /r/ variants.

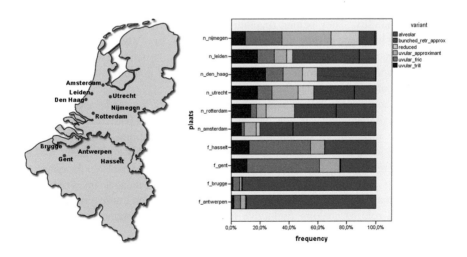

Alveolar /r/ variants are realized with the tongue tip, the uvular variants with the back of the tongue. A trill gives a regular impression, a fricative is marked by friction. An approximant is a underachieved realization. The bunched variant is the vowel-like realization, which came to be part of standard Dutch in the Netherlands as a post vocalic variant ever since the 1960s. The bars in the figure above show how different the pronunciations are between towns, but even within a number of towns.

These two intersections show that variation between (type 1 variability) and within (type 2 and 3 variability) share the same linguistic characteristics and uses the same sources of linguistic elements or components. That is an important conclusion, which also means that chance and variation in language are not as such a source for language evolution. Language can be related, as family members, and inheritance is an important aspect in the historical development of languages, but specific types of languages or language structures are not better equipped to deal with social life, thinking, or culture. Language change does not result in the selection of a best language. The only filter that seems relevant is learnability of a language. The language has to be transmitted from one generation to the next. Children need to shape their own language on the basis of the input of their parents and other speakers.

The intersection of the three types of variability seem to help us in understanding how quickly languages may change, although we admittedly do not understand completely how specific structures may originate from other ones. We have to investigate further the elements involved in making human languages.

We have explicitly formulated this as 'human language making', to emphasize the active role of humans in creating communication through language. They use their mouths and ears, their primary biological sources, which give them almost infinite possibilities to shape sound structures to communicate. The adaptiveness and flexibility of human sound systems at the same time create probabilistic patterns of variability. They belong together. What is the form of this link on other levels of the language system, like words, morphemes, or syntactic structures? The link seems recursive. We see again many possibilities, structures, with open ends, ready to adapt to the communicative needs. That means that probabilistic patterns are fundamental in human language.

In many applications in language research probabilistic properties are becoming part of the (computational) models developed. That applies in particular to language and speech technology. In automated computer translations several software methods are used, among which probabilistic approaches play a prominent role in establishing relationships between the languages involved and between the concrete constructions and linguistic schemes belonging to the languages involved. One could say that probabilistic grammars take over, but even more crucial is the fundamental role of analogy (inductive patterns, based on frequency patterns in language use and matching patterns of language use).

This development runs counter to the basic assumption of the conception of language and language structure as a rule system. This approach was dominant over the last decades in language sciences, in which Chomskyan linguistics handled rules (or concepts related to rules, like movement) as absolute entities, excluding whatever probabilistic mechanisms. Variability was excluded by defining the research object of the language sciences as the competence of the ideal speaker/hearer, all variability being excluded and related to performance factors.

Assuming homogeneity deprives chance from being a conundrum. This is a wrong point of view that deprives linguists from the proper drive to explain the enormous amount of variation in languages. It is the very task of linguistics to solve the conundrum of variability, the curse of Babel. In doing so, we need to involve cognition (the brains), but also the way we construct social reality and the social group(s) we belong to. Cognition is not only an inside property of the brains, it is the outcome of social interaction. Language is, as Labov states, outward bound.

What does it mean to say that chance is an inherent property of human language? It means that language has infinite ways of expressing meaning, often careful ways, but not always. At the same time it means that so many different subsystems are being involved that their interactions can be understood in the end, hopefully, but not predicted. In understanding language variability it remains fundamental to solve the link between the individual and her/his group. Many patterns of variation are defined by the level of aggregation and that certainly applies to language and social behavior.

Open Access This chapter is distributed under the terms of the Creative Commons Attribution-Noncommercial 2.5 License (http://creativecommons.org/licenses/by-nc/2.5/) which permits any noncommercial use, distribution, and reproduction in any medium, provided the original author(s) and source are credited. The images or other third party material in this chapter are included in the work's Creative Commons license, unless indicated otherwise in the credit line; if such material is not included in the work's Creative Commons license and the respective action is not permitted by statutory regulation, users will need to obtain permission from the license holder to duplicate, adapt or reproduce the material.

References

Blevins, J. (2004). *Evolutionary phonology: The emergence of sound patterns*. Cambridge: Cambridge University Press.

Browman, C., & Goldstein, L. (1989). Articulatory gestures as phonological units. *Phonology, 6*, 201–251.

Bybee, J. (2002). Word frequency and context of use in the lexical diffusion of phonetically conditioned sound change. *Language Variation and Change, 14*, 261–290.

Dryer, M. S., & Haspelmath, M. (Eds.). (2013). *The World Atlas of language structures online*. Leipzig: Max Planck Institute for Evolutionary Anthropology.

Durkheim, D. É. (1897). *Le suicide*. Alcan: Étude de sociologie. Paris.

Dunn, M., Greenhill, S. J., Levinson, S. C., & Gray, R. D. (2011). Evolved structure of language shows lineage-specific trends in word-order universals. *Nature, 473*, 79–82.

Giles, H. (1973). Accent mobility: A model and some data. *Anthropological Linguistics, 15*, 87–105.

Givón, T. (1985). Iconicity, isomorphism, and non-arbitrary coding in syntax. In John Haiman (Ed.), *Iconicity in Syntax* (pp. 187–220). Amsterdam: Benjamins.

Greenberg, J. H. (1963). Some universals of grammar, with particular reference to the order of meaningful elements. In J. H. Greenberg (Ed.) *Universals of Language* (pp. 73–113). London: MIT Press.

Gussenhoven, C. (2000). On the origin and development of the central franconian tone contrast. In: A. Lahiri (Ed.), *Analogy, Levelling, Markedness. Principles of change in phonology and morphology* (pp. 215–260). Berlin/New York: Mouton de Gruyter.

Hinton, L., Nichols, J., & Ohala, J. J. (Eds.). (1994). *Sound symbolism*. Cambridge: Cambridge University Press.

Jacobi, I. (2009). *On variation and change in diphthongs and long vowels of spoken Dutch.* Amsterdam: University of Amsterdam.

Jakobson, R. (1960). Why 'Mama' and 'Papa'? In B. Kaplan & S. Wapner (Eds.), *Perspectives in psychological theory: Essays in honor of Heinz Werner* (pp. 124–134). New York: International Universities Press.

Köhler, W. (1929). *Gestalt psychology*. New York: Liveright.

Labov, W. (1972). *Sociolinguistic patterns*. Philadelphia: University of Pennsylvanias Press.

Ladefoged, P., & Maddieson, I. (1996). *The sounds of the world's languages*. Oxford: Blackwell.

Luján, M., Minaya, L., & Sankoff, D. (1984). The universal consistency hypothesis and the prediction of word order acquisition stages in the speech of Bilingual children. *Language, 60*, 343–371.

Maddieson, I. (1984). *Patterns of sound*. Cambridge: Cambridge University Press.

Nicholas, E., & Levinson, S. C. (2009). The myth of language universals: Language diversity and its importance for cognitive science. *Behavioral and Brain Sciences, 32*, 429–492.

Muysken, P. (2008). *Functional categories*. Cambridge: Cambridge University Press.

Ohala, J. (1983). The origin of sound patterns in vocal tract constraints. In P. F. MacNeilage (Ed.), *The production of speech* (pp. 189–216). New York: Springer.

Ohala, J. (1981). The listener as a source of sound change. In C. S. Masek, R. A. Hendrick & M. F. Miller (Eds.). In *Proceedings of the Chicago Linguistics Society 17: Papers from the Parasession on Language and Behaviour* (pp. 178–203).

Ohala, J. J. (1997). Sound Symbolism. *In Proceedings of 4th Seoul International Conference on Linguistics [SICOL]* (pp. 98–103). August 11–15, 1997.

Pierrehumbert, J. (2001). Exemplar dynamics: Word frequency, lenition, and contrast. In J. Bybee & P. Hopper (Eds.), *Frequency effects and the emergence of lexical structure* (pp. 137–157). Amsterdam: John Benjamins.

Ramachandran, V. S., & Hubbard, E. M. (2001). Synesthesia—A window into perception, thought, and language. *Journal of Consciousness Studies, 8*, 3–34.

Saussure, F. de. (1916). *Cours de linguistique générale*. In C. Bally & A. Sechehaye, with the collaboration of A. Riedlinger. Lausanne and Paris: Payot; (W. Baskin, Trans.). *Course in General Linguistics*. Glasgow: Fontana/Collins, 1977.

Schwartz, Jean-Luc, Boë, Louis-Jean, Vallée, Nathalie, & Abry, Christian. (1997). Major trends in vowel system inventories. *Journal of Phonetics, 25*, 233–253.

Sebregts, K. (2015). The sociophonetics and phonology of Dutch r [PhD thesis]. Utrecht: LOT.

Stroop, J. (1998). *Poldernederlands: Waardoor het ABN verdwijnt*. Amsterdam: Bert Bakker.

van Heuven, V. J., van Bezooijen, R., & Edelman, L. (2005). Pronunciation of/Ei/in avant-garde Dutch: A cross-sex acoustic study. In M. Filppula, J. Klemola, M. Planet & E. Penttila (Eds.), *Dialects across borders* (pp. 185–210). Amsterdam: John Benjamins.

Wiese, R. (2011). The representation of rhotics. In M. van Oostendorp, C. Ewen, E. Hume & K. Rice (Eds.), *The Blackwell Companion to Phonology* (pp. 711–729). Malden, MA & Oxford: Wiley-Blackwell.

Wohlgemuth, J., & Cysouw, M. (Eds.). (2010). *Rara & Rarissima: Documenting the fringes of linguistic diversity*. Berlin: De Gruyter.

Yu, A. (2013). Individual differences in socio-cognitive processing and the actuation of sound change. In A. C. L. Yu (Ed.), *Origins of sound change: Approaches to phonologization* (pp. 201–227). Oxford: Oxford University Press.

Zsiga, E. C. (1997). Features, gestures, and Igbo vowels: An approach to the phonology/phonetics interface. *Language, 73*(2), 227–274.

Biographies

Harold Bekkering (1965) has been Professor of Cognitive Psychology at Radboud University since 2002. Before, he worked at universities in Maastricht, St. Louis, and Groningen, and he was a senior scientist at the Max Planck Institute for Psychological Research in Munich. At the moment, he is member of the Board of Directors of the Donders Institute for Brain, Cognition and Behaviour. Harold's research interests cover the broad field of Cognitive Neuroscience including Cognitive, Social and Developmental Psychology, Cognitive Robotics and Educational Neuroscience. His main interest is to unravel learning mechanisms in the Brain, e.g., in Social Interaction. The idea that learning is to improve probabilistic inference of the observable world is the basis for his contribution to the topic of coincidence.

Noortje ter Berg (1980) studied Communication Management at the University of Applied Sciences in Utrecht and subsequently Religious Studies at Radboud University. She is currently Programme Director at the Radboud Honours Academy and served as Project Manager for this book. Her interest in chance derives from an underlying search for the various ways people give meaning to life.

Han Brunner (1956) studied Medicine in Groningen. He specialized in Clinical Genetics at Radboud University, where he obtained his Ph.D. in 1993. He has been head of the Department of Human Genetics at Radboud University Medical Center (Radboud UMC) since 1998, and also of the Department of Clinical Genetics at Maastrichtumc from 2014 onwards. Han is a member of the Academia Europea as well as of the Royal Netherlands Academy of Arts and Sciences (KNAW). He is a Knight in the Order of the Dutch Lion. His research begins with patient observations as the starting point for molecular investigation of intellectual disability, and human behaviour.

Michiel van Elk (1980) studied Philosophy, Biological Psychology, and Psychology of Religion in Utrecht, Amsterdam, and Nijmegen. He obtained his Ph.D. in Cognitive Neuroscience at the Donders Institute for Brain, Cognition and Behaviour, followed by a post-doctoral position at the École Polytechnique Fédérale de Lausanne, Switzerland. He is currently working as a researcher at the University of

© The Author(s) 2016
K. Landsman and E. van Wolde (eds.), *The Challenge of Chance*,
The Frontiers Collection, DOI 10.1007/978-3-319-26300-7

Amsterdam. Michiel's research focuses on the neurocognitive and psychological basis of religious and spiritual beliefs and experiences.

Karl Friston (1959) studied Natural Sciences (Physics and Psychology) at the University of Cambridge and went on to complete his medical studies at King's College Hospital, London. He is a theoretical neuroscientist and an authority on brain imaging. Currently he is a Professor of Neuroscience at University College London. Karl received the first Young Investigators Award in Human Brain Mapping (1996) and was elected a Fellow of the Academy of Medical Sciences (1999). In 2000 he was President of the Organization of Human Brain Mapping. In 2003 he was awarded the Minerva Golden Brain Award and was elected a Fellow of the Royal Society in 2006. In 2008 he received a Medal, Collège de France and an Honorary Doctorate from the University of York in 2011. He became of Fellow of the Society of Biology in 2012, received the Weldon Memorial prize and Medal in 2013 for contributions to Mathematical Biology and was elected as a member of EMBO (excellence in the life sciences) in 2014.

Jelle Goeman (1976) is mathematician and historian. He obtained his Ph.D. in Statistics from Leiden University in 2006. He worked at Leiden UMC and Imperial College, London, before joining Radboud UMC as a Professor of Biostatistics in 2013. He obtained Veni and Vidi grants from the Netherlands Organisation for Scientific Research (NWO). His research focuses on statistical inference in high-dimensional data. Variation, chance, and probability are at the core of his field, his motto being: "Once we really understand the question, the answer is often surprisingly simple."

Olivier Hekster (1974) obtained his Ph.D. in History from Radboud University in 2002. He was a lecturer in Ancient History at Wadham College, Oxford (2001–2002) and Fellow and Tutor in Ancient History at Merton College, Oxford (2002–2004) before taking up the Chair in Ancient History at the Radboud University in 2004. From 2005 to 2010, he was a member of The Young Academy. Olivier's research focuses on the role of ideology in ancient Rome, specifically on Roman imperial representation. He is particularly interested in the ways Roman emperors employed different 'media' to broadcast their image, and in the reception of that image by the heterogeneous population of the Roman Empire.

Roeland van Hout (1952) obtained his Ph.D. in Linguistics from Radboud University on Sociolinguistic Variation in the Dialect of Nijmegen. He specialized in the methodology and statistics of empirical language research, and he was Professor on Methodology of Empirical Linguistics from 1995–2002 at Tilburg University. For a long period he has been Research Director of the language research institute CLS, both in Tilburg and later on in Nijmegen, between 1995 and 2005. Roeland has been a Professor of Applied and Variation Linguistics at the Department of Linguistics of Radboud University since 2002. Language variation, both from a sociolinguistic and geographical angle, has always been at the core of his research interests, including the role of chance and probability in the configurations of linguistic distributions.

Corjo (C.J.H.) Jansen (1961) obtained his Ph.D. in Law from the University of Utrecht in 1987. He was Associate Professor at the University of Groningen from 1990–1998. He has been a Professor of Legal History and Civil Law at Radboud University since 1998. He was extra-ordinary Professor of Roman Law at the University of Amsterdam from 2001 to 2012. He was Dean of the Faculty of Law of the Radboud University from 2003 to 2005 and from 2008 to 2010. Corjo has been Chairman of the Business and Law Research Centre since 2007. His research is mainly concerned with Justinian Roman Law, the history of Civil Law in the 19th century, and the administration of justice in the Second World War.

Eelke Jongejans (1975) obtained his Ph.D. in Biology from Wageningen University in 2004. He was subsequently a postdoc at Pennsylvania State University and Radboud University, where he now works as a tenured researcher at the Animal Ecology group. Eelke studies the impact of environmental drivers on the spatial demography of animals and plants. He wants to understand how ecological and evolutionary processes at the individual level integrate and scale-up to population dynamics. His focus is mainly on ecological frameworks and models that can augment the scientific underpinning of management of invasive or endangered species. Within the Centre for Avian Population Studies, for instance, he aims to understand why certain bird species are declining and to develop tools to detect critical declines as soon as possible.

Hans de Kroon (1959) received his Ph.D. from Utrecht University in 1990. After a postdoc at Indiana University (USA) he returned to the Netherlands with a Royal Academy Research Fellowship at Utrecht University. In 1994 he accepted a position as Assistant and subsequently Associate Professor at Wageningen University. Since 2000 Hans has been aProfessor of Plant Ecology at Radboud University. His work focuses on plant traits, plant interactions, plant populations and questions of biodiversity maintenance. He combines experimental approaches with population modelling. In recent years, his modelling expertise was also applied to birds, together with partners at Radboud University campus.

Klaas Landsman (1963) obtained his Ph.D. in Theoretical High-Energy Physics from the University of Amsterdam in 1989. He was a research fellow at the University of Cambridge from 1989 to 1997, interrupted by an Alexander von Humboldt Fellowship at Hamburg in 1993–94. He was subsequently a Royal Academy Research Fellow at the University of Amsterdam from 1997 to 2002, and obtained a Pioneer Grant from NWO in 2002. Klaas has held the Chairs in Analysis and subsequently in Mathematical Physics at the Radboud University since 2004, and in 2011 was awarded the Bronze Medal of this university for his outreach work in mathematics. His research is mainly concerned with non-commutative geometry and with the mathematical foundations of quantum theory. The latter topic lies behind his interest in (pure) chance and probability.

David R. Loy (1947) obtained his Ph.D. in comparative philosophy from the National University of Singapore in 1985. He taught at Bunkyo University in Japan 1991–2005 and Xavier University in Cincinnati, Ohio 2006–2010. He has been a

Zen practitioner since 1971 and is a teacher in the Sanbo Zen tradition. He writes and lectures on contemporary Buddhism and is especially interested in the social and ecological implications of Buddhist teachings.

Christoph Lüthy (1964) studied Philosophy and Modern Languages in Oxford, Physics in Basel, and History of Science at Harvard where he obtained his Ph.D. in 1995. After some postdoc years in Italy and Germany, he ended up at Radboud University where he holds a chair in the History of Philosophy and Science. A specialist in the early modern period, Christoph works particularly on natural philosophy, matter theory, ontology, and the logic of scientific imagery. He also nurtures a passion for the implications of evolutionary biology for the philosophy of mind.

Pieter Muysken (1950) is Professor of Linguistics at Radboud University and at Stellenbosch University, having previously taught at Amsterdam and Leiden. He did his undergraduate work at Yale University and obtained his Ph.D. at the University of Amsterdam. He is a member of the KNAW and the Max-Planck-Gesellschaft. Prizes awarded to him include the Bernhard Prize, the Prix des Ambassadeurs, and the Spinoza Prize (1998, from NWO). He was decorated with a Knighthood in 2008. His main research interests are language contact, Andean linguistics, and Creole studies.

Carla Rita Palmerino (1969) obtained her Ph.D. in the History of Science from the University of Florence in 1998. Subsequently she has been affiliated to the Center of the History of Philosophy and Science of Radboud University, where she was appointed Professor in the History of Modern Philosophy in 2014. She is also part-time Professor of Philosophy at the Open University of the Netherlands in Heerlen. Her research focuses on early modern science and philosophy, notably on theories of matter and motion, on the debate concerning the ontological status of mathematical entities, and on the heuristic and polemical function of thought experiments.

Stephanie Rosenkranz (1965) is Professor of Multidisciplinary Microeconomics at Utrecht University School of Economics. She obtained her Ph.D. in Economics from the Humboldt University in Berlin and was a postdoc at the J.L. Kellogg Graduate School of Management, Northwestern University. She subsequently obtained her Habilitation in Economic Theory at the University of Bonn. Her research topics include theoretical and experimental economics, industrial organization, social networks and behavioural economics.

Sebastiaan Terwijn (1969) studied Mathematics in Amsterdam and Heidelberg. He obtained his Ph.D. from the University of Amsterdam in 1998, and his Habilitation from the Technical University of Vienna in 2008. He held temporary positions in Munich, Vienna, and Amsterdam before moving to Radboud University in 2010. His research is in mathematical logic, in particular computability and complexity theory, but he has also worked in proof theory and on topics in theoretical computer science such as information theory and computational learning theory. A particular emphasis of his work has been on the combination of logic and probability theory, to wit, computable measure theory, Kolmogorov complexity, and algorithmic randomness.

Johannes M.M.H. Thijssen (1957) is Professor of History of Philosophy and Dean of the Faculty of Philosophy, Theology and Religious Studies at Radboud University. In the past, he was a post-doc at Harvard University and UC Santa Barbara and the recipient of a KNAW research fellowship and various grants from NWO, among which a Pioneer Grant. He is currently working on an essay about the connection between the rise of scientific thinking and the decline of philosophy as a way of life.

Utz Weitzel (1967) is Professor of Finance at Radboud University and at Utrecht University School of Economics. He obtained his Ph.D. in Economics from Humboldt University in Berlin. Before his appointment in Nijmegen, he was affiliated with Utrecht University and the Max Planck Institute of Economics in Jena. His research topics include experimental and behavioural finance/economics, corporate finance, and decision making under uncertainty.

Ellen van Wolde (1954) obtained her Ph.D. in Biblical Studies from Radboud University in 1989. She was a professor at the Faculty of Theology of University of Tilburg from 1992–2008, and has held the chair in Textual Sources of Judaism and Christianity at Radboud University since 2009. In 2005 she was appointed a member of the KNAW, becoming a member of its Executive Board in 2011. Ellen's research is mainly concerned with the Old Testament Books of Genesis and Job, and with methodological approaches that acknowledge the role culture and language plays in the formation of biblical texts. A related field of interest of hers is the question how chance, bad luck, or coincidence were explained in ancient cultures and religions, especially in so far as these explanations still influence our present views.

Titles in this Series

Quantum Mechanics and Gravity
By Mendel Sachs

Quantum-Classical Correspondence
Dynamical Quantization and the Classical Limit by Dr. A.O. Bolivar

Knowledge and the World: Challenges Beyond the Science Wars
Ed. by M. Carrier, J. Roggenhofer, G. Küppers and P. Blanchard

Quantum-Classical Analogies
By Daniela Dragoman and Mircea Dragoman

Life—As a Matter of Fat
The Emerging Science of Lipidomics by Ole G. Mouritsen

Quo Vadis Quantum Mechanics?
Ed. by Avshalom C. Elitzur, Shahar Dolev and Nancy Kolenda

Information and Its Role in Nature by Juan G. Roederer

Extreme Events in Nature and Society
Ed. by Sergio Albeverio, Volker Jentsch and Holger Kantz

The Thermodynamic Machinery of Life
By Michal Kurzynski

Weak Links
The Universal Key to the Stability of Networks and Complex Systems by Csermely Peter

The Emerging Physics of Consciousness
Ed. by Jack A. Tuszynski

Quantum Mechanics at the Crossroads
New Perspectives from History, Philosophy and Physics Ed. by James Evans and Alan S. Thorndike

Mind, Matter and the Implicate Order
By Paavo T.I. Pylkkanen

Particle Metaphysics
A Critical Account of Subatomic Reality by Brigitte Falkenburg

The Physical Basis of the Direction of Time
By H. Dieter Zeh

Asymmetry: The Foundation of Information
By Scott J. Muller

Decoherence and the Quantum-To-Classical Transition
By Maximilian A. Schlosshauer

The Nonlinear Universe
Chaos, Emergence, Life by Alwyn C. Scott

Quantum Superposition
Counterintuitive Consequences of Coherence, Entanglement, and Interference
by Mark P. Silverman

Symmetry Rules
How Science and Nature are Founded on Symmetry by Joseph Rosen

Mind, Matter and Quantum Mechanics
By Henry P. Stapp

Entanglement, Information, and the Interpretation of Quantum Mechanics
By Gregg Jaeger

Relativity and the Nature of Spacetime
By Vesselin Petkov

The Biological Evolution of Religious Mind and Behavior
Ed. by Eckart Voland and Wulf Schiefenhovel

Homo Novus-A Human without Illusions
Ed. by Ulrich J. Frey, Charlotte Stormer and Kai P. Willfiihr

Brain-Computer Interfaces
Revolutionizing Human-Computer Interaction
Ed. by Bernhard Graimann, Brendan Allison and Gert Pfurtscheller

Extreme States of Matter
On Earth and in the Cosmos by Vladimir E. Fortov

Searching for Extraterrestrial Intelligence
SETI Past, Present, and Future Ed. by H. Paul Shuch

Essential Building Blocks of Human Nature
Ed. by Ulrich J. Frey, Charlotte Störmer and Kai P. Willführ

Mindful Universe
Quantum Mechanics and the Participating Observer by Henry P. Stapp

Principles of Evolution
From the Planck Epoch to Complex Multicellular Life Ed. by Hildegard Meyer-Ortmanns and Stefan Thurner

The Second Law of Economics
Energy, Entropy, and the Origins of Wealth by Reiner Köummel

States of Consciousness
Experimental Insights into Meditation, Waking, Sleep and Dreams Ed. by Dean Cvetkovic and Irena Cosic

Elegance and Enigma
The Quantum Interviews The Quantum Interviews Ed. by Maximilian Schlosshauer

Humans on Earth
From Origins to Possible Futures by Filipe Duarte Santos

Evolution 2.0
Implications of Darwinism in Philosophy and the Social and Natural Sciences Ed. by Martin Brinkworth and Friedel Weinert

Probability in Physics
Ed. by Yemima Ben-Menahem and Meir Hemmo

Chips 2020
A Guide to the Future of Nanoelectronics Ed. by Bernd Hoefflinger

From the Web to the Grid and Beyond
Computing Paradigms Driven by High-Energy Physics Ed. by Rene Brun, Federico Carminati and Giuliana Galli Carminati

The Language Phenomenon
Human Communication from Milliseconds to Millennia Ed. by P.-M. Binder and K. Smith

The Dual Nature of Life
By Gennadiy Zhegunov

Natural Fabrications
By William Seager

Ultimate Horizons
By Helmut Satz

Physics, Nature and Society
By Joaquín Marro

Extraterrestrial Altruism
Ed. by Douglas A. Vakoch

The Beginning and the End
By Clément Vidal

A Brief History of String Theory
By Dean Rickles

Singularity Hypotheses
Ed. by Amnon H. Eden, James H. Moor, Johnny H. Søraker and Eric Steinhart

Why More Is Different
Philosophical Issues in Condensed Matter Physics and Complex Systems
Ed. by Brigitte Falkenburg and Margaret Morrison

Questioning the Foundations of Physics
Ed. by Anthony Aguirre, Brendan Foster and Zeeya Merali

It From Bit or Bit From It?
Ed. by Anthony Aguirre, Brendan Foster and Zeeya Merali

The Unknown as an Engine for Science
By Hans J. Pirner

How Should Humanity Steer the Future
Ed. by Anthony Aguirre, Brendan Foster and Zeeya Merali

Trick or Truth
Ed. by Anthony Aguirre, Brendan Foster and Zeeya Merali

How Can Physics Underlie the Mind
By George Ellis

The Challenge of Chance
Ed. By Klaas Landsman, Ellen van Wolde